Dynamic Formal Epistemology

SYNTHESE LIBRARY

STUDIES IN EPISTEMOLOGY, LOGIC, METHODOLOGY, AND PHILOSOPHY OF SCIENCE

VOLUME 351

For further volumes:
http://www.springer.com/series/6607

Dynamic Formal Epistemology

Edited by

Patrick Girard
University of Auckland, New Zealand
Olivier Roy
University of Groningen, The Netherlands
and
Mathieu Marion
Université du Québec à Montréal, Canada

Springer

Editors

Prof. Patrick Girard
Department of Philosophy
University of Auckland
Private Bag 92019
1142 Auckland
New Zealand
p.girard@auckland.ac.nz

Dr. Olivier Roy
University of Groningen
Faculty of Philosophy
Oude Boteringestraat 52
9712 GL Groningen
Netherlands
o.roy@rug.nl

Mathieu Marion
Département de Philosophie
Université du Québec à Montréal
C.P. 8888, Succursale Centre-Ville
Montréal, Québec
Canada H3C 3P8
marion.mathieu@uqam.ca

ISBN 978-94-007-0073-4 e-ISBN 978-94-007-0074-1
DOI 10.1007/978-94-007-0074-1
Springer Dordrecht Heidelberg London New York

Printed on acid-free paper

Springer is part of Springer Science+Business Media (www.springer.com)

Acknowledgements

This volume is born out of the workshop *Dynamic Logic Montreal*, held in Montreal in June 2007. As both the editors of this volume and organizers of the workshop, we would like to thank all the participants, attendees, local staff and students, who made this event such a success. The conference had been made possible by a grant from the *Canadian Social Sciences and Humanities Research Council*, which we gratefully acknowledge.

We would like to thank the authors in this volume for their contribution and also for their patience during the lengthy process that led to this publication. We are also grateful to the referees for careful reading and constructive suggestions on each chapter and on the book as a whole.

Finally, we would like to thank Neil Newton, Imogene Paterson, Kym Pram and Vincent Motard-Côté, who helped with the production of this volume, as well as the editors and production team at *Synthese Library*. The editing work has been supported by the University of Auckland's Faculty Research Development Fund.

July 2010 Patrick Girard, Olivier Roy and Mathieu Marion

Contents

Contributors

Horacio Arló-Costa Department of Philosophy, Carnegie Mellon University, Baker Hall 135, Pittsburgh, PA 15213-3890, USA, hcosta@andrew.cmu.edu

Guillaume Aucher Faculty of Sciences, Technology and Communication (FSTC), University of Luxembourg, 6 rue Richard Coudenhove – Kalergi L-1359, Luxembourg, guillaume.aucher@uni.lu

Johan van Benthem Institute for Logic, Language & Computation (ILLC), University of Amsterdam, PO Box 94242, 1090 GE, Amsterdam, The Netherlands; Department of Philosophy, Stanford University, Stanford, CA 94305, USA, johan@science.uva.nl; johan.vanbenthem@uva.nl

Denis Bonnay Département de Philosophie, Université Paris Ouest Nanterre, 200 avenue de la Rpublique, 92001 Nanterre CEDEX, France; Département d'Etudes Cognitives de l'ENS, IREPH/IHPST, 29 rue d'Ulm, 75005 Paris, France, denis.bonnay@u-paris10.fr; denis.bonnay@ens.fr

Paul Egré Département d'Etudes Cognitives, Institut Jean-Nicod, Ecole Normale Supérieure, 29 rue d'Ulm, 75005 Paris, France, paulegre@gmail.com

Patrick Girard Department of Philosophy, University of Auckland, 18 Symonds Street, Auckland, New Zealand, p.girard@auckland.ac.nz

Andreas Herzig IRIT-LILaC, 118 route de Narbonne, F-31062, Toulouse Cedex 9, France, herzig@irit.fr

John F. Horty Philosophy Department, Institute of Advanced Computer Studies, University of Maryland, College Park, MD 20742, USA, horty@umiacs.umd.edu

Barteld Kooi Faculty of Philosophy, University of Groningen, Oude Boteringestraat 52, 9712 GL Groningen, The Netherlands, B.P.Kooi@rug.nl

François Lepage Département de Philosophie, Université du Québec à Montréal, C.P. 6128, succursale Centre-ville, Montréal, QC H3C 3J7, Canada, francois.lepage@umontreal.ca

Mathieu Marion Département de Philosophie, Université du Québec à Montréal, C.P. 8888, Succursale Centre-Ville, Montréal, Québec, Canada H3C 3P8, marion.mathieu@uqam.ca

Charles Morgan Department of Mathematics, University College London, Gower Street, London WC1E 6BT, UK, charles.morgan@ucl.ac.uk

Eric Pacuit Center for Logic and Philosophy of Science, Tilburg University, 5000 LE Tilburg, The Netherlands, e.j.pacuit@uvt.nl

Bryan Renne Faculty of Philosophy, University of Groningen, Oude Boteringestraat 52, 9712 GL Groningen, The Netherlands, bryan@renne.org

Olivier Roy Faculty of Philosophy, University of Groningen, Oude Boteringestraat 52, 9712 GL Groningen, The Netherlands, o.roy@rug.nl

Darko Sarenac Department of Philosophy, Colorado State University, Fort Collins, CO 80523-1781, USA, darko@colostate.edu

Krister Segerberg Filosofiska institutionen, Uppsala Universitet, 751 26 Uppsala, Sweden, krister.segerberg@filosofi.uu.se

Daniel Vanderveken Département de Philosophie, Université du Québec à Trois-Rivières, Trois-Rivières, QC G9A 5H7, Canada, daniel.vanderveken@uqtr.ca

Audrey Yap Department of Philosophy, University of Victoria, Victoria, BC V8W 3P4, Canada, ayap@uvic.ca

Chapter 1
Introduction

Patrick Girard, Mathieu Marion, and Olivier Roy

The frontier of contemporary epistemology is dynamic. Shifting from purely conceptual analysis, the theory of individual knowledge, belief and justification now includes an increasing amount of formal work, utilizing either logic or probabilities. Epistemology has also moved to questions regarding information change, its flow among groups, and its place within interaction, whereas for a long time it was centered mainly on the question of individual knowledge and its acquisition in a static environment. Epistemology is thus expanding beyond the conceptual analysis of justified true belief, allowing for a broad and formal philosophical inquiry into the notion of information and how it is acquired, changed, passed on, and aggregated. By doing so it provides new insights and methods relevant not only to the theory of knowledge but also for our understanding of interaction, obligations, and scientific discovery.

Dynamic epistemology played a part in the renewal of formal analytical philosophy. Numerous seminal contributions in the middle of the twentieth century were closely connected with formal investigations: A. Prior on time and determinism, G. H. von Wright on preferences and obligations, S. Kripke on direct reference, J. Hintikka on knowledge and beliefs, D. Lewis on conditionals and conventions, and R. Jeffrey on Bayesian rationality, to name but a few. Since the 1970s, however, formal work within the field of logic has moved slowly away from analytic philosophy towards mathematics, computer science and linguistics. These different areas have provided logic with a plethora of new mathematical tools, and have aided the development of techniques that can analyze information flow, belief revision, preference change, and strategic interaction. These tools proved to be much more than just fruitful applications; philosophers now realize that they also shed new light on foundational questions.

This book brings together original contributions from the actors of this dynamic turn in epistemology. It aims to bring their work under a single umbrella by highlighting the coherence of their current research themes, and by establishing connec-

---

O. Roy (✉)
Faculty of Philosophy, University of Groningen, Oude Boteringestraat 52, 9712 GL Groningen, The Netherlands
e-mail: o.roy@rug.nl

P. Girard et al. (eds.), *Dynamic Formal Epistemology*, Synthese Library 351,
DOI 10.1007/978-94-007-0074-1_1,

tions between topics that up until now have been investigated independently. It will also be a helpful red for any analytic philosopher who is not yet acquainted with the dynamic perspective, as it illustrates how the new analytical toolbox unveils fresh questions about the theory of knowledge, belief, preference, action, and rationality.

The contributions in this book explore a number of central axes in dynamic epistemology: temporal, social, probabilistic and even deontic dynamics. This diversity is by no means a sign of disunity; rather, the dynamic way of thinking sheds light on a broad array of topics. The following is a short overview of these contributions.

Temporal information change is an obvious subject to arise from the study of dynamic epistemology. An interest in time and events is certainly not new in philosophy, especially in the analytic or the logical tradition, but the emergence of propositional dynamic logic (PDL), epistemic temporal logic (ETL) and dynamic epistemic logic (DEL) has revealed a number of crucial issues involving temporal reasoning. In Chapter 2, Eric Pacuit and Barteld Kooi provide the unaccustomed reader with an introduction to these formal frameworks. For the expert they demonstrate that PDL, ETL and DEL are not competing systems; their complementarity supersedes their differences. The reader will also find a comprehensive bibliography of the themes covered in this book.

Dynamic epistemic logic, in its standard form, suffers from a widespread limitation, which Chapters 3 and 4 endeavor to tackle. In their most common forms, these logical languages only refer to current information states and their transformations, and do not refer back to how things were before certain events occurred. In Chapter 3, Audrey Yap avoids this limitation by extending the standard dynamic epistemic language to include a past operator. She provides a sound and complete axiomatization for this new inclusion and investigates its expressive power. In Chapter 4, Guillaume Aucher and Andreas Herzig investigate past operators in propositional dynamic logic. Both contributions illustrate the interconnections, demonstrated by Pacuit and Kooi in Chapter 2, between the different logical systems used to talk about temporal information change. Yaps models for DEL are similar to ETL structures, while Aucher and Herzig show that standard DEL can be faithfully embedded into their extended PDL.

In Chapter 5, Sarenac investigates a general approach to dynamic systems, using a notion borrowed from computer science, namely iterative function systems (IFS). Dynamic epistemic logic is one among many categories of dynamic problems that can be analyzed in this setting. Indeed, Sarenac traces his analysis back to Poincare's work on the three-body problem, and contrasts dynamical analyses in mathematical physics with those more common in computer science, which are central to the present book.

Dennis Bonnay and Paul Egre show in Chapter 6 that, when one considers temporal dynamics, elegant solutions may be found to questions regarding imprecise knowledge. They provide a dynamic analysis of Timothy Willamson's Margin of Error Paradox. The analysis shows that Willamsons paradox stems from a rather simplistic assumption about the rigidity of margins of error through time.

In Chapter 7, Bryan Renne addresses the topic of evidence, and in particular, evidence elimination, which is another issue involved in the study of temporal change.

Using justification logic, a logic that can analyze the evidence put forward to support a proof, Renne provides a sound and complete axiomatic system for the logic of evidence change in a way that, again, bridges different logical systems this time between justification and dynamic epistemic logics.

In Chapter 8, Johan van Benthem bridges the gap between the single-agent perspective and the dynamics of social interaction. He shows that well-known dynamic epistemic logical systems, originally designed to analyze information updates after epistemic events, can be seen as systems of preference aggregation commonly used in Social Choice Theory. He does so by providing a characterization of the standard DEL update rule in terms of a priority update, which opens up a whole new perspective on dynamic epistemic logical systems.

The study of belief change already has a long history within Bayesian and probabilistic approaches to epistemology. Despite this, the following two contributions demonstrate that if we take information dynamics seriously, there are still major challenges that the existing approaches face.

In Chapter 9, François Lepage and Charles Morgan take Lewis well-known triviality result for the conditional probability of a counterfactual and generalize it to any two-place probability function satisfying minimal requirements. This result applies to a wide range of belief change operations from classical conditioning to imaging. Alternative probabilistic views on counterfactual, reasoning and belief change are called for if one is to pursue this approach.

In Chapter 10, Horacio Arló Costa puts forward another important challenge to contemporary probabilistic approaches to belief change, namely, how one may account for such phenomenon in situations where attitudes are incomplete or indeterminate. Arló Costa sketches a two-tier theory of belief change and presents the various challenges that its formalization poses. He also considers the application of this theory to Philosophy of Science, thus illustrating how dynamic epistemology can contribute to traditional debates about the theory of knowledge and scientific inquiry.

The next two chapters in this collection show that notions of obligation can be analyzed dynamically, thus extending the scope of dynamic studies to the deontic realm. John Horty's paper in Chapter 11, which nicely complements the temporal logic of the first chapters, looks at deontic modalities in branching time structures. In previous works Horty has demonstrated that an elegant formalization of act utilitarianism can be made by extending Belnap's stit (see to it that) logic with deontic operators. In this contribution, he argues that various interpretations of these operators are possible in stit models, interpretations that seem equally plausible but which turn out to make contradicting assessments of actions in certain contexts. He shows, however, that by borrowing the idea of double time references from temporal logic, a very general account of act utilitarianism based on different perspectives can be given.

In Chapter 12, Krister Segerberg addresses issues pertaining to deontic logic over temporal structures. He proposes a formalism to deal with factual and normative changes, and the interaction between the two, thus improving on some of his earlier work.

Finally, in Chapter 13, Daniel Vanderveken proposes a general logical typology of propositional attitudes, ranging from the basic notions of belief, desire and intention, to sophisticated notions such as regret and expectation. In his chapter Vanderveken shows that his approach avoids the pitfalls of logical omniscience by providing a fine-grained intensional semantic. He also provides a general theory of attitude revision that can deal with all kinds of attitude changes.

* * * *

The contributions in this volume result from the workshop *Dynamic Logic Montreal*/, held at the Université du Québec à Montréal (UQAM) in June 2007. We would like to thank the Social Sciences and Humanities Research Council of Canada (SSHRC), UQAM, the Université de Montréal, the Canada Research Chair in the Philosophy of Logic and Mathematics and the Institute for Logic, Language and Computation for their financial support. We are also grateful to: Imogene Paterson, Neil Newton, Kym Pram and Vincent Motard-Côté for their help during the workshop and/or during the preparation of this volume; the referees of the various chapters, and of the volume as a whole, for their useful comments and suggestions; and Vincent Hendricks Ingrid van Laarhoven at Synthese Library for their enthusiasm and support. Finally, thanks to all contributors of this book for their efforts, and for their patience, during the rather lengthy process that lead to this publication.

Chapter 2
Logics of Rational Interaction

Barteld Kooi and Eric Pacuit

2.1 Introduction

There is a growing literature focused on using logical methods to reason about communities of agents engaged in some form of social interaction. Much of the work builds upon existing logical frameworks developed by philosophers and computer scientists incorporating insights and ideas from philosophy (especially epistemology and philosophy of action), game theory, decision theory and social choice theory. The result is a web of logical systems each addressing different aspects of rational agency and social interaction. Rather than providing an encyclopedic account of these different logical systems,[1] this chapter focuses on issues surrounding the modeling of informational attitudes in social interactive situations. The main objective is to introduce the two main approaches to modeling "rational interaction" and provide pointers to the current literature.

Of course, there is no single approach that can address *all* of the complex phenomena that arise when rational agents interact with one another and the environment. Thus it is important to understand how the different analyses from within and across the disciplines mentioned above can fit together. This suggests the following three general questions:

1. How can we *compare* different logical frameworks addressing similar aspects of rational agency and social interaction (i.e., how information evolves through social interaction)?
2. How should we *combine* logical systems which address *different* aspects of social interaction towards the goal of a comprehensive (formal) theory of rational agency?

B. Kooi (✉)
Faculty of Philosophy, University of Groningen, Oude Boteringestraat 52, 9712 GL, Groningen, The Netherlands
e-mail: B.P.Kooi@rug.nl

[1]The interested reader can consult Meyer and Veltman (2007), van der Hoek and Wooldridge (2003), van Benthem (2008a) and references therein.

P. Girard et al. (eds.), *Dynamic Formal Epistemology*, Synthese Library 351,
DOI 10.1007/978-94-007-0074-1_2, © Springer Science+Business Media B.V. 2011

3. How do the logical frameworks discussed in this literature contribute to the broader discussion of rational agency and social interaction within philosophy and the social sciences?

Certainly, the first two questions raise numerous methodological issues and technical problems. However, they also make explicit certain foundational and philosophical issues surrounding rational interaction (cf. van Benthem et al. 2009a). In particular, viewing the various logical systems found in the literature as (sometimes competing) accounts of rational agency forces us to carefully examine what we even mean by a "rational agent" (see van Benthem 2005, for an extensive discussion). Of course, the nature of rationality and human agency is a central concern of many philosophers from Aristotle to David Hume to present-day philosophers (cf. Bratman 2007, Searle 1985, Hyman and Steward 2004). The point here is that there are many different types of reasoning and dynamic processes that agents use when interacting with other agents. Comparing and combining the different logical systems forces us to consider how these different processes interact.

In this survey, the modeling of informational attitudes of a group of *rational* agents engaged in some form of social interaction (e.g. having a conversation or playing a card game) takes center stage. Indeed, there are many logical systems today that describe how an agent's information changes over time. Sometimes the differences between two competing logical systems are technical in nature reflecting different conventions used by different research communities. And so, with a certain amount of technical work, such frameworks are seen to be equivalent up to model transformations (cf. Halpern 1999, Lomuscio and Ryan 1997, Pacuit 2007a, van Benthem et al. 2009a). Other differences point to key conceptual issues about rational interaction. We will introduce the two main logical accounts of rational interaction and highlight such similarities and differences.

2.2 Reasoning About Rational Interaction

This section introduces two logical frameworks that describe the dynamics of information over time in a multiagent situation. The first is *epistemic temporal logic* (ETL, Fagin et al. 1995, Parikh and Ramanujam 1985) which uses linear or branching time models with added epistemic structure induced by the agents' different capabilities for observing events. These models provide a "grand stage" where histories of some social interaction unfold constrained by a **protocol**. Here a **protocol** is intended to represent the rules or conventions that govern many of our social interactions. For example, in a conversation, it is typically not polite to "blurt everything out at the beginning", as we must speak in small chunks. Other natural conversational protocol rules include "do not repeat yourself", "let others speak in turn", and "be honest". Imposing such rules *restricts* the legitimate sequences of possible statements.

The other framework is *dynamic epistemic logic* (DEL, Gerbrandy 1999a; Baltag et al. 1998b; van Ditmarsch et al. 2007) that describes social interactions in terms of epistemic **event models** (which may occur inside modalities of the language). Similar to the way Kripke structures are used to capture the information the agents have about a *fixed* social situation,[2] an **event model** describes the agents' information about which actual events are currently taking place. The temporal evolution of the situation is then computed from some initial epistemic model through a process of successive "product updates". Details of both frameworks are provided in the subsequent sections.

Often DEL and ETL are presented as *competing* ways of adding dynamics to multi-agent epistemic models. Based on van Benthem et al. (2009, 2006) and van Benthem and Pacuit (2006), we will see how DEL and ETL should rather be viewed as *complementary* accounts of social interaction. The focus is on conceptual issues leaving some of the more technical details and proofs to the relevant papers. The following running example will help guide intuitions (also discussed in Pacuit and Parikh 2006).

Example 2.1 Suppose that Ann would like Bob to attend her talk; however, she only wants Bob to attend if he is interested in the subject of her talk, not because he is just being polite. There is a very simple procedure to solve Ann's problem: Have a (trusted) friend tell Bob the time and subject of her talk.

Taking a cue from computer science, perhaps we can *prove* that this simple procedure correctly solves Ann's problem. However, it is not so clear how to define a correct solution to Ann's problem. If Bob is actually present during Ann's talk, can we conclude that Ann's procedure succeeded? Not really. Bob may have figured out that Ann wanted him to attend, and so is there only out of politeness. Thus for Ann's procedure to succeed, she must achieve a certain "level of knowledge" (cf. Parikh, 2003) between her and Bob. Besides both Ann and Bob knowing about the talk and Ann knowing that Bob knows about

Bob *does not know* that Ann knows about the talk.

This last point is important, since, if Bob knows that Ann knows that he knows about the talk, he may feel social pressure to attend.[3] Thus, the procedure *to have a friend tell Bob about the talk, but not reveal that it is at Ann's suggestion*, will satisfy all the conditions. Telling Bob directly will satisfy the first three, but not the essential last condition.

[2] A Kripke structure is a set of states with relations on this set for each agent. The states, or possible worlds, represent different ways the social situation could have evolved and the relations describe the agents' (current) information. See, for example, Fagin et al. (1995) for details.

[3] Of course, this is not meant to be a complete analysis of "social politeness".

2.2.1 Epistemic Temporal Logic

Fix a finite set of agents $\mathcal{A}$ and a (possibly infinite) set of events[4] Σ. A **history** is a finite sequence of events[5] from Σ. We write Σ^* for the set of histories built from elements of Σ. For a history h, we write he for the history h followed by the event e. Given $h, h' \in \Sigma^*$, we write $h \preceq h'$ if h is a prefix of h', and $h \prec_e h'$ if $h' = he$ for some event e.

For example, consider the social interaction described in Example 2.1. There are three relevant participants: Ann (A), Bob (B) and Ann's friend (call him Charles (C)). What are the relevant primitive events? To keep things simple, assume that Ann's talk is either at 2PM or 3PM and initially none of the agents know this. Say, that Ann receives a message stating that her talk is at 2PM (denote this event – Ann receiving a private message saying that her talk is at 2PM – by $\mathsf{e}_{\mathsf{A}}^{\mathsf{2PM}}$). Now, after Ann receives the message that the talk is at 2PM, she proceeds to tell her trusted friend Charles that the talk is at 2PM (and that she wants him to inform Bob of the time of the talk without acknowledging that the information can from her – call this event $\mathsf{e}_{\mathsf{C}}^{\mathsf{A}}$), then Charles tells Bob this information (call this event $\mathsf{e}_{\mathsf{B}}^{\mathsf{C}}$). Thus, the history

$$\mathsf{e}_{\mathsf{A}}^{\mathsf{2PM}}\ \mathsf{e}_{\mathsf{C}}^{\mathsf{A}}\ \mathsf{e}_{\mathsf{B}}^{\mathsf{C}}$$

represents the sequence of events where "Ann receives a (private) message stating that the talk is at 2PM, Ann tells Charles the talk is at 2PM, then Charles tells Bob the talk is at 2PM". Of course, there are other events that are also relevant to this situation. For one thing, Ann could have received a message stating that her talk is at 3PM (denote this event by $\mathsf{e}_{\mathsf{A}}^{\mathsf{3PM}}$). This will be important to capture Bob's uncertainty about whether Ann knows that he knows about the talk. Furthermore, Charles may learn about the time of the talk independently of Ann (denote these two events by $\mathsf{e}_{\mathsf{C}}^{\mathsf{2PM}}$, $\mathsf{e}_{\mathsf{C}}^{\mathsf{3PM}}$). So, for example, the history

$$\mathsf{e}_{\mathsf{A}}^{\mathsf{2PM}}\ \mathsf{e}_{\mathsf{C}}^{\mathsf{2PM}}\ \mathsf{e}_{\mathsf{B}}^{\mathsf{C}}$$

represents the situation where Charles independently learns about the time of the talk and informs Bob.

There are a number of simplifying assumptions that we adopt in this section. They are not crucial for the analysis of Example 2.1, but do simplify the some of

---

[4]There is a large literature addressing the many subtleties surrounding the very notion of an *event* and when one event *causes* another event (see, for example, Cartwright 2007). However, for this chapter we take the notion of event as primitive. What is needed is that if an event takes place at some time t, then the fact that the event took place can be observed by a relevant set of agents at t. Compare this with the notion of an event from probability theory. If we assume that at each clock tick a coin is flipped exactly once, then "the coin landed heads" is a possible event. However, "the coin landed heads more than tails" would not be an event, since it cannot be observed at any one moment. As we will see, the second statement will be considered a *property* of histories, or sequences of events.

[5]To be precise, elements of Σ should, perhaps, be thought of as event *types* whereas elements of a history are event *tokens*.

the formal details. Since, histories are sequences of (discrete) events, we assume the existence of a global discrete clock (whether the agents have access to this clock is another issue that will be discussed shortly). The length of the history then represents the amount of time that has passed. Note that this implies that we are assuming a finite past with a possibly infinite future. Furthermore, we assume that at each clock tick, or moment, *some* event takes place (which need not be an event that any agent directly observes). Thus, we can include an event $\mathsf{e_t}$ (for "clock tick") which can represent that "Charles does *not* tell Bob that the talk is at 2PM." So the history

$$\mathsf{e_A^{2PM}\ e_C^A\ e_t}$$

describes the sequence of events where, after learning about the time of the talk, Ann informs Charles, but Charles does *not* go on to tell Bob that the talk is at 2PM. Once a set of events Σ is fixed, the temporal evolution and moment-by-moment uncertainty of the agents can be described.

Definition 2.1 (ETL Frames) Let Σ be a set of events. A **protocol** is a set $\mathsf{H} \subseteq \Sigma^*$ closed under non-empty prefixes. An **ETL frame** is a tuple $\langle \Sigma, \mathsf{H}, \{\sim_i\}_{i \in \mathcal{A}} \rangle$ with H a protocol, and for each $i \in \mathcal{A}$, a binary relation $\sim_i$ on[6] H.

An ETL frame describes how the agents' *hard* information[7] evolves over time in some social situation. The protocol describes (among other things) the temporal structure, with h' such that $h \prec_e h'$ representing the point in time after e has happened in h. The relations $\sim_i$ represent the uncertainty of the agents about how the current history has evolved. Thus, $h \sim_i h'$ means that from agent i's point of view, the history h' looks the same as the history h.

Note that the protocol in an ETL frame captures not only the temporal structure of the social situation being modeled but also assumptions about the nature of the participants. For example, the following is a possible protocol built from the events described above:

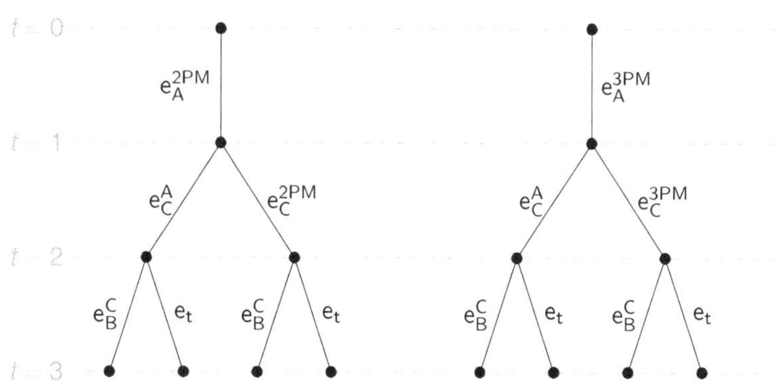

[6] Although we will not do so here, typically it is assumed that $\sim_i$ is an equivalence relation.
[7] As opposed to *soft* information which may be revised. See van Benthem (2007) for a general discussion of hard and soft information.

While this protocol does describe possible ways the situation described in Example 2.1 could evolve, it does not account for the *motivation* of the agents. For example, the history

$$\mathsf{e}_A^{3PM}\ \mathsf{e}_C^{A}\ \mathsf{e}_B^{C}$$

describes the sequence of events where Ann learns the talk is at 3PM but tells Charles (who goes on to inform Bob) that the talk is at *2PM*. Of course, given the assumption that Ann *wants* Bob to attend her talk, this should not be part of (Ann's) protocol. Similarly, since we assume Charles is trustworthy, we should not include any histories where e_t follows the event e_C^A. Taking into account these underlying assumptions about the motivations (e.g. Ann wants Bob to attend the talk) and dispositions (e.g. Charles tells the truth and lives up to his promises) of the agents we can drop a number of histories from the protocol shown above. Note that we keep the history

$$\mathsf{e}_A^{2PM}\ \mathsf{e}_C^{2PM}\ \mathsf{e}_t$$

in the protocol, since if Charles learns independently about the time of the talk, then he is under no obligation to inform Bob. In the picture below, we also add some of the uncertainty relations for Ann and Bob (to keep the picture simple, we do not draw the full ETL frame). The solid line represents Bob's uncertainty while the dashed line represents Ann's uncertainty. The main assumption is that Bob can only observe the event (e_B^C). So, for example, the histories $h = \mathsf{e}_A^{2PM}\ \mathsf{e}_C^{A}\ \mathsf{e}_B^{C}$ and $h' = \mathsf{e}_A^{2PM}\ \mathsf{e}_C^{2PM}\ \mathsf{e}_B^{C}$ look the same to Bob (i.e., $h \sim_B h'$).[8]

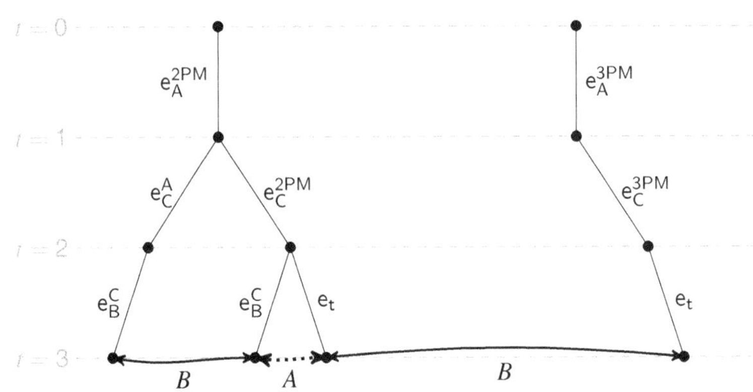

Assumptions about the underlying protocol in an ETL frame corresponds to "fixing the playground" where the agents will interact. As we have seen, the protocol not only describes the temporal structure of the situation being modeled, but also any *causal* relationships between events (e.g., sending a message must always proceed receiving that message) plus the motivations and dispositions of the participants (e.g., liars send messages that they *know* – or believe – to be false). Thus

[8] Note that we do not include any reflexive arrows in the picture in order to keep things simple.

the "knowledge" of agent i at a history h in some ETL frame is derived from both i's observational powers (via the $\sim_i$ relation) and i's information about the (fixed) protocol.

Remark 2.1 (Three Equivalent Approaches) There are at least two further approaches to uncertainty in the literature. The first, discussed by Parikh and Ramanujam (1985), explicitly describes the agents' "observational" power. That is, each agent i has a set E_i of events she *can* observe.[9] For simplicity, we assume $E_i \subseteq \Sigma$ but this is not necessary. A **local view** function is a map $\lambda_i : \mathsf{H} \to E_i^*$. Given a finite history $h \in \mathsf{H}$, the intended interpretation of $\lambda_i(h)$ is "the sequence of events observed by agent i at h". The second approach comes from Fagin et al. (1995). Each agent has a set L_i of **local states** (if necessary, one can also assume a set L_e of environment states). Events e are tuples of local states (one for each agent) $\langle l_1, \ldots, l_n \rangle$ where for each $i = 1, \ldots, n$, $l_i \in L_i$. Then two finite histories h and h' are i-equivalent provided the local state of the last of event on h and h' is the same for agent i. From a technical point of view, the three approaches (uncertainty relations, local view functions and local states) to modeling uncertainty are equivalent (Pacuit 2007a, van Benthem and Pacuit 2006, provide the relevant discussions).

Although, syntactic issues do not play an important role in this chapter, we give the bare necessities to facilitate a comparison between ETL and DEL. Different modal languages describe ETL frames (see, for example, Hodkinson and Reynolds 2006, Fagin et al. 1995), with "branching" or "linear" variants. Let At be a countable set of atomic propositions. The language $\mathcal{L}_{ETL}$ is generated by the following grammar:

$$P \mid \neg\varphi \mid \varphi \wedge \psi \mid K_i\varphi \mid \langle e \rangle \varphi$$

where $i \in \mathcal{A}$, $e \in \Sigma$ and $P \in \mathsf{At}$. The usual boolean connectives ($\vee, \to, \leftrightarrow$) and the dual modal operators (L_i, $[e]$) are defined as usual. The pure epistemic language, denoted $\mathcal{L}_{EL}$, is the fragment of $\mathcal{L}_{ETL}$ with only epistemic modalities (which we will refer to both as the "language of epistemic logic" and the "epistemic fragment" of $\mathcal{L}_{ETL}$ or the language $\mathcal{L}_{\mathrm{DEL}}$ defined below). The intended interpretation of "$K_i\varphi$" is "according to agent i's current information, φ is true." The intended interpretation of "$\langle e \rangle \varphi$" is "after event e (does) take place, φ is true." Formulas are interpreted at histories in an **ETL model**:

Definition 2.2 (ETL Model) An **ETL model** is a tuple $\langle \Sigma, \mathsf{H}, \{\sim_i\}_{i \in \mathcal{A}}, V \rangle$ with $\langle \Sigma, \mathsf{H}, \{\sim_i\}_{i \in \mathcal{A}} \rangle$ an ETL frame and V a valuation function (V: $\mathsf{At} \to 2^{\mathsf{H}}$).

Definition 2.3 (Truth of $\mathcal{L}_{\mathrm{ETL}}$ Formulas) Let $\mathcal{H} = \langle \Sigma, \mathsf{H}, \{\sim_i\}_{i \in \mathcal{A}}, V \rangle$ be an ETL model. The truth of a formula φ at a history $h \in \mathsf{H}$, denoted $\mathcal{H}, h \models \varphi$, is defined inductively as follows:

1. $\mathcal{H}, h \models P$ iff $h \in V(P)$
2. $\mathcal{H}, h \models \neg\varphi$ iff $\mathcal{H}, h \not\models \varphi$
3. $\mathcal{H}, h \models \varphi \wedge \psi$ iff $\mathcal{H}, h \models \varphi$ and $\mathcal{H}, h \models \psi$

[9] This may be different from what the agent *does* observe in a given situation.

4. $\mathcal{H}, h \models K_i\varphi$ iff for each $h' \in \mathsf{H}$, if $h \sim_i h'$ then $\mathcal{H}, h' \models \varphi$
5. $\mathcal{H}, h \models \langle e \rangle\varphi$ iff there exists an $h' \in \mathsf{H}$ such that $h \prec_e h'$ and $\mathcal{H}, h' \models \varphi$

It is often natural to extend the language $\mathcal{L}_{ETL}$ with group knowledge operators (e.g., common or distributed knowledge) and more expressive temporal operators (e.g., arbitrary future or past modalities).

2.2.2 Dynamic Epistemic Logic

An alternative account of interactive dynamics was elaborated by Gerbrandy (1999a), Baltag et al. (1998b), van Benthem (2006), van Benthem et al. (2006) and others. From an initial epistemic model, temporal structure evolves as explicitly triggered by complex informative events.

Definition 2.4 (Epistemic Model) Let $\mathcal{A}$ be a finite set of agents and At a set of atomic propositions. An **epistemic model** is a tuple $\langle W, \{R_i\}_{i \in \mathcal{A}}, V\rangle$ where W is a non-empty set, for each $i \in \mathcal{A}$, R_i is a relation[10] on W ($R_i \subseteq W \times W$) and V a valuation function ($V : \mathsf{At} \rightarrow 2^W$). We call the set W the domain of $\mathcal{M}$, denoted by $D(\mathcal{M})$. A pair $\mathcal{M}, w$ where $\mathcal{M}$ is an epistemic model and $w \in D(\mathcal{M})$ is called a **pointed epistemic model**.

We can interpret the epistemic language, $\mathcal{L}_{\text{EL}}$, defined above at states in an epistemic model. Truth is defined as usual. We only recall the definition of the knowledge operators:

$$\mathcal{M}, w \models K_i\varphi \;\text{ iff for each }\; w' \in W, \text{if } \; wR_iw' \;\text{ then } \mathcal{M}, w' \models \varphi$$

Returning to our running example (Example 2.1), initially we assume that none of the agents knows the time of Ann's talk. Let P be the proposition "Ann's talk is at 2PM." Then this initial model can be pictured as follows: there are two states w and v with P true at w ($w \in V(P)$). The agent's uncertainty relations is the universal relation (since all agents have the same information, we do not label the arrows). Note that the convention followed in this section is that a solid line around a state means that state is the *actual* or current state (i.e., where the formulas are to be evaluated):

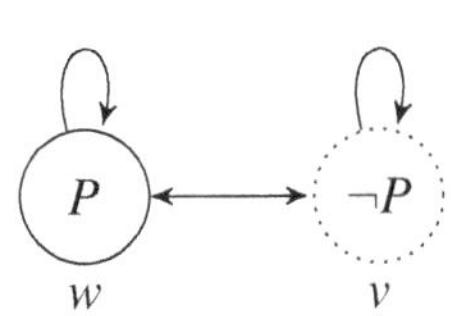

Whereas an ETL frame describes the agents' information at all moments, **event models** are used to build new epistemic models as needed.

Definition 2.5 (Event Model) An **event model** is a tuple $\langle S, \{\longrightarrow_i\}_{i \in \mathcal{A}}, \mathsf{pre}\rangle$, where S is a nonempty set of **primitive events**, for each $i \in \mathcal{A}$, $\longrightarrow_i \subseteq S \times S$

[10] Again, the R_i are often taken to be equivalence relations on W – but we do not commit.

and $\mathsf{pre}\ S \to \mathcal{L}_{EL}$ is the **pre-condition function**. The set S in an event model $\mathcal{E}$ is called the domain of $\mathcal{E}$, denoted $D(\mathcal{E})$.

Given two primitive events e and f, $e \longrightarrow_i f$ means that "according to agent i, event e looks like event f." Event models then describe an "epistemic event". In Example 2.1 the first event is Ann receiving a private message that the talk is at 2PM. This can be described by a simple event model: there are two primitive events e and f. The precondition of e is P ($\mathsf{pre}(e) = P$) and the precondition of f is $\top$ (i.e., f is the "skip event").

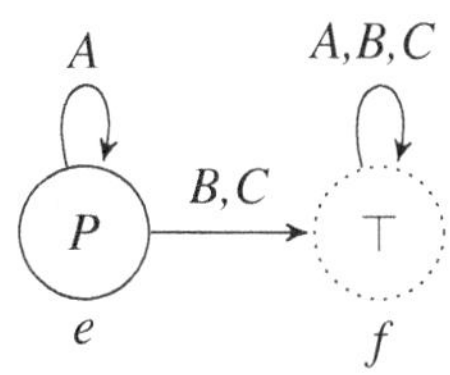

Thus, initially Ann observes the actual event e (and so, learning that P is true) while Bob an Charles observe a skip event (and so, their information does not change). What is the effect of this event on the initial model pictured above? Intuitively, it is not hard to see that after the initial event, Ann knows that P is true while Bob and Charles are still ignorant of P *and the fact that Ann knows* P. That is, combining the initial epistemic model with the above event model should yield the following epistemic model (for simplicity we only draw Ann and Bob's uncertainty relations):

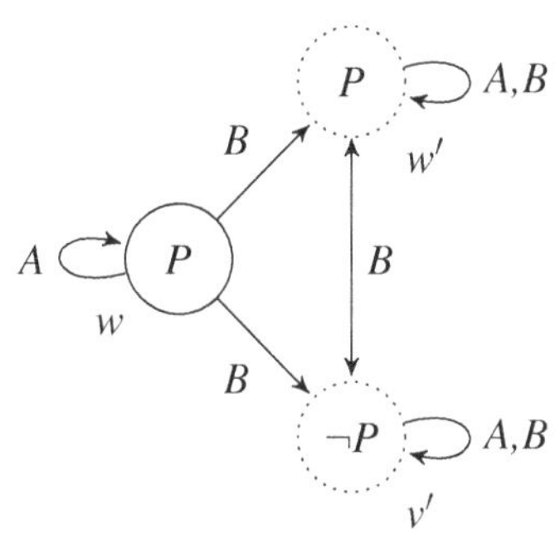

The following definition gives a general procedure for constructing a new epistemic model from a given epistemic model and an event model.

Definition 2.6 (Product Update) The **product update** $\mathcal{M} \otimes \mathcal{E}$ of an epistemic model $\mathcal{M} = \langle W, \{R_i\}_{i \in \mathcal{A}}, V \rangle$ and event model $\mathcal{E} = \langle S, \{\longrightarrow_i\}_{i \in \mathcal{A}}, \mathsf{pre} \rangle$ is the epistemic model $\langle W', R_i', V' \rangle$ with

1. $W' = \{(w, e) \mid w \in W, e \in S \text{ and } \mathcal{M}, w \models \mathsf{pre}(e)\}$,
2. $(w, e) R_i' (w', e')$ iff $w R_i w'$ in $\mathcal{M}$ and $e \longrightarrow_i e'$ in $\mathcal{E}$, and
3. For all $P \in \mathsf{At}$, $(s, e) \in V'(P)$ iff $s \in V(P)$

We illustrate this construction using our running example. The main event in Example 2.1 is "Charles telling Bob (without Ann present) that Ann's talk is at 2PM". This can be described using the following event model (again only the Ann and Bob relations will be drawn): Ann is aware of the actual event taking place while Bob thinks the event is a private message to himself.

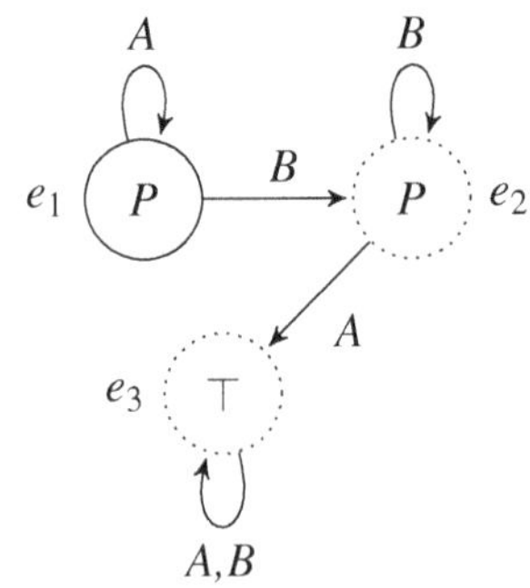

As in the previous section, there are implicit assumptions here about the motivations and dispositions of the agents. Thus, even though Ann is not present during the actual event,[11] she *trusts* that Charles will honestly tell Bob that the talk is at 2PM (without revealing he received the information from her). This explains why in the above event model, $e_1 \longrightarrow_A e_1$. Starting from a slightly modified epistemic model from the one given above (where Bob now knows that Ann knows *whether* the talk is at 2PM), using Definition 2.6, we can calculate the effect of the above event model as follows (again focusing only on Ann and Bob's information):

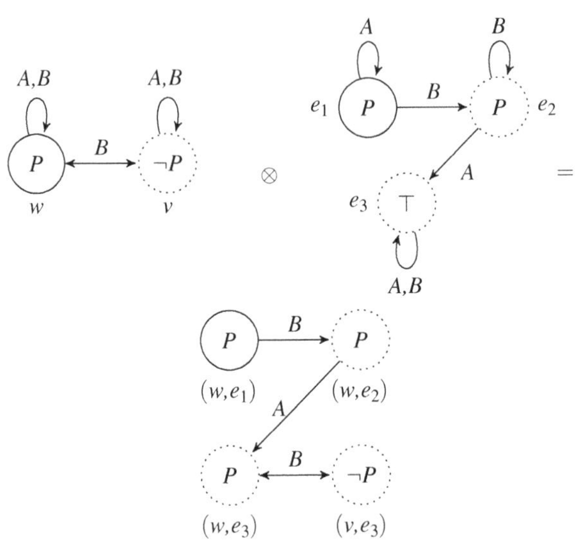

---

[11] Of course, we must assume that she knows precisely *when* Charles will meet with Bob.

Note that, in the epistemic model on the right, for simplicity, the reflexive arrows are not drawn.

Finally, a few comments about syntactic issues. The language $\mathcal{L}_{\mathrm{DEL}}$ extends $\mathcal{L}_{\mathrm{EL}}$ with operators $\langle \mathcal{E}, e\rangle$ for each pair of event models $\mathcal{E}$ and event e in the domain of $\mathcal{E}$. Truth for $\mathcal{L}_{\mathrm{DEL}}$ is defined as usual. We only define the typical DEL modalities:

$$\mathcal{M}, w \models \langle \mathcal{E}, e\rangle\varphi \text{ iff } \mathcal{M}, w \models \mathsf{pre}(e) \text{ and } \mathcal{M} \otimes \mathcal{E}, (w, e) \models \varphi$$

Example 2.2 Public Announcement Logic The **public announcement** of a formula $\varphi \in \mathcal{L}_{\mathrm{EL}}$ is the event model $\mathcal{E}_\varphi = \langle \{e\}, \{\longrightarrow_i\}_{i\in\mathcal{A}}, \mathsf{pre}\rangle$ where for each $i \in \mathcal{A}$, $e \longrightarrow_i e$ and $\mathsf{pre}(e) = \varphi$ (see Plaza 2007, Gerbrandy 1999a). As the reader is invited to verify, the product update of an epistemic model $\mathcal{M}$ with a public announcement model $\mathcal{E}_\varphi$ is the submodel of $\mathcal{M}$ containing all the states that satisfy φ. In this case, the DEL modality $\langle \mathcal{E}_\varphi, e\rangle$ will be denoted $\langle\varphi\rangle$. Henceforth, $\mathcal{L}_{PAL}$ will denote this language.

2.2.3 Comparing DEL and ETL

Both ETL and DEL are logical frameworks that describe the flow of information in a social interactive situation. For instance the *broadcasts* studied by van der Meyden (1996) and Lomuscio et al. (2000) are essentially the *public announcements* of Example 2.2. So, it is natural to ask how these two frameworks are related (cf. question 1 from the Introduction). Different logical frameworks, such as DEL and ETL, can be compared along many different dimensions. One key way to compare two different logical frameworks focuses on their *expressivity*. In order to show that one logic is at least as expressive as another logic, there are two main tasks to be carried out:

1. One has to establish a relation between the models of the two logics so that if we are given a model from the one logic, we can construct a corresponding model for the other logic;
2. One has to provide a formal translation so that if we are given a formula in the one formal language, we can produce a formula in the other with the same meaning.

Connections[12] between DEL and ETL along these lines have been worked out in detail by van Benthem and Pacuit (2006) and van Benthem et al. (2009).

The key observation is that by repeatedly updating an epistemic model with event models, the machinery of DEL (i.e., Definition 2.6) in effect creates ETL models. Note that an ETL model contains not only a description of how the agents' information changes over time, but also "protocol information" describing *when* each

[12] The first formal connection was established by Gerbrandy (1999a, Section 5.3).

event *can* be performed.[13] Details of this comparison can be found in van Benthem et al. (2009). Instead we identify the properties present in all DEL-generated ETL models. These properties have been discussed elsewhere (cf. Fagin et al. 1995, Bonanno 2004), but can also be seen as coming out of the definition of product update (Definition 2.6).

Definition 2.7 (Synchronicity, Perfect Recall, Uniform No Miracles) Let $\mathcal{H} = \langle \Sigma, \mathsf{H}, \{\sim_i\}_{i \in \mathcal{A}}, V \rangle$ be an ETL model. $\mathcal{H}$ satisfies:

- **Synchronicity** iff for all $h, h' \in \mathsf{H}$, if $h \sim_i h'$ then $\mathsf{len}(h) = \mathsf{len}(h')$ ($\mathsf{len}(h)$ is the number of events in h).
- **Perfect Recall** iff for all $h, h' \in \mathsf{H}$, $e, e' \in \Sigma$ with $he, h'e' \in \mathsf{H}$, if $he \sim_i h'e'$, then $h \sim_i h'$
- **Uniform No Miracles** iff for all $h, h' \in \mathsf{H}$, $e, e' \in \Sigma$ with $he, h'e' \in \mathsf{H}$, if there are $h'', h''' \in \mathsf{H}$ with $h''e, h'''e' \in \mathsf{H}$ such that $h''e \sim_i h'''e'$ and $h \sim_i h'$, then $he \sim_i h'e'$.

Note that Definition 2.7 are properties of ETL *frames*. Already with these properties we can say something about how to relate the two frameworks. Suppose that $\mathcal{H}$ is an ETL frame satisfying the properties in Definition 2.7. We can easily read off an epistemic *frame* (i.e., a set of states W and relations R_i for each agent $i \in \mathcal{A}$ on W) to serve as the initial model (let the histories of length 1 be the states and simply copy the uncertainty relations). Furthermore, we can define a "DEL-like" protocol $\mathsf{P}_{\mathcal{H}}$ consisting of sequences of event models where the precondition function assigns to the primitive events *sets of finite histories*. Intuitively, if e is a primitive event (i.e., a state in an event model), then $\mathsf{pre}(e)$ is the set of histories where e can "be performed". Thus, we have a comparison of the two frameworks at the level of frames provided we work with a modified definition of an event model. However, the comparison is between *models*, so we need additional properties. In particular, at each level of the ETL model we will need to specify a *formula* of $\mathcal{L}_{\mathrm{EL}}$ as a pre-condition for each primitive event e (cf. Definition 2.6). As usual, this requires that the set of histories preceding an event e be *bisimulation-closed* (cf. Blackburn et al. 2002, for a discussion of the notion of bisimulation). One final assumption that propositional variables do not change their truth value along a fixed history is needed since we are assuming that product update does not change the ground facts (although see the discussion in the next section about *factual change*). Consult (van Benthem et al. 2009, Theorem 1) for the details of the proof that any ETL model with the properties discussed above is generated from an initial epistemic model by a *DEL protocol* (i.e., a sequence of event models).

This technical result and discussion illustrates how DEL product update (Definition 2.6) may be used to generate interesting ETL frames and describes the observational powers of the agents presupposed in the DEL setting. Of course, this is not the only way to compare DEL and ETL. We can also we can also draw distinctions

[13]The *preconditions of DEL* also encode protocol information of a "local" character, and hence they can do some of the work of global protocols, as has been pointed out by van Benthem (2006).

and comparisons by focusing on technical properties such as axiomatization and/or complexity results.

2.2.3.1 Axiomatizations

Axiomatizations in both DEL and ETL frameworks have been extensively studied. Both take as a starting point standard axiomatizations of epistemic logic (cf. Fagin et al. 1995). This short section reports on some of these results and highlights some of the important technical issues.

A sound and complete axiomatization of a number of different classes of ETL frames under the assumptions discussed in the previous section can be found in Halpern et al. (2004). Without assumptions about the observational powers of the agents (cf. Definition 2.7), such axiomatizations involve a straightforward *fusion* of appropriate axiomatizations of epistemic logic and temporal logic (see Kurucz 2006, Section 3.2, for an extended discussion of this). It becomes much more interesting when there are assumptions connecting knowledge and time. For example, assuming an ETL frame satisfies perfect recall validates the following axiom scheme:

$$K_i\langle e\rangle\varphi \to \langle e\rangle K_i\varphi$$

For if agent i knows (at the current moment) that φ will be true at the next moment (i.e., after event e) then, since i has perfect recall, i cannot lose this piece of information. Therefore, at the next moment (after event e) agent i will know φ.[14]

There are three parameters that govern axiomatization results in the ETL framework. The first is the expressiveness of the language (i.e., does the language include a common knowledge operator? an *arbitrary future* operator? a past operator?). The second is structural conditions on the ETL frames (i.e., is the ETL frame a single tree with a unique root? finitely branching?). Finally, the third parameter is the assumptions made about the observational powers of the agents (i.e., do the agents have perfect recall? do the agents agree on the time? do the agents satisfy the properties from Definition 2.7?). At one extreme, with at least two agents and languages containing common knowledge operators knowledge and arbitrary future operators, the validity problem over classes of ETL frames that satisfy perfect recall is Π^1_1-*complete* (see Halpern and Vardi 1989, van Benthem and Pacuit 2006, for proofs). Nonetheless, many classes of ETL frames (under different combinations of assumptions about the observational power of the agents) in a variety of modal languages (typically without a common knowledge operator or in a restricted temporal language) can be found in Halpern et al. (2004), French et al. (2004), and van der Meyden and Wong

[14]Interestingly, van der Meyden (1994) showed in languages with an "until" operator ($\varphi U\psi$ meaning there is a point in the future satisfying ψ and that φ is true at every moment until that point) adding only this axiom to an epistemic and temporal logic is *not* complete for ETL frames with perfect recall. What is needed is the more complex axiom scheme: $K_i\varphi_1 \wedge N(K_i\varphi_2 \wedge \neg K_i\varphi_3) \to L_i((K_i\varphi_1)\, U[(K_i\varphi_2)U\neg\varphi_3])$, where "$N$" is the *next-time operator* – after any event e (cf. Halpern et al. 2004).

(2003). Despite many different axiomatization and non-axiomatization results, it is fair to say that no general picture has yet emerged (although see the discussion by van Benthem and Pacuit (2006) and Kurucz (2006) for some first steps in this direction).

In contrast, the so-called *reduction axioms* have proven an invaluable method for providing sound and complete axiomatizations in the DEL framework. They were first used by Plaza to prove completeness for *public announcement logic* (see Example 2.1). This is the logic where the only event models are those where there is one primitive event, and the uncertainty relation for all agents is the universal relation. A public announcement can then be referred to simply by its precondition, resulting in formulas of the form $\langle\varphi\rangle\psi$. The following are reduction axioms for PAL:

$$\langle\varphi\rangle p \leftrightarrow (\varphi \wedge p)$$
$$\langle\varphi\rangle\neg\psi \leftrightarrow (\varphi \wedge \neg\langle\varphi\rangle\psi)$$
$$\langle\varphi\rangle(\psi \wedge \chi) \leftrightarrow (\langle\varphi\rangle\psi \wedge \langle\varphi\rangle\chi)$$
$$\langle\varphi\rangle K_i\psi \leftrightarrow (\varphi \wedge K_i\langle\varphi\rangle\psi)$$

These are reduction axioms in the sense that going from left to right either the number of announcement operators is reduced or the complexity of the formulas within the scope of announcement operators is reduced. In the first axiom we see that an announcement has been eliminated. In the second axiom we see that the announcement operator and the negation have switched place. In the third we see an announcement of a conjunction on the left and a conjunction of announcements on the right. In the fourth axiom we also see that the announcement and the epistemic operator have switched place. The reduction axioms for event models in general are a straightforward generalization of the axioms above.

The reduction axioms for PAL provide an insightful syntactic analysis of announcements which complements the semantic analysis. In a sense, the reduction axioms describe the effect of an announcement in terms of what is true before the announcement. By relating pre- and postconditions for each logical operator, the reduction axioms completely characterize the announcement operator.

In the completeness proof for PAL the reduction axioms play an essential role. Given a formula containing an announcement operator, one can completely eliminate the announcement by repeatedly applying the reduction axioms. In this way one produces a formula of epistemic logic. By adding the appropriate reduction axioms to a complete axiomatization for epistemic logic, it is straightforward to show the resulting proof system is complete in the following manner. Suppose a formula φ is a semantic tautology. By applying the reduction axioms one obtains a *provably* equivalent formula φ'. This is a semantic tautology in the language of epistemic logic. By completeness of the proof system for epistemic logic, there must be a proof of φ', and since φ and φ' are provably equivalent one can construct a proof of φ. This technique for proving completeness is considered so elegant that many have adopted it (Plaza 2007, Gerbrandy 1999a, Baltag et al. 1998b, Herzig et al. 2000b, Kooi 2003a, Renardel de Lavalette 2004, Kooi and van Benthem 2004, van Eijck

2004b, a, Ruan 2004, van Benthem 2007, van Benthem and Liu 2007, Kooi 2007, van Benthem and Ikegami 2008).

Reduction axioms are not only useful in providing a syntactic analysis of updates and for proving completeness, they also show that the language containing the update is just as expressive as the language without it. So the results mentioned above are also expressivity results showing the language of PAL is no more expressive than the language of epistemic logic. Yet Lutz (2006) has shown that in the case of PAL at least, the language is more succinct than the language of epistemic logic (there is a formula scheme in PAL such that every equivalent formula scheme in epistemic logic is exponentially longer). This suggests that PAL describes announcements at an appropriate level of abstraction.

When a logical language becomes strictly more expressive by adding dynamic operators, reduction axioms are not available. Adding public announcement operators to epistemic logic with common knowledge is such a case. It was shown by Baltag et al. (1998b) that the language of epistemic logic with common knowledge and public announcements is more expressive than epistemic logic with common knowledge. Therefore a reduction axiom for formulas of the form $[\varphi]C_{\mathcal{B}}\psi$ does not exist. Baltag, Moss and Solecki also showed that adding private announcements to epistemic logic with common knowledge adds expressivity. Renne (2007b) showed that the expressivity of these two logics is incomparable. In cases where adding dynamic operators strictly increases the expressivity of the language a completeness proof using reduction axioms is not available and a complete proof system is harder to obtain.

2.3 Extensions, Connections and Applications

The previous section introduced two different logical frameworks that describe how an agent's information evolves through observation when interacting with other agents. The results discussed in Section 2.2.3 provide a concrete answer to question 1 from the Introduction (how should we compare two logical frameworks addressing the same aspect of rational agency). But what about the other two questions? Here, especially regarding question 3 (how the logics of rational interaction contribute to broader discussions on rational agency), we cannot point to any concrete results as answers to these questions. Rather, this section turns to several extensions of these logics of rational interaction, as well as connections with other fields, and some applications.

To keep this survey at a manageable length we will not be able to provide anything approaching a complete survey of all extensions and applications of ETL and DEL. See Fagin et al. (1995) for a textbook presentation of a number of extensions and applications of ETL, and van Ditmarsch et al. (2007), van Benthem (2008a) for applications and extensions of DEL. The topics discussed below where chosen because they are representative of current research directions and issues addressed in this volume. We start by briefly discussing a few other logics frameworks that can broadly be categorized as "logics of rational interaction".

2.3.1 Propositional Dynamic Logic

The language of DEL is set up similarly to the languages of *propositional dynamic logic* (PDL). Two distinct classes of formulas and programs (i.e. updates) are defined. Therefore it would be natural to include the Kleene star for iteration as well, as it allows one to express things such as no matter how matter how many times it is announced that φ, it will not become common knowledge that ψ as $[\varphi^*]\neg C_{\mathcal{B}}\psi$. Miller and Moss (2005) showed that the satisfiability problem for PAL with iteration is undecidable.[15] Hence, DEL with iteration has the same problem.

Still, DEL can be embedded in PDL, i.e. for each formula in DEL, there is a formula in PDL which is equivalent to it van Eijck (2004b). In van Eijck's approach, PDL formulas are read epistemically, for instance $[i]\varphi$ is read as agent i knows that φ. Another link between DEL and PDL is developed by Aucher and Herzig (2007) where $[e]\varphi$ is read as after event e it is the case that φ, i.e. e is taken to be an event from an event model. Separate modalities for agents are added to PDL (just as was done by van Benthem 2001). With the addition of a converse operator, this logic can express properties of event models.

2.3.2 Belief Revision

The ground breaking paper by Alchourrón et al. (1985) put information change prominently on the agenda of philosophical logic. Their approach, abbreviated as AGM, focuses on what to do when receiving (and accepting) information not in accordance with the agent's theory of the current state of affairs. This led to a stream of publications in an area nowadays called *belief revision*.

Indeed, there are by now many different approaches to modeling how agents (should) change their beliefs in the presence of new (and trusted) information. Shoham and Leyton-Brown (2009, Section 14.2) discuss many of these different approaches (including *nonmonotonic* consequence relations, default logics and probabilistic frameworks). Rather than discussing this expanding literature, we point to contributions that are most relevant to the logics we discussed in Section 2.2. These can be roughly divided into two categories. The first are ETL-style logics that describe how an agent's beliefs change through time (see, for example, Friedman and Halpern (1997, 1999), Bonanno (2007) and references therein). The second category can be described as dynamic modal logics of belief revision. Building on a suggestion of van Benthem (1989), de Rijke (1994) took some first steps to develop a dynamic modal logic of belief revision. This led to the development of

---

[15] In fact, they show that the validity problem for public announcement logic with iteration is *highly undecidable* (Π^1_1-complete). In light of the translation between the DEL framework and the ETL framework discussed in Section 2.2.3, this is related to classic results of Halpern and Vardi (1989) showing that the validity problem for ETL frames that satisfy perfect recall and no miracles in certain modal languages is Π^1_1-complete. See van Benthem et al. (2009, Section 6.1) for an extended discussion of this relationship.

dynamic doxastic logic (see Segerberg's contribution to this volume and Lindström and Segerberg (2007) for an overview of this approach). Recent work has provided a multi-agent perspective with a number of DEL-style logics of belief revision[16] (see, for example, Aucher 2003, van Ditmarsch 2005, Cantwell 2006, van Benthem 2007, Baltag and Smets 2008a). Building on the results discussed in Section 2.2.3, van Benthem and Dégremont (2009) formally compare the ETL-style and DEL-style logics of belief change.

2.3.3 Probability Logic

Probability theory provides a quantitative analysis of information. Rather than a proposition being known or unknown, its degree of certainty is represented by a number. For instance, the chance that the queen of hearts is drawn from a shuffled ordinary deck of cards is 1/52. The are many connections between probability and logic, including epistemic logic (see Halpern (2003) for a textbook presentation). The connection with logics discussed in this survey becomes apparent by noting that a Bayesian update resembles the public announcement of Example 2.2. It is therefore is quite natural to combine probability logics and dynamic epistemic logics (cf. van Benthem 2003, Kooi 2003b, Aucher 2007, van Benthem et al. 2009b, Sack 2008a).

2.3.4 Situation Calculus

Reasoning about actions is an important area of research in artificial intelligence and the *situation calculus* of McCarthy and Hayes (1969) is one of the most influential approaches (see Reiter (2001) for a textbook on the subject). The situation calculus is a fragment of second order logic that can describe many situations and how situations change due to actions. Typical examples involve robots moving blocks. Comparisons with ETL-style logics is relatively straightforward since the situation calculus can express most epistemic and temporal modalities. The comparison with DEL is more subtle. The link between the two formalisms was established by van Ditmarsch et al. (2007), who use situation calculus and DEL to approach the *frame problem*.[17]

The comparisons between the different logical frameworks discussed above and in Section 2.2.3 suggest a number of *extensions* to the basic DEL and ETL frameworks. For example, the results of van Benthem et al. (2009a) suggest adding temporal operators, such as a past-time operator, to DEL (cf. Sack 2008b, Hoshi and Yap 2009). Again we do not have the space to cover all extensions to the logics of rational interaction (see van Benthem 2008a, for an extended discussion), so we focus on a few key research avenues.

---

[16] Closely related are the dynamic logics of preferences discussed by van Benthem and Liu (2007).

[17] Both Reiter (2001) and McCarthy and Hayes (1969) discuss this classic problem of AI.

2.3.5 Factual Change

Although DEL is mainly used to model information change due to communication, comparisons with ETL and the situation calculus suggest that it may be convenient to model situations where the bare facts of the world do change. This was already foreshadowed in the CWI technical report version of the paper by Baltag et al. (1998b). The first DEL with factual change was proposed by Bleeker and van Eijck (2000), where multiple propositional letters can change simultaneously. Baltag (2002) considers DEL with "flip" actions, which changes the extension of a propositional letter p to its complement. Renardel de Lavalette (2004) uses operators $p := \varphi$ which changes the extension of p to the extension of φ and applies the same idea to agents where $i := \pi$ changes the accessibility relation of i to that of π. van Ditmarsch et al. (2005) provide a logic with such operators and public announcements. Van Eijck (2004a) showed that DEL with simultaneous factual changes in event models can be reduced to PDL. Factual change has been further studied in van Benthem et al. (2006), Herzig and de Lima (2006), Kooi (2007), and van Ditmarsch and Kooi (2008).

2.3.6 Logics of Rational Agency

The logics discussed in this survey focus primarily on information change. But logics have also been developed to reason more broadly about rational agency. Indeed, there are now many different "logics of rational agency" (see van der Hoek and Wooldridge 2003, Meyer and Veltman 2007, Horty 2001, for a discussion and pointers to the relevant literature) that not only focus on describing various informational and/or motivational attitudes but also explicating their relationships. An overarching theme in many of these papers is that during a social interaction, an agent's "knowledge" and "beliefs" both influence *and* shape the *social* events. The following example (taken from Pacuit et al. 2006) illustrates this point.

Example 1: Uma is a physician whose neighbour is ill. Uma does not know and has not been informed. Uma has no obligation (as yet) to treat the neighbour.

Example 2: Uma is a physician whose neighbour Sam is ill. The neighbour's daughter Ann comes to Uma's house and tells her. Now Uma does have an obligation to treat Sam, or perhaps call in an ambulance or a specialist.

Example 3: Mary is a patient in St. Gibson's hospital. Mary is having a heart attack. The caveat which applied in case 1) does not apply here. The hospital cannot plead ignorance, but rather it has an obligation to *be aware* of Mary's condition at all times and to provide emergency treatment as appropriate.

In all the cases we mentioned above, the issue of an obligation arises. This obligation is circumstantial in the sense that in other situations, the obligation might not apply.

If Sam is ill, Uma needs to know that he is ill, and the nature of the illness, but not where Sam went to school. Thus an agent's obligations are often dependent on what the agent knows, and indeed one cannot reasonably be expected to respond to a problem if one is not aware of its existence. This, in turn, creates a secondary obligation on Ann to inform Uma that her father is ill.

Based on the logical framework discussed in Section 2.2.1 and Horty (2001), Pacuit et al. (2006) develop a logical framework that formalizes the reasoning of Uma and Ann in the above examples. It is argued that this reasoning is shaped by the assumption that Uma and Ann's preferences are aligned (i.e., both want Sam to get better). For example, Ann will not be under any obligation to tell Uma that her father is ill, if Ann justifiably believes that Uma would not treat her father even if she knew of his illness. Thus, in order for Ann to *know* that she has an obligation to tell Uma about her father's illness, Ann must *know* that "Uma will, in fact, treat her father (in a reasonable amount of time) upon learning of his illness". More formally, in all the histories that Ann currently considers possible, the event where her father is treated for his illness is always preceded by the event where she tells Uma about his illness. That is, the histories where Uma learns of Sam's illness but does not treat him are not part of the protocol. Similar reasoning is needed for Uma to derive that she has an obligation to treat Sam. Obviously, if Uma has a good reason to believe that Ann always lies about her father being ill, then she is under no obligation to treat Sam. See Pacuit et al. (2006) for a formal treatment of these examples.

2.3.7 Inference Logic

Besides information about the world and the discourse information, there is a third kind of information that plays a role in interaction, namely information derived from (logical) *inference*. What conclusions is one allowed to draw from a set of premises, and how is the process of inference carried out? A number of logical frameworks have been developed that explicitly reason about such inferential steps (Duc 1997, 2001, Jago 2006). Frameworks that extend ETL style logics include (Ågotnes and Alechina 2007, Alechina et al. 2009). Combinations of "inference logics" and DEL have been put forward by van Benthem (2008b) and Velazquez-Quesada (2009).

2.3.8 Justification Logic

Justification logic is an epistemic logic where explicit reasons for knowledge are represented. A formula $t\colon \varphi$ is intended to mean "the agent knows φ for reason t". It was introduced by Artemov and Nogina (2005) based on their work on explicit provability logic. Renne (2010) added public announcements to this logic and proved a number of expressivity results. See also Renne's contribution to this volume for an extended discussion.

Finally, we conclude this section with a brief discussion of a number of key applications of the logics of rational interaction discussed in Section 2.2.

2.3.9 Puzzles and Paradoxes

The development of DEL in particular was fueled by a number of puzzles and paradoxes. These did not only function as an inspiration, but also as a touchstone for DEL. Both Plaza and Gerbrandy analyzed the Muddy Children puzzle using PAL. Plaza also treats the Sum and Product puzzle. Another example of a puzzle where a specification of the solution in DEL offers a method of evaluating solutions suggested in the literature is the Russian cards problems van Ditmarsch (2003).

Although some of these puzzles are also found in recreational mathematics, some have serious philosophical repercussions. The hangman paradox, or unexpected examination paradox was first analyzed using PAL by Gerbrandy (1999a), Gerbrandy (2007). A judge sentences a prisoner to death and says that he will be hanged next week but that the day of the execution will come as a surprise. The prisoner then reasons as follows. If the execution were on Friday, then I would know on Thursday evening that this is so, and the day of the execution would not be a surprise. Therefore the execution cannot take place on Friday. So, Thursday is the last possible day for the execution. By the same reasoning as before the prisoner concludes that the execution cannot take place on Thursday either, and so he continues eliminating all days of the week. The prisoner cheerfully infers that the execution cannot take place at all. To his great surprise he is executed on Tuesday.

The central point of Gerbrandy's analysis is that the announcement of the judge may be an *unsuccessful update*. That is, a formula that becomes false by its announcement. This phenomenon also occurs in Update Semantics, when an update system does not satisfy the condition of *idempotence* (cf. Veltman 1996). A literary example of this phenomenon is found in the fairy tale *Rumpelstilzchen* Grimm and Grimm (1857), where a goblin who sings the following song (our translation below[18]):

> Heute back ich, morgen brau ich,
> übermorgen hol ich der Königin ihr Kind;
> ach, wie gut dass niemand weiß,
> dass ich Rumpelstilzchen heiß!

> Today I bake, tomorrow I brew,
> The day after tomorrow I will fetch the queen's child;
> Oh, it's good that nobody knows,
> that I'm called Rumpelstilzchen.

In the fairy tale his song is overheard and therefore it is no longer true that nobody knows the goblin's name. Thus uttering a true statement, can make that statement itself false! In PAL such statements are called unsuccessful updates. A successful update is a formula φ such that $[\varphi]\varphi$ is a tautology. An update is unsuccessful if it is not successful. The announcement of the judge is an unsuccessful update, i.e. the judge may ruin the surprise by saying that the day of the execution will come as a

[18] Regrettably the English translations we consulted do not contain this phenomenon.

surprise. Van Ditmarsch and Kooi (2006) discuss this phenomenon in a number of contexts.

The formula $p \wedge \neg K_a p$ is a typical example of an unsuccessful update, which play a role in the *Fitch paradox* or *knowability paradox*. If one accepts that all truths are knowable, then if p is true but unknown it should be knowable that p is true but unknown. This leads to a contradiction. Using DEL the paradox was analyzed by van Benthem (2004a). This led to the development of arbitrary public announcement logic, where formulas $\Diamond\varphi$ occur, which are read as there is some announcement such that afterwards φ is true (Balbiani et al. 2007).

2.3.10 Game Theory

Any (formal) model that addresses issues of (practical) rationality needs to account for the possibility of conflicting *goals* of the different agents. Starting from the work of Ramsey (1926), de Finetti (1937), von Neumann and Morgenstern (1944) and Savage (1954), the mathematical analyses provided by decision and game theorists have generated many important insights about such *strategic* interactive situations. Indeed, in their classic text, von Neumann and Morgenstern explain that they want "to find the mathematically complete principles which define "rational behavior" for the participants" (von Neumann and Morgenstern 1944, p. 31). Nonetheless, many foundational questions remain open. These questions are not mathematical in nature but involve the meaning of the fundamental concepts employed in the mathematical analyses.

Building on seminal work by John Harsanyi (1967) on incomplete information games[19] and Robert Aumann (1999, 1976) introducing common knowledge to game theory, many researchers have forcefully argued that the basic mathematical model of a "game-theoretic situation" should be extended with an explicit representation of the players' relevant informational attitudes[20] (following Harsanyi (1967), this parameter is called a player's *type*. See, for example, Brandenburger (2007), Bonanno and Battigalli (1999), for an extended discussion). A central concern in this literature is the players' attitude towards statements about the *rationality of the other players* and whether such statements can be revised during the course of a (dynamic) game[21]. Although there is considerable disagreement over the precise

---

[19]That is, situations in which the *structure* of the game is not common knowledge. For example, games where players may be uncertain about their own available actions and preferences and/or the available actions and preferences of the other players. This should be contrasted with *imperfect information* games where players may receive different information during the course of the game. See Myerson (2004) for a recent discussion of Harsanyi's classic paper.

[20]Typically this means the players *first-order* beliefs about the available choices of the other players, the players beliefs about the other players beliefs about these first-order beliefs, and so on *ad infinitum*.

[21]Or, in the case of a one-shot strategic game, whether such statements can be revised during the players' initial period of deliberation.

formulation, it is generally assumed that such statements about the rationality of the other players are more *entrenched* than, for example, higher-order change of beliefs about the *types* of the other players.

One lesson to take away from this discussion is that game-theoretic analyses of multiagent strategic situations should be embedded in a larger framework that describes how the players' (hard and soft) information evolves over time. The logical systems discussed in this chapter focus on precisely this issue (cf. Section 2.2). Thus, these frameworks complement the game theoretic models described above by focusing on how a player's type may evolve over time and how a player may change types during the course of a game. Much more can be said on this general topic, but we will not go into this here (see van der Hoek and Pauly 2006, for discussion along these lines and pointers to the relevant literature).

2.3.11 Security

One of the recent application areas of logics of rational interaction is security, especially authentication and privacy. Both DEL and ETL frameworks have been used to verify that security protocols meet their specification Bleeker and van Eijck (2000), Hommersom et al. (2004), Dechesne and Wang (2007), Halpern and Pucella (2003), Ramanujam and Suresh (2005) and van der Meyden and Wilke (2007). A topic of special interest is so-called *zero-knowledge protocols*. These are security protocols where security does not depend on the bounded computational resources of the participants in the protocol. An example is the solution of the *Russian cards problem* (see van Ditmarsch 2003). This problem has been modeled in DEL and formal checked by the model checker DEMO, developed by van Eijck (2005), and van Ditmarsch et al. (2006). Other typical problems found in the literature on security have been analyzed, including the dining cryptographers problem (van Eijck and Orzan 2005, van der Meyden 2007).

2.4 Conclusion: Towards a Unified Account of Rational Interaction

There is a multitude of logics that aim to model different aspects of rational interaction. Often, for one and the same aspect there are numerous approaches. Do these alternative approaches represent radically different conceptual frameworks, or are the same concepts represented in different guises? In our case, how should we compare two different frameworks that model how an agent's information changes through interaction with other agents and the environment. This was exactly what we described in Section 2.2 for ETL and DEL. So these observations point to one "coherent" account of rational interaction. Yet this is not the whole story of rational interaction.

Agents are faced with many diverse tasks as they interact with the environment and one another. At certain moments, agents must *react* to the (perhaps surprising) events they observe while at other moments they must be *proactive* and choose to perform a specific action. One central underlying assumption is that rational agents obtain what they want via the implementation of (successful) *plans* (cf. Bratman 1987). And this implementation often requires, among other things, representation of various informational attitudes of the other agents involved in the social interaction. As illustrated by the discussion of Example 2.1 in Section 2.2, in social situations there are many (sometimes competing) *sources* of information for these attitudes: for example, the type of "communicatory event" (public announcement, private announcement, etc), the disposition of the other participants (liars, truth tellers, etc.) and other implicit assumptions about the protocol information (reducing the number of possible histories). This naturally leads to different notions of "knowledge" and "belief" that drive social interaction.

The conclusion is that a comprehensive account of rational interaction cannot be isolated from other aspects of rational agency and social interaction. This chapter presented some recent work which points to such a comprehensive account. Once the technical results of Section 2.2.3 are in place, the two major current views of how information evolves through social interaction can be seen as *complementary*. This opens the door to merging these two perspectives (cf. van Benthem et al. 2009, Section 4) which will, in turn, lead to a more diversified account of the reasoning and dynamic processes that govern social interaction.

References

Ågotnes T, Alechina N (2007) The dynamics of syntactic knowledge. Journal of Logic and Computation 17(1):83–116

Alchourrón C, Gärdenfors P, Makinson D (1985) On the logic of theory change: Partial meet contraction and revision functions. Journal of Symbolic Logic 50:510–530

Alechina N, Logan B, Nguyen HN, Rakibl A (2009) Verifying time, memory and communication bounds in systems of reasoning agents. Synthese (Knowledge, Rationality & Action) 169(2):385–403

Artemov S, Nogina E (2005) Introducing justification into epistemic logic. Journal of Logic and Computation 15(6):1059–1073

Aucher G (2003) A combined system for update logic and belief revision. Master's thesis, ILLC, University of Amsterdam, Amsterdam

Aucher G (2007) Interpreting an action from what we perceive and what we expect. Journal of Applied Non-Classical Logics 17(1):9–38

Aucher G, Herzig A (2007a) From DEL to EDL: exploring the power of converse events. In: Mellouli K (ed) Proceedings of the 9th European Conference on Symbolic and Quantitative Approaches to Reasoning with Uncertainty (ECSQARU 2007), Springer Verlag Berlin, LNCS, vol 4724, pp 199–209, ftp://ftp.irit.fr/IRIT/LILAC/Ecsqaru2007.pdf

Aumann R (1976) Agreeing to disagree. Annals of Statistics 4:1236–1239

Aumann R (1999) Interactive epistemology I: Knowledge. International Journal of Game Theory 28:263–300

Balbiani P, Baltag A, van Ditmarsch H, Herzig A, Hoshi T, de Lima T (2007) What can we achieve by arbitrary announcements? – A dynamic take on Fitch's knowability. In: Samet D (ed) Proceedings of TARK 2007, pp 42–51

Baltag A (2002) A logic for suspicious players: epistemic actions and belief-updates in games. Bulletin of Economic Research 54(1):1–45

Baltag A, Moss LS (2004) Logics for epistemic programs. Synthese 139(2):165–224

Baltag A, Smets S (2008a) A qualitative theory of dynamic interactive belief revision. In: Bonanno G, van der Hoek W, Wooldridge M (eds) Proceedings of LOFT 2007, Amsterdam University Press, Amsterdam, Texts in Logic and Games, vol 3, pp 9–58

Baltag A, Moss L, Solecki S (1998a) The logic of public announcements, common knowledge and private suspicions. In: Gilboa I (ed) Proceedings TARK 1998, Morgan Kaufmann Publishers Inc., pp 43–56

Baltag A, Moss L, Solecki S (1998b) The logic of public announcements, common knowledge, and private suspicions. In: Gilboa I (ed) Proceedings of the 7th conference on Theoretical aspects of rationality and knowledge, Morgan Kaufmann Publishers Inc., p 56

van Benthem J (1989) Semantic parallels in natural language and computation. In: Ebbinghaus HD, Fernandez-Prida J, Garrido M, Lascar D, Artalejo MR (eds) Logic Colloquium '87, North-Holland, Amsterdam

van Benthem J (2001) Games in dynamic-epistemic logic. Bulletin of Economic Research 53(4):219–248

van Benthem J (2003) Conditional probability meets update logic. Journal of Logic, Language and Information 12(4):409–421

van Benthem J (2004) What one may come to know. Analysis 64(2):95–105

van Benthem J (2005) Rational animals: What is 'KRA'?, http://staff.science.uva.nl/johan/publications.html, invited lecture Malaga ESSLLI Summer School 2006

van Benthem J (2006) One is a lonely number: on the logic of communication. In: Chatzidakis Z, Koepke P, Pohlers W (eds) Logic Colloquium'02, Lecture Notes in Logic ASL & A.K. Peters, Wellesley, pp 96–129, Tech. Rep. PP-2003-07, ILLC, Amsterdam (2002)

van Benthem J (2007) Dynamic logic for belief revision. Journal of Applied Non-Classical Logics 17(2):129–155

van Benthem J (2008a) Logical dynamics of information and interaction. Book manuscript

van Benthem J (2008b) Merging observation and access in dynamic logic. Manuscript

van Benthem J (2010) Logical dynamics of information and interaction. ILLC Manuscript, to appear with Cambridge University Press

van Benthem J, Dégremont C (2009) Building bridges between dynamic and temporal doxastic logics. In: Proceedings of LOFT 8

van Benthem J, Ikegami D (2008) Modal fixed-point logic and changing models. In: Avron A, Dershowitz N, Rabinovich A (eds) Pillars of computer science, essays dedicated to Boris (Boaz) Trakhtenbrot on the occasion of his 85th birthday, LNCS, vol 4800, Springer-Verlag, Berlin, pp 146–165

van Benthem J, Liu F (2007) Dynamic logic of preference upgrade. Journal of Applied Non-Classical Logics 17(2):157–182

van Benthem J, Pacuit E (2006) The tree of knowledge in action: Towards a common perspective. In: Governatori G, Hodkinson I, Venema Y (eds) Advances in modal logic volume 6, King's College Press, London pp 87–106

van Benthem J, van Eijck J, Kooi B (2006) Logics of communication and change. Information and Computation 204(11):1620–1662

van Benthem J, Gerbrandy J, Hoshi T, Pacuit E (2009a) Merging frameworks of interaction. Journal of Philosophical Logic 38:491–526

van Benthem J, Gerbrandy J, Kooi B (2009b) Dynamic update with probabilities. Studia Logica 93(1):67–96

Blackburn P, de Rijke M, Venema Y (2002) Modal logic. Cambridge University Press, Cambridge

Bleeker A, van Eijck J (2000) The epistemics of encryption. Tech. Rep. INS-R0019, CWI, Amsterdam, available from http://db.cwi.nl/rapporten/

Bonanno G (2004) Memory and perfect recall in extensive games. Games and Economic Behaviour 47:237–256

Bonanno G (2007) Axiomatic characterization of AGM belief revision in a temporal logic. Artificial Intelligence 171:144–160

Bonanno G, Battigalli P (1999) Recent results on belief, knowledge and the epistemic foundations of game theory. Research in Economics 53(2):149–225
Brandenburger A (2007) The power of paradox: some recent developments in interactive epistemology. International Journal of Game Theory 35:465–492
Bratman M (1987) Intention, plans and practical reason. Harvard University Press, London
Bratman M (2007) Structures of agency. Oxford University Press, Oxford
Cantwell J (2006) A formal model of multi-agent belief-interaction. Journal of Logic, Language and Information 15(4):303–329
Cartwright N (2007) Hunting causes and using them: Approaches in philosophy and economics. Cambridge University Press, Cambridge
Dechesne F, Wang Y (2007) Dynamic epistemic verification of security protocols: framework and case study. In: van Benthem J, Ju S, Veltman F (eds) A meeting of the minds, Proceedings of the Workshop on Logic, Rationality and Interaction, Beijing, 2007, College Publications, London, Texts in Computer Science, vol 8, pp 129–143
van Ditmarsch H, Kooi B (2008) Semantic results for ontic and epistemic change. In: Bonanno G, van der Hoek W, Wooldridge M (eds) Logic and the foundations of game and decision theory. Amsterdam University Press, Amsterdam, pp 87–117
van Ditmarsch H, van der Hoek W, Kooi B (2007) Dynamic epistemic logic. Synthese library, Springer, New York
van Ditmarsch HP (2003) The Russian cards problem. Studia Logica 75:31–62
van Ditmarsch HP (2005) Prolegomena to dynamic logic for belief revision. Synthese (Knowledge, Rationality & Action) 147:229–275
van Ditmarsch HP, Kooi B (2006) The secret of my success. Synthese 151(2):201–232
van Ditmarsch HP, van der Hoek W, Kooi BP (2005) Dynamic epistemic logic with assignment. In: Dignum F, Dignum V, Koenig S, Kraus S, Singh MP, Wooldridge M (eds) Proceedings of the Fourth International Joint Conference on Autonomous Agents and Multi-Agent Systems (AAMAS 05), ACM Inc, New York, vol 1, pp 141–148
van Ditmarsch HP, van der Hoek W, van der Meyden R, Ruan J (2006) Model checking russian cards. Electronic Notes in Theoretical Computer Science 149:790–795
Duc HN (1997) Reasoning about rational, but not logically omniscient, agents. Journal of Logic and Computation 7(5):633–648
Duc HN (2001) Resource-bounded reasoning about knowledge. PhD thesis, Universität Leipzig
van Eijck J (2004a) Guarded actions. Tech. Rep. SEN-E0425, CWI, Amsterdam, available from http://db.cwi.nl/rapporten/
van Eijck J (2004b) Reducing dynamic epistemic logic to PDL by program transformation. Tech. Rep. SEN-E0423, CWI, Amsterdam, available from http://db.cwi.nl/rapporten/
van Eijck J (2005) DEMO: a system for dynamic epistemic modeling. http://homepages.cwi.nl/jve/demo
van Eijck J, Orzan SM (2005) Modelling the epistemics of communication with functional programming. In: van Eekelen M (ed) Sixth Symposium on Trends in Functional Programming TFP'05, pp 44–59
Fagin R, Halpern J, Moses Y, Vardi M (1995) Reasoning about knowledge. The MIT Press, Boston
de Finetti B (1937) La prevision: Ses lois logiques, ses sources subjectives. In: Annales de l'Institut Henri Poincare 7, Paris, pp 1–68, translated into English by Henry E. Kyburg Jr., Foresight: Its logical laws, its subjective sources. In Henry E. Kyburg Jr. and Howard E. Smokler (1964, eds), Studies in subjective probability, 53–118, Wiley, New York
Fitting M (2009) Reasoning with justifications. In: Makinson D, Malinowski J, Wansing H (eds) Towards mathematical philosophy, trends in logic, vol 28, Springer, Dordrecht, Netherlands, pp 107–123
French T, van der Meyden R, Reynolds M (2004) Axioms for logics of knowledge and past time: Synchrony and unique initial states. In: Advances in Modal Logic, pp 53–72
Friedman N, Halpern J (1997) Modeling belief in dynamic systems. Part I: Foundations. Artificial Intelligence 95(2):257–316

Friedman N, Halpern J (1999) Modeling belief in dynamic systems. Part II: Revision and update. Journal of AI Research 10:117–167

Gerbrandy J (1999a) Bisimulations on planet kripke. PhD thesis, Institute for Logic, Language and Computation (DS-1999-01)

Gerbrandy J, Groeneveld W (1997) Reasoning about information change. Journal of Logic, Language and Information 6(2):147–169

Gerbrandy JD (2007) The surprise examination in dynamic epistemic logic. Synthese 155(1): 21–33

Grimm J, Grimm W (1857) Kinder- und Hausmärchen. Göttingen

Halpern J (1999) Set-theoretic completeness for epistemic and conditional logic. Annals of Mathematics and Artificial Intelligence 26:1–27

Halpern J, Pucella R (2003) Modeling adversaries in a logic for security protocol analysis. In: Formal Aspects of Security

Halpern J, Vardi M (1989) The complexity of reasoning about knowledge and time. Journal of Computer and System Sciences 38:195–237

Halpern J, van der Meyden R, Vardi M (2004) Complete axiomatizations for reasoning about knowledge and time. SIAM Journal of Computing 33(2):674–703

Halpern J (2003) Reasoning about Uncertainty. The MIT Press, Cambridge, MA

Harsanyi J (1967) Games with incomplete informations played by 'bayesian' players. Management Science 14:159–182, 320–334, 486–502

Herzig A, de Lima T (2006) Epistemic actions and ontic actions: a unified logical framework. In: Simão Sichman J, Coelho H, Oliveira Rezende S (eds) Advances in Artificial Intelligence – IBERAMIA-SBIA 2006, Springer-Verlag, Berlin, Lecture Notes in Computer Science, vol 4140, pp 409–418

Herzig A, Lang J, Polacsek T (2000b) A modal logic for epistemic tests. In: Horn W (ed) Proceedings of ECAI 2000, pp 553–557

Hodkinson I, Reynolds M (2006) Temporal logic. In: Blackburn P, van Benthem J, Wolter F (eds) Handbook of modal logic, studies in logic, vol 3, Elsevier, Amsterdam, pp 655–720

van der Hoek W, Pauly M (2006) Modal logic for games and information. In: Blackburn P, van Benthem J, Wolter F (eds) Handbook of modal logic, studies in logic, vol 3, Elsevier, Amsterdam pp 1077–1148

van der Hoek W, Wooldridge M (2003) Towards a logic of rational agency. Logic Journal of the IGPL 11(2):135–160

Hommersom A, Meyer JJC, de Vink E (2004) Update semantics of security protocols. Synthese (Knowledge, Rationality & Action) 142:229–267

Horty J (2001a) Agency and deontic logic. Oxford University Press, Oxford

Hoshi T, Yap A (2009) Dynamic epistemic logic with branching temporal structures. Synthese (Knowledge, Rationality & Action)

Hyman J, Steward H (eds) (2004) Agency and action. Cambridge University Press, Cambridge

Jago M (2006) Logics for resource-bounded agents. PhD thesis, University of Nottingham

Kooi B (2003a) Knowledge, chance, and change. PhD thesis, University of Groningen, iLLC Dissertation Series DS-2003-01

Kooi B (2003b) Probabilistic dynamic epistemic logic. Journal of Logic, Language and Information 12(4):381–408

Kooi B (2007) Expressivity and completeness for public update logics via reduction axioms. Journal of Applied Non-Classical Logics 17(2):231–253

Kooi B, van Benthem J (2004) Reduction axioms for epistemic actions. In: Schmidt R, Pratt-Hartmann I, Reynolds M, Wansing H (eds) AiML-2004: Advances in Modal Logic, Department of Computer Science, University of Manchester, Technical report series, UMCS-04-9-1, pp 197–211

Kurucz A (2006) Combining modal logics. In: Blackburn P, van Benthem J, Wolter F (eds) Handbook of modal logic, studies in logic, vol 3, Elsevier, Amsterdam, pp 869–924

Lindström S, Segerberg K (2007) Modal logic and philosophy. In: Blackburn P, van Benthem J, Wolter F (eds) Handbook of modal logic, studies in logic and practical reasoning, vol 3, Elsevier, Amsterdam, pp 1149–1214

Lomuscio A, Ryan M (1997) On the relation between interpreted systems and kripke models. In: Proceedings of the AI97 Workshop on Theoretical and Practical Foundation of Intelligent Agents and Agent-Oriented Systems, vol LNCS p 1441

Lomuscio AR, van der Meyden R, Ryan M (2000) Knowledge in multiagent systems: initial configurations and broadcasts. ACM Transactions on Computational Logic 1(2):247–284

Luce RD, Raiffa H (1957) Games and decisions. John Wiley and Sons, New York

Lutz C (2006) Complexity and succinctness of public announcement logic. In: Proceedings AAMAS-06: Fifth International Joint Conference on Autonomous Agents and Multiagent Systems, Hakodate, Japan

McCarthy J, Hayes PJ (1969) Some philosophical problems from the standpoint of artificial intelligence. Machine Intelligence 4:463–502

van der Meyden R (1994) Axioms for knowledge and time in distributed systems with perfect recall. In: Proceedings of the IEEE Symposium on Logic in Computer Science, pp 448–457

van der Meyden R (1996) Finite state implementations of knowledge-based programs. In: Chandru V, Vinay V (eds) Proceedings of the Conference on the Foundations of Software Technology and Theoretical Computer Science, Springer-Verlag, Berlin, Lecture Notes in Computer Science, vol 1180, pp 31–50

van der Meyden R (2007) Two applications of epistemic logic in computer security. In: Logic at the Crossroads, pp 207–222

van der Meyden R, Wilke T (2007) Preservation of epistemic properties in security protocol implementations. In: Proceedings of Theoretical Aspects of Rationality and Knowledge

van der Meyden R, Wong K (2003) Complete axiomatizations for reasoning about knowledge and branching time. Studia Logica 75(1):93–123

Meyer JJ, Veltman F (2007) Intelligent agents and common sense reasoning. In: Blackburn P, van Benthem J, Wolter F (eds) Handbook of modal logic, Elsevier, Amsterdam

Miller J, Moss L (April 2005a) The undecidability of iterated modal relativization. Studia Logica 79:373–407

Myerson R (2004) Harsanyi's games with incomplete information. Management Science 50(12):1818–1824

von Neumann J, Morgenstern O (1944) A theory of games and economic behaviour. Princeton University Press, Princeton, NJ

Pacuit E (2007a) Some comments on history based structures. Journal of Applied Logic 5(4):613–624

Pacuit E, Parikh R (2006) Social interaction, knowledge and social software. In: Interactive Computation: The New Paradigm, Springer, pp 441–462

Pacuit E, Parikh R, Cogan E (2006) The logic of knowledge based obligation. Knowledge, Rationality and Action: A Subjournal of Synthese 149(2):311–341

Parikh R (2003) Levels of knowledge, games, and group action. Research in Economics 57:267–281

Parikh R, Ramanujam R (1985) Distributed processes and the logic of knowledge. In: Logic of programs, Springer, Lecture Notes in Computer Science, vol 193, pp 256–268

Plaza JA (2007) Logics of public communications. Synthese 158(2):165–179 (this paper was originally published as Plaza, J. A. (1989). Logics of public communications. In M. L. Emrich, M. S. Pfeifer, M. Hadzikadic, and Z.W. Ras (Eds.), *Proceedings of the fourth international symposium on methodologies for intelligent systems: Poster session program* (pp. 201–216). Publisher: Oak Ridge National Laboratory, ORNL/DSRD-24).

Ramanujam R, Suresh S (2005) Deciding knowledge properties of security protocols. In: Proceedings of Theoretical Aspects of Rationality and Knowledge, pp 219–235

Ramsey F (1926) Truth and probability. In: Braithwaite R (ed) The foundations of mathematics and other logical essays, Routledge, London
Reiter R (2001) Knowledge in action: Logical foundations for specifying and implementing dynamical systems. The MIT Press, Cambridge
Renardel de Lavalette GR (2004) Changing modalities. Journal of Logic and Computation 14(2):253–278
Renne B (2010) Public Communication in Justification Logic. Journal of Logic and Computation. doi:10.1093/logcom/exp/026
Renne B (2007b) The relative expressivity of public and private communication in BMS logic. In: van Benthem J, Ju S, Veltman F (eds) A Meeting of the Minds: Proceedings of the Workshop on Logic, Rationality, and Interaction, Beijing, 2007, College Publications, London, Texts in Computer Science, vol 8, pp 213–229
de Rijke M (1994) Meeting some neighbours. In: van Eijck J, Visser A (eds) Logic and information flow, The MIT Press, Cambridge, pp 170–195
Ruan J (2004) Exploring the update universe. Master's thesis, ILLC, Amsterdam
Sack J (2008a) Extending probabilistic dynamic epistemic logic. In: Proceedings of the Workshop on Logic and Intelligent Interaction
Sack J (2008b) Temporal languages for epistemic programs. Journal of Logic, Language and Information 17(2):183–216
Savage L (1954a) The foundations of statistics. John Wiley and Sons, New York, second revised edition published by Dover Publications, 1972
Searle J (1985) The construction of social reality. The Free Press, New York
Shoham Y, Leyton-Brown K (2009) Multiagent systems: Algorithmic, game-theoretic, and logical foundations. Cambridge University Press, Cambridge
Velazquez-Quesada FR (2009) Inference and update. Synthese (Knowledge, Rationality & Action) 169(2):283–300
Veltman F (1996) Defaults in update semantics. Journal of Philosophical Logic 25(3):221–261

Chapter 3
Dynamic Epistemic Logic and Temporal Modality

Audrey Yap

3.1 Dynamic Epistemic Logic

Dynamic epistemic logic (DEL) allows us to model agents who learn new information about the world from events which they observe. However, the language of DEL has only forward-looking modal operators, which allow us to talk about an agent's informational state after an event takes place. This chapter describes a method for supplementing DEL with a backward-looking modal operator, which allows for a rich increase in expressive power without losing completeness.

3.1.1 Language and Models

As the framework of DEL is the one which will be extended with a past modal operator, we will first present the (static) system of epistemic logic and its extension to DEL. Such a system is standard, and more detailed presentations can be found, for instance in van Ditmarsch et al. (2007).

Definition 3.1 A static epistemic model M is a tuple

$$M = (W, \{\sim_j : j \in G\}, V, w_0).$$

1. W is a set of possible worlds, or states of the model.
2. G is a set of agents.
3. $\sim_j$ is an equivalence relation defined on W for each agent j. The intended interpretation is that $w \sim_j v$ whenever j cannot differentiate between worlds w and v.
4. V is a valuation on worlds. The intended interpretation is that $w \in V(p)$ whenever p holds at w.
5. w_0 is the actual world.

A. Yap (✉)
Department of Philosophy, University of Victoria, P.O. Box 3045, Victoria, BC VBW 3p4, Canada
e-mail: ayap@uvic.ca

P. Girard et al. (eds.), *Dynamic Formal Epistemology*, Synthese Library 351,
DOI 10.1007/978-94-007-0074-1_3,

Definition 3.2 The language for the static epistemic models is the language of epistemic logic:

$$\mathcal{L}_{St}\varphi := p|\neg\varphi|\varphi \wedge \varphi|K_j\varphi$$

The semantics for the propositional part are standard. And for an epistemic model M and a world w, the semantics for $K_j\varphi$ are as follows:

$$M, w \models K_j\varphi \text{ iff for all } v \text{ s.t.} w \sim_j v, M, v \models \varphi.$$

Definition 3.3 Now, we can define an epistemic action model

$$A = (\Sigma, \{\sim_j : j \in G\}, \{\text{Pre}_\sigma : \sigma \in \Sigma\}, \sigma_0).$$

1. Σ is the set of simple actions. These can also be interpreted as events.
2. $\sim_j$ is an equivalence relation which is defined on Σ for each agent j. The intended interpretation is that $\sigma \sim_j \tau$ whenever j cannot differentiate between actions σ and τ.
3. For each simple action σ, Pre_σ defines the preconditions which must be true at a world in order for σ to be performed at that world. Pre_σ is a formula in $\mathcal{L}_{St}$, not the dynamic language, so we cannot define paradoxical preconditions.[1]
4. σ_0 is the actual action performed in our update.

Having introduced action models, we can define the models of dynamic epistemic logic itself.

Definition 3.4 We define $M \times A$ as the product model

$$M \times A = (W \times \Sigma, \{\sim'_j : j \in G\}, V', w'_0).$$

1. $W \times \Sigma = \{(w, \sigma) : M, w \models \text{Pre}_\sigma\}$. The updated model is the product of the two previous models, restricted only by the condition that a world must satisfy the preconditions for an action for that action to be performed there.
2. We define $\sim'_j$ such that $(w_1, \sigma_1) \sim'_j (w_2, \sigma_2)$ iff $w_1 \sim_j w_2$ and $\sigma_1 \sim_j \sigma_2$. So j is only uncertain between two updated states if he could not previously tell the difference between the worlds, and the actions performed are also indistinguishable.
3. V' is essentially the old valuation on worlds, such that $(w, \sigma) \in V'(p)$ iff $w \in V(p)$.
4. $w'_0 = (w_0, \sigma_0)$. The new actual world is the product of the previous actual world and the actual action.

[1]As shown in Baltag et al. (1998), it is possible to allow formulas in the action dynamic language if we are careful, but for our purposes here, it will be simpler to confine things to the static language.

Then $M \times A$ is a new static model, which can produce further updated models, when we take the product with any action model whose preconditions are in the same language, with the same set of agents.

Definition 3.5 The dynamic language extends the static language with an update operator:

$$\mathcal{L}_{DEL}\varphi := p|\neg\varphi|\varphi \wedge \varphi|K_j\varphi|[A,\sigma]\varphi$$

The semantics for $[A,\sigma]\varphi$ are as follows:

$$M, w \models [A,\sigma]\varphi \text{ iff } M, w \models \text{Pre}_\sigma \text{ implies } M \times A, (w, a) \models \varphi.$$

Where the action model is understood, $[A,\sigma]$ can simply be written as $[\sigma]$.

3.1.1.1 Public Announcement Logic

Further, the logic of public announcements from Baltag et al. (1998) could be seen as a special case of product update.

Example 3.1 Given an epistemic model M, we can construct an action model representing the announcement of a formula $\varphi \in \mathcal{L}_{St}$.

$$A = (\{!\varphi\}, \{\langle !\varphi, !\varphi\rangle\}_{j\in G}, \{\text{Pre}_{!\varphi} \equiv \varphi\}, !\varphi).$$

1. In this case, there is just one formula announced, φ. But we could alternately have had a set of distinct formulas which could be announced.
2. Agents can differentiate between different announcements, so the $\sim_j$ relation is just a reflexive loop for each agent $j \in G$.
3. The only precondition for announcing φ at a world is that φ holds at that world.
4. We simply designate one announcement as the actual one. In this case, our action model has only one announcement, but if there were more, one would simply be distinguished.

Then the public announcement operator $[!\varphi]$ is just a special case of the $[\sigma]$ operator in dynamic epistemic logic .

3.2 Dynamic Epistemic Logic with History

Having introduced the DEL framework in Section 3.1, we will now present its extension to DEL+H, which supplements $\mathcal{L}_{\text{DEL}}$ with a past modal operator P_σ that will allow us to express what was the case before the event σ occurred.

3.2.1 Language and Models

To some extent, worlds in $M \times A$ already encode their history. For instance, after we take the product of an initial epistemic model M by an action model A n many times, the worlds in the resulting model can be seen as $n+1$-tuples, such that each world is of the form $(w, \sigma_1, \sigma_2, \ldots, \sigma_n)$, where w is a world in the original state model M, and each σ_i is an action in Σ. But this only tracks the events that led to a particular world, and not the uncertainties of agents in previous worlds. For that reason, we will modify the models of DEL.

Definition 3.6 Define the *length* of a world $\text{len}(w)$ to be the number of updates it encodes, so $\text{len}((w, \sigma_1, \ldots \sigma_n)) = n$.

Now we need to redefine the product model $M \times A$ in such a way that it retains the structure of original model M.

Definition 3.7 Redefine $M \times A$ to be the epistemic action model

$$M \times A = (W \cup (W \times \Sigma), \{\sim_j \cup \sim'_j : j \in G\}, \{R_\sigma \cup R'_\sigma : \sigma \in A\}, V \cup V', w'_0).$$

1. Let n be the maximum length of a world in W. Let $W \times \Sigma = \{(w, \sigma) : M, w \models \text{Pre}_\sigma \text{ and } \text{len}(w) = n\}$, as before. The set of worlds in the new model is the union of those in the original, together with the product of the epistemic and action models. The restriction on the product is that a world must be a leaf, and satisfy the preconditions for an action for that pair to be included.
2. Let $(w_1, \sigma_1) \sim'_j (w_2, \sigma_2)$ iff $w_1 \sim_j w_2$ and $\sigma_1 \sim_j \sigma_2$, as before. And since we keep the worlds from M, we also keep the uncertainties from M as well.
3. The R_σ relations are a new addition. We let $R'_\sigma = \{\langle (w, \sigma), w \rangle : (w, \sigma) \in W \times \Sigma\}$. Each world in a product model points to its ancestor. So $w R_\sigma v$ implies that $w = (v, \sigma)$, and $M, v \models \text{Pre}_\sigma$.
   So when we take a product, we keep all the old R-relations, and add a new arrow for every world in $W \times \Sigma$, pointing to its unique ancestor. If our epistemic model M was not a product model before, then its $R_\sigma = \emptyset$.
4. Let $(w, \sigma) \in V'(p)$ iff $w \in V(p)$, as before. For the valuation as well, we keep the old valuation, and add valuations for the new worlds as expected.
5. $w'_0 = (w_0, \sigma_0)$, as before.

Action models remain the same, in spite of the new update mechanism. One way to picture the effect of updates is as adding subsequent layers to a tree (or more precisely, a forest). The models do in fact have a forest structure under the R_σ relations. Also, due to the way in which the R_j relations are defined, agents are only uncertain between worlds which have the same length. This builds in a synchronicity assumption, for the sake of simplicity.

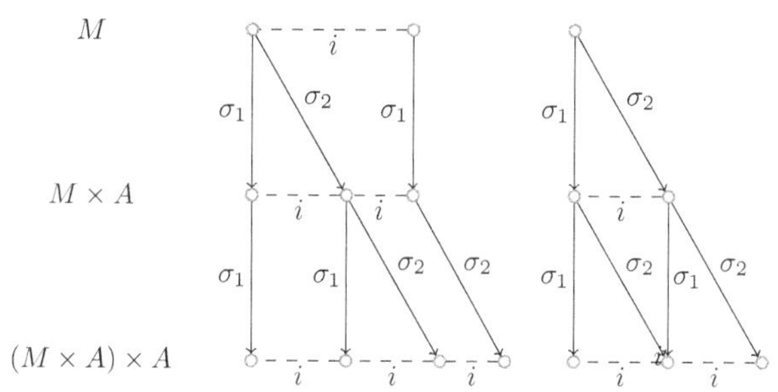

Fig. 3.1 The expansion of product models under updates

Definition 3.8 We now can extend the language once more in order to add a past temporal operator.

$$\mathcal{L}_{\text{DEL}+H} \quad \varphi := p|\neg\varphi|\varphi \wedge \varphi|K_i\varphi|[A,\sigma]\varphi|P_\sigma\varphi$$

The semantics for P_σ are as follows:

$$M, w \models P_\sigma\varphi \text{ iff } \exists v \text{ such that } w = (v, \sigma) \text{ and } M, v \models \varphi.$$

P_σ is defined as a diamond modality, whereas $[\sigma]$ is defined as a box. The motivation for this choice of semantics is one of naturalness. The idea is that the claim that $P_\sigma\varphi$ holds only makes sense if there was in fact a past in which σ did occur. So we should not allow $P_\sigma\varphi$ to hold at a world which was not in fact the result of a σ action.

3.2.2 About the Logic

Having introduced the models and semantics under which DEL is to be expanded, in this section, we will discuss the logic of the extended system. A system of axioms will be given, extending those of the original DEL system to include axioms governing the behavior of the past modal operator. Further, it will be shown that the resulting system is sound and complete.

3.2.2.1 Axiomatizing the Language

The Baltag et al. (1998) axiomatization of product update is complete, and gives a reduction to epistemic logic. We to a certain degree, we can follow this treatment for our past modality, adding axioms for our new operator. Below are the basic axioms and those for the forward-looking portion of our language.

Basic Axioms and Modal Rules

All propositional tautologies

S5 axioms for K_i

(Modus Ponens)	From $\models \varphi$ and $\models \varphi \to \psi$, infer $\models \psi$
(K-necessitation)	From $\models \varphi$, infer $\models K_i\varphi$
($[\sigma]$-necessitation)	From $\models \varphi$, infer $\models [\sigma]\varphi$
(P_σ-necessitation)	From $\models \varphi$, infer $\models P_\sigma \text{Pre}_\sigma \to P_\sigma\varphi$

Action Axioms

(Atomic Permanence)	$[\sigma]p \leftrightarrow (\text{Pre}_\sigma \to p)$
($[\sigma]$-normality)	$[\sigma](\varphi \to \psi) \to ([\sigma]\varphi \to [\sigma]\psi)$
(Partial Functionality)	$[\sigma]\neg\varphi \leftrightarrow (\text{Pre}_\sigma \to \neg[\sigma]\varphi)$
(Action-Knowledge)	$[\sigma]K_i\varphi \leftrightarrow (\text{Pre}_\sigma \to \bigwedge\{K_i[\tau]\varphi : \sigma \sim_i \tau\})$
(Interaction with Past)	$[\sigma]P_\sigma\varphi \leftrightarrow (\text{Pre}_\sigma \to \varphi)$

There are five reduction axioms for $[\sigma]$, and we might want to look for their counterparts with respect to P_σ. First note that we need something playing the same role as Pre_σ, since these define when it is possible to perform an action at a world. This means we want something like a Post_σ, which defines when an world has been attained by the performance of an action. For this, we can simply use $P_\sigma\text{Pre}_\sigma$. This ensures that there was a previous world in which it was possible to perform σ. Given this, the first few axioms are straightforward:

Past Looking Axioms

(Atomic Permanence)	$P_\sigma q \leftrightarrow (P_\sigma\top \wedge q)$
(Unique Arrows)	$\neg(P_\sigma\top \wedge P_\tau\top)$ when $\sigma \neq \tau$
($\neg$-Reduction)	$P_\sigma\neg\varphi \leftrightarrow (P_\sigma\text{Pre}_\sigma \wedge \neg P_\sigma\varphi)$
($\wedge$-Reduction)	$P_\sigma(\varphi \wedge \psi) \leftrightarrow (P_\sigma\varphi \wedge P_\sigma\psi)$
(Interaction with $[\sigma]$)	$P_\sigma[\sigma]\varphi \leftrightarrow (P_\sigma\text{Pre}_\sigma \wedge \varphi)$
(Interaction with K_i)	$P_\sigma K_i\varphi \leftrightarrow P_\sigma\text{Pre}_\sigma \wedge K_i \bigvee\{P_\tau\varphi : \sigma \sim_i \tau\} \wedge$ $P_\sigma K_i(\bigwedge\{\neg\text{Pre}_\tau : \sigma \sim_i \tau\} \to \varphi)$

- Atomic Permanence: This ensures that the truth values of atomic formulas are not changed after events occur.
- Unique Arrows: This is an axiom schema, ensuring that it is inconsistent to state that two different events led to the same world. Since every world except those at the root can be uniquely written as $w = (v, \sigma)$, if it is also the case that $w = (v, \tau)$, then $\sigma = \tau$. So there is only one action which leads to any given world.
- $\neg$-Reduction: The reduction axiom for $\neg$ ensures that each world will have a unique σ-ancestor.
- $\wedge$-Reduction: This axiom (together with the Unique Arrows axiom) ensures that there is a unique path back to the root at each world. This corresponds to showing that there is only one order in which we can correctly nest past operators. To see this, note the following instance of this axiom:

$$P_\sigma(P_{\tau_1}\top \wedge P_{\tau_2}\top) \leftrightarrow (P_\sigma P_{\tau_1}\top \wedge P_\sigma P_{\tau_2}\top).$$

By Unique Arrows,

$$(P_{\tau_1}\top \wedge P_{\tau_2}\top) \rightarrow \tau_1 = \tau_2.$$

So

$$(P_\sigma P_{\tau_1}\top \wedge P_\sigma P_{\tau_2}\top) \rightarrow \tau_1 = \tau_2.$$

- Interaction with $[\sigma]$: This axiom tells us that if we take a σ-step back and then a σ-step forward, we end up in the same world.
- The Interaction with K_i axiom will be discussed in the following section.
- We do not need an axiom to deal with the formula $P_\tau[\sigma]\varphi$, since that involves taking a step back in the history, and then a step forward, but when $\sigma \neq \tau$, the two steps do not lead back to the same world. Further, when φ does not contain any P-operators, we can apply the forward-looking reduction axioms to reduce $[\sigma]\varphi$ to a formula in the static language.

Furthermore, we can prove a soundness theorem for our new axioms. The soundness of Atomic Permanence and Unique Arrows is straightforward from the definition of our models, but we can prove the result for the remaining axioms.

Theorem 3.1 (***Soundness***) *The axioms are valid in all DEL+H models.*

Proof

- **¬-Reduction**: ($\Rightarrow$) Suppose $M, w \models P_\sigma\neg\varphi$. Then there exists v such that $w = (v, \sigma)$, and $M, v \models \neg\varphi$. Since v is the unique predecessor of w, $M, w \not\models P_\sigma\varphi$ and $M, w \models \text{Pre}_\sigma$. So $M, w \models P_\sigma\text{Pre}_\sigma \wedge \neg P_\sigma\varphi$.
  ($\Leftarrow$) Suppose $M, w \models P_\sigma\text{Pre}_\sigma \wedge \neg P_\sigma\varphi$. Then there exists v such that $w = (v, \sigma)$. Also, $M, v \not\models \varphi$, so $M, v \models \neg\varphi$. That implies $M, w \models P_\sigma\neg\varphi$.
- **∧-Reduction**: ($\Rightarrow$) Suppose $M, w \models P_\sigma(\varphi \wedge \psi)$. Then there exists v such that $w = (v, \sigma)$, and $M, v \models \varphi \wedge \psi$. So there exists v such that $M, v \models \varphi$, which implies $M, w \models P_\sigma\varphi$, and such that $M, v \models \psi$, which implies $M, w \models P_\sigma\psi$, so $M, w \models P_\sigma\varphi \wedge P_\sigma\psi$.
  ($\Leftarrow$) Suppose $M, w \models P_\sigma\varphi \wedge P_\sigma\psi$. Then there exists v_1 such that $w = (v_1, \sigma)$, and $M, v \models \varphi$, and there exists v_2 such that $w = (v_2, \sigma)$, and $M, v \models \psi$. But then clearly $v_1 = v_2$, so we have $M, v_1 \models \varphi \wedge \psi$, which implies $M, v \models P_\sigma(\varphi \wedge \psi)$.
- **Interaction with** $[\sigma]$: ($\Rightarrow$) Suppose $M, w \models P_\sigma[\sigma]\varphi$. Then there exists v such that $w = (v, \sigma)$, and $M, v \models [\sigma]\varphi$. Then $M, (v, \sigma) \models \varphi$. However, $w = (v, \sigma)$, so $M, w \models P_\sigma\text{Pre}_\sigma \wedge \varphi$.
  ($\Leftarrow$) Suppose $M, w \models P_\sigma\text{Pre}_\sigma \wedge \varphi$. Then there exists v such that $w = (v, \sigma)$. Since $M, (v, \sigma) \models \varphi$, we know $M, v \models [\sigma]\varphi$. So $M, w \models P_\sigma[\sigma]\varphi$.
- **Interaction with** K_i: ($\Rightarrow$) Suppose $M, w \models P_\sigma K_i\varphi$. Then there is some v such that $w = (v, \sigma)$ and $M, v \models K_i\varphi$. Thus, $M, w \models P_\sigma\text{Pre}_\sigma$. Now let u be any world such that $w \sim_i u$. By construction of our models, this implies that u is of

the form (t, τ), with $v \sim_i t$, and $\sigma \sim_i \tau$. Since $M, v \models K_i\varphi$, we can conclude $M, t \models \varphi$, and thus $M, u \models P_\tau\varphi$. Therefore, $M, u \models \bigvee\{P_\tau\varphi : \sigma \sim_i \tau\}$. So $M, w \models K_i \bigvee\{P_\tau\varphi : \sigma \sim_i \tau\}$. Finally, let t be any world such that $v \sim_i t$. Since $M, v \models K_i\varphi$, we know that $M, t \models \bigwedge\{\neg\text{Pre}_\tau : \sigma \sim_i \tau\} \rightarrow \varphi$. So $M, w \models P_\sigma K_i(\bigwedge\{\neg\text{Pre}_\tau : \sigma \sim_i \tau\} \rightarrow \varphi)$.
($\Leftarrow$) Suppose $M, w \models P_\sigma\text{Pre}_\sigma \wedge K_i \bigvee\{P_\tau\varphi : \sigma \sim_i \tau\} \wedge P_\sigma K_i(\bigwedge\{\neg\text{Pre}_\tau : \sigma \sim_i \tau\} \rightarrow \varphi)$. Then there is some v such that $w = (v, \sigma)$. Let t be any world such that $v \sim_i t$. There are two cases: (i) for some τ such that $\sigma \sim_i \tau$, $M, t \models \text{Pre}_\tau$, or (ii) there is no such τ.
Case (i): By construction of our models, $u = (t, \tau)$ is such that $w \sim_i u$. That implies $M, u \models \bigvee\{P_\tau\varphi : \sigma \sim_i \tau\}$. Since $u = (t, \tau)$, it can only be that $M, u \models P_\tau\varphi$. Thus, $M, t \models \varphi$.
Case (ii): Then $M, t \models \bigwedge\{\neg\text{Pre}_\tau : \sigma \sim_i \tau\}$. However, $M, v \models K_i(\bigwedge\{\neg\text{Pre}_\tau : \sigma \sim_i \tau\} \rightarrow \varphi)$. So since $v \sim_i t$, this implies $M, t \models \varphi$.
Thus since in both cases, $M, t \models \varphi$, we know $M, v \models K_i\varphi$, and thus $M, w \models P_\sigma K_i\varphi$.

3.2.2.2 Reduction Axioms for Knowledge

It is clear, however, that the interaction with K_i axiom is not a reduction axiom as the others are, as the last conjunct is still a formula of the form $P_\sigma K_i\psi$. The first two conjuncts correspond to a principle of perfect recall, but they do not imply $P_\sigma K_i\varphi$ by themselves. In fact, an axiom such as this is the best we can do.

Theorem 3.2 *No reduction axioms are possible for the P_σ operator.*

Proof Consider the following two models:

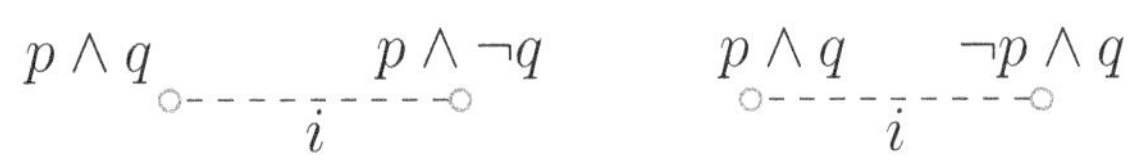

Fig. 3.2 Two models before an update

After a public announcement of $p \wedge q$ in the actual world, we obtain the following two updated models shown in Fig. 3.3.

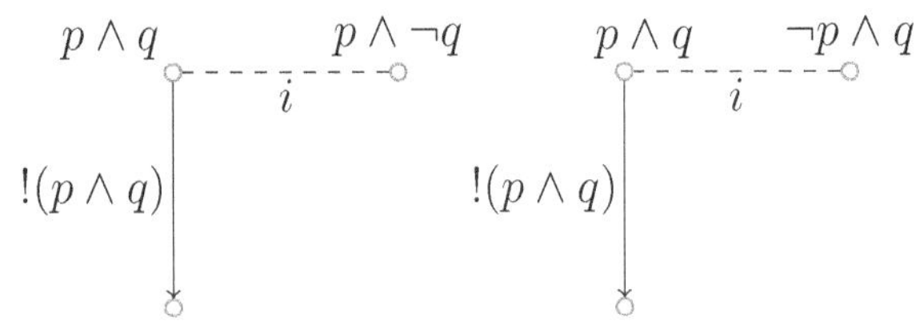

Fig. 3.3 The updated models

It is clear that no formula in $\mathcal{L}_{\mathrm{DEL}}$ can distinguish the updated worlds, since the same propositional formulas are true in each and there are no uncertainties. Now, consider the $\mathcal{L}_{\mathrm{DEL}+H}$ formula

$$P_{!(p \wedge q)} K_i p.$$

In other words, i knew p was true before the announcement. This clearly holds in the updated world on the left, but not in the world on the right. Since $\mathcal{L}_{\mathrm{DEL}+H}$ can distinguish the worlds, but $\mathcal{L}_{\mathrm{DEL}}$ cannot, there can be no reduction axioms for $\mathcal{L}_{\mathrm{DEL}+H}$, since the latter is clearly more expressive.

The interaction with K_i axiom is at best a partial reduction axiom, which does not entirely allow us to eliminate the P_σ modality, but isolates the cases in which it is not possible to eliminate it. Since updates allow agents to eliminate uncertainties between worlds, they can learn about their situation. So they can discover that certain worlds which were previously indistinguishable from the actual world, are actually impossible past states of affairs. The intuitive reading of the axiom is that we split the past into two cases: worlds the agent still holds possible, and worlds the agent now knows are impossible. The first conjunct deals with the former class of worlds, and the second deals with the worlds that have been ruled out, stating that φ was still true in those.

Example 3.2 The way in which the axiom deals with its different cases can be seen in Fig. 3.4.

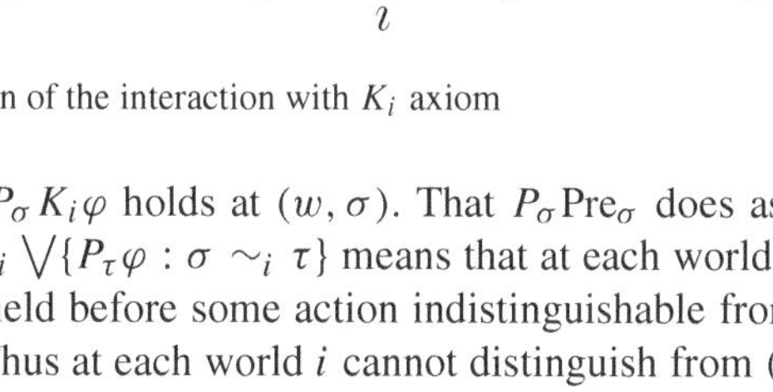

Fig. 3.4 An illustration of the interaction with K_i axiom

In this model, $P_\sigma K_i \varphi$ holds at (w, σ). That $P_\sigma \mathrm{Pre}_\sigma$ does as well is clear. The second conjunct $K_i \bigvee\{P_\tau \varphi : \sigma \sim_i \tau\}$ means that at each world i holds possible, it is the case that φ held before some action indistinguishable from σ. In our model, we have $\sigma \sim_i \tau$. Thus at each world i cannot distinguish from (w, σ), $P_\sigma \varphi \vee P_\tau \varphi$ must hold. And this is clearly the case. In other words, φ was true in all the histories i still considers possible. The third conjunct $P_\sigma K_i \bigwedge(\{\neg \mathrm{Pre}_\tau : \sigma \sim_i \tau\} \rightarrow \varphi)$ deals with the histories i no longer considers possible. These are exactly the past worlds which did not satisfy the preconditions for any action that i thinks could have taken place. So at every world i cannot distinguish from w, $(\neg \mathrm{Pre}_\sigma \wedge \neg \mathrm{Pre}_\tau) \rightarrow \varphi$ must hold. The only world satisfying the antecedent of this conditional is v, and φ holds at v. So the third conjunct also holds at (w, σ).

We can see this also in Figs. 3.2 and 3.3. Since the right hand worlds do not satisfy the precondition for the announcement of $(p \wedge q)$, after the announcement, they are ruled out as possibilities. But when we state $P_{!(p \wedge q)} K_i p$, we want to say that p was true in all the past worlds still considered possible (the ones which satisfied the preconditions for $(p \wedge q)$ to be announced), as well as in the past worlds now known to be impossible (those which did not satisfy those preconditions).

Given these considerations, the fact that only a partial reduction is possible should be seen as a good thing. If there were a full reduction of $\mathcal{L}_{\mathrm{DEL}+H}$ to $\mathcal{L}_{St}$, then information about the past is reducible to information about the present. But an agent's current state of knowledge does not let us deduce anything about her past knowledge states, so the latter *should not* be reducible to the former. Otherwise, we would not be able to talk about agents' learning. The ability to talk about what agents learn in an update was the point of introducing the new modal operator to the language; so the lack of a full reduction is an encouraging result. In fact, the past modality allows us to express an agent's having learned something, by what is almost a converse Moore sentence: $P_\sigma \neg K_i \varphi \wedge K_i \varphi$, or "Before σ occurred, i did not know that φ, but she knows it now." However, the fact that no reduction to dynamic epistemic logic is possible means that we cannot use reduction axioms to provide a completeness result for these kinds of models.

3.2.2.3 Completeness

A variation on the usual canonical model construction can provide us with a weak completeness result. However, instead of taking the worlds in the canonical model simply to be maximal consistent sets, we will have to order the maximal consistent sets temporally, so that they do not refer too far into the past or future than is permitted by the construction of the model. The completeness proof relies on the fact that any particular formula will only require looking finitely far into the past or future in order to be evaluated, so a forest of only finite depth will be required to satisfy it.

Definition 3.9 Let the *past depth* of a formula φ, $d_p(\varphi)$, measure how many steps into the past φ looks. We define it with the following inductive definition:

- $d_p(p) = 0$
- $d_p(\neg\varphi) = d_p(\varphi)$
- $d_p(\varphi \wedge \psi) = \max(d_p(\varphi), d_p(\psi))$
- $d_p(K_i\varphi) = d_p(\varphi)$
- $d_p([\sigma]\varphi) = d_p(\varphi) - 1$
- $d_p(P_\sigma\varphi) = \max(d_p(\varphi), 0) + 1$

The reason why we do not simply add 1 in the last clause is because it is possible for $d_p(\varphi)$ to be negative, but even if it is, evaluating $P_\sigma\varphi$ requires looking at a past world. For instance: evaluating $P_\sigma[\sigma][\tau]p$ still requires looking one step into the past, even though $d_p([\sigma][\tau]p) = -2$. So we cannot simply add 1 in this case, or we will obtain the wrong result.

Let the *future depth* for a formula $d_f(\varphi)$ be defined symmetrically, with the clauses for past and future switched, representing how far into the future we have to look to evaluate φ.

The canonical model for a formula φ will have to be constructed in stages, and the number of stages depends on how far we have to look in both directions: $d_p(\varphi)+d_f(\varphi)$.

Definition 3.10 Let $d_p(\varphi) + d_f(\varphi) = N$. We define the n-depth *closure* of φ for $n \leq N$.

φ_0 is the minimal set such that:

- $\top \in \varphi_0$,
- Every subformula of φ in $\mathcal{L}_{\text{DEL}}$ is in φ_0,
- If $\psi \in \varphi_0$, then $\neg\psi \in \varphi_0$ for ψ not a negation,
- If $\psi, \chi \in \varphi_0$, then $\psi \wedge \chi \in \varphi_0$,
- If $\psi \in \varphi_0$, then $K_i\psi \in \varphi_0$ for every agent i,
- $\text{Pre}_\sigma \in \varphi_0$ for every action $[\sigma]$,
- If $d_f(\psi) < N$, then $[\sigma]\psi \in \varphi_0$.

Thus φ_0 is a set of formulas which only looks forward. We can look at it as the original model, since every formula in it has past depth 0. Now we can define φ_{n+1} such that:

- If $\psi \in \varphi_n$ and $d_f(\psi) + n < N$, then $\psi \in \varphi_{n+1}$,
- If $\psi \in \varphi_{n+1}$, then $\neg\psi \in \varphi_{n+1}$, for ψ not a negation,
- If $\psi, \chi \in \varphi_{n+1}$, then $\psi \wedge \chi \in \varphi_{n+1}$,
- If $\psi \in \varphi_{n+1}$, then $K_i\psi \in \varphi_{n+1}$ for every agent i,
- If $\psi \in \varphi_n$, then $P_\sigma\psi \in \varphi_{n+1}$ for every action σ.
- If $\psi \in \varphi_{n+1}$ and $d_f(\psi) + n + 1 < N$, then $[\sigma]\psi \in \varphi_n$ for every action σ.

The last two clauses add the forward and backward-looking operators . The main thing we need to worry about with respect to the past-looking operators is that, at the n-th stage, we do not look further into the past than n steps, so we must add a bound based on the temporal depth of the formula. Similarly, we must restrict the addition of forward-looking operators , so that we do not look further forward into the future than φ does.

Lemma 3.1 (***Maximality Lemma***) *$\psi \in \varphi_k$ iff $d_p(\psi) \leq k$ and $d_f(\psi) + k \leq N$, for any ψ containing only actions σ, agents i and propositional subformulas mentioned in φ.*

Proof By induction on k and on the complexity of ψ.

For $k = 0$**.** The propositional case is immediate, as are the cases for K_i, $\neg$, and $\wedge$. The $P_\sigma\psi$ case is not applicable here, so we only consider $[\sigma]\psi$. However, the only way to have $[\sigma]\psi \in \varphi_0$ is for $d_f(\psi) < N$, so it is clear that $d_f([\sigma]\psi) \leq N$.

For $k+1$**.** The propositional case here is again immediate, as are the cases for K_i, $\neg$, and $\wedge$.

Case P_σ. ($\Rightarrow$). Let $P_\sigma\psi \in \varphi_{k+1}$. Then either $P_\sigma\psi \in \varphi_k$, in which case we are done, or $\psi \in \varphi_k$, which by the IH implies $d_p(\psi) \leq k$. Thus, since $d_p(P_\sigma\psi) = \max(d_p(\psi), 0) + 1$, we know $d_p(P_\sigma\psi) \leq k + 1$. Also, since $d_f(\psi) + k \leq N$ by the IH, and $d_f(P_\sigma\psi) = d_f(\psi) - 1$, we have $d_f(\psi) + k - 1 \leq N$. Clearly, $d_f(P_\sigma\psi) + k + 1 \leq N$.

($\Leftarrow$). Let $d_p(P_\sigma\psi) \leq k + 1$ and $d_f(P_\sigma\psi) + k + 1 \leq N$. By definition of P, we know that $d_p(\psi) \leq k$. And by definition of d_f, we know that $d_f(\psi) + k \leq N$. Thus by the IH, $\psi \in \varphi_k$. But then by definition of φ_{k+1}, $P_\sigma\psi \in \varphi_{k+1}$.

Case $[\sigma]$. ($\Rightarrow$). Let $[\sigma]\psi \in \varphi_{k+1}$. Then either $[\sigma]\psi \in \varphi_k$, in which case we are done, or $\psi \in \varphi_{k+1}$ and $d_f(\psi)+k+1 < N$. But then $d_f([\sigma]\psi)+k+1 \leq N$. Also, since $d_p([\sigma]\psi) = d_p(\psi) - 1$, since by the IH, $d_p(\psi) \leq k+1$, $d_p([\sigma]\psi) \leq k+1$.

($\Leftarrow$). Let $d_p([\sigma]\psi) \leq k + 1$ and $d_f([\sigma]\psi) + k + 1 \leq N$. By definition of d_p, we know that $d_p(\psi) \leq k$, and by definition of d_f, we know that $d_f(\psi) + k \leq N$. Thus by the IH, $\psi \in \varphi_k$. However, since $d_f([\sigma]\psi) = \max(d_f(\psi), 0) + 1$, we know that $d_f(\psi) + 1 + k + 1 \leq N$, so $d_f(\psi) + k + 1 < N$. Then by definition of φ_{k+1}, we know that $\psi \in \varphi_{k+1}$, and also that $[\sigma]\psi \in \varphi_{k+1}$. □

In order to construct the canonical model M_φ for a formula φ, we will need to use a two-stage construction. First, the tree must be built from the leaves to the root, in order to determine the histories, and second, the uncertainties must be set from the root to the leaves, to ensure that the update definitions are satisfied.

Lemma 3.2 (***Predecessor Lemma***) *For every Γ which is an MCS in φ_{k+1}, there is a unique Δ, which is an MCS in φ_k such that $\psi \in \Delta$ for all $P_\sigma\psi \in \Gamma$.*

Proof Let Γ be an MCS in φ_{k+1}. Consider

$$\Delta = \{\psi : P_\sigma\psi \in \Gamma\}.$$

It is fairly immediate that Δ is a set in φ_k. Note also that $\mathrm{Pre}_\sigma \in \Delta$. So it remains to show that it is maximal and consistent. Its consistency follows from the consistency of Γ, for if $\varphi \in \Delta$ and $\neg\varphi \in \Delta$, then $P_\sigma\varphi, P_\sigma\neg\varphi \in \Gamma$. Then since $P_\sigma\neg\varphi \leftrightarrow P_\sigma\mathrm{Pre}_\sigma \wedge \neg P_\sigma\varphi$ is an axiom, this implies $\neg P_\sigma\varphi \in \Gamma$, so Γ would have been inconsistent as well.

So now we show it is maximal in φ_k. Suppose for some $\psi \in \varphi_k$, that neither ψ nor $\neg\psi \in \Delta$. Since $\psi \notin \Delta$, $P_\sigma\psi \notin \Gamma$. By construction of φ_{k+1}, we know $P_\sigma\psi \in \varphi_{k+1}$. So since Γ is an MCS in φ_{k+1}, we know that $\neg P_\sigma\psi \in \Gamma$. Note also that $P_\sigma\mathrm{Pre}_\sigma \in \Gamma$. However,

$$P_\sigma\neg\psi \leftrightarrow P_\sigma\mathrm{Pre}_\sigma \wedge \neg P_\sigma\psi$$

is an axiom – so for Γ to be an MCS, $P_\sigma\neg\psi \in \Gamma$, which means $\neg\psi \in \Delta$. The converse also holds. If $\neg\psi \notin \Delta$, then $P_\sigma\neg\psi \notin \Gamma$. So $\neg P_\sigma\neg\psi \in \Gamma$, which implies by the same axiom that $P_\sigma\psi \in \Gamma$. Thus, $\psi \in \Delta$.

Last, we show that Δ is the unique predecessor of Γ. Suppose there were distinct Δ_1, Δ_2 such that $\Gamma R_\sigma \Delta_1$ and $\Gamma R_\sigma \Delta_2$. Since both are maximal, there is some ψ with modal depth $\leq k$ such that $\psi \in \Delta_1$ and $\neg\psi \in \Delta_2$. This implies that neither $P_\sigma \psi$ nor $P_\sigma \neg\psi \in \Gamma$. But as above, if $P_\sigma \psi \notin \Gamma$, then $\neg P_\sigma \psi \in \Gamma$, by maximality, since these formulas have modal depth $k + 1$. And $P_\sigma \mathrm{Pre}_\sigma \in \Gamma$, so $P_\sigma \neg\psi \in \Gamma$. Thus each world must have a unique predecessor.

Definition 3.11 (*Canonical Model*) The *canonical model* M_φ for a formula φ with depth n is a quadruple $(W_\varphi, R_{\sigma,\varphi}, \sim_{i,\varphi} V_\varphi)$ which we will define in two stages.

First, let $W_\varphi = \bigcup_{i \leq n}\{\Gamma \subseteq \varphi_i : \Gamma$ is maximally consistent in $\varphi_i\}$ and $V_\varphi(p) = \{\Gamma : p \in \Gamma\}$. Now we let $(\Gamma, \Delta) \in R_{\sigma,\varphi}$ iff Γ is an MCS in φ_{k+1} and Δ is an MCS in φ_k, $\mathrm{Pre}_\sigma \in \Delta$, and $\psi \in \Delta$ for all ψ with $P_\sigma \psi \in \Gamma$. This sets the $R_{\sigma,\varphi}$ relations from the leaves up to the root, giving us the vertical structure of the model. Next, we need to construct the uncertainties horizontally.

Define the $\sim_{i,\varphi}$ relations of the *canonical model* from the top down, starting with the MCS's in φ_0. For Γ, Δ as MCS's in φ_0, let $(\Gamma, \Delta) \in \sim_{i,\varphi}$ iff for every φ with $K_i\varphi \in \Gamma$, $\varphi \in \Delta$. The Predecessor Lemma entitles us to refer to the unique predecessor of any world, provided it has past depth > 0. So for Γ, Δ as MCS's in φ_{k+1}, let Γ', Δ' be the unique predecessors of Γ and Δ respectively. Then $(\Gamma, \Delta) \in \sim_{i,\varphi}$ iff $\Gamma \sim_{i,\varphi} \Delta$, and where $P_\sigma\varphi \in \Gamma$ and $P_\tau\psi \in \Delta$, $\sigma \sim_i \tau$. This sets the uncertainties according to the usual update definition, from the top down, and completes the construction of M_φ.

The construction of the canonical model ensures that it will structurally be a forest with horizontal uncertainties, such as the model in Fig. 3.1. One thing it does not provide, however, is a guarantee that we will be able to recover the individual action models giving rise to each update.

Lemma 3.3 (*Truth Lemma*) *For all* $\psi \in \varphi_k$, $\psi \in \Gamma$ *iff* $M_\varphi, \Gamma \models \psi$.

Proof By induction on k and ψ. We will only do the $K_i\psi$ and $P_\sigma\psi$ cases at each level, since the $[\sigma]\psi$ case follows from the reduction axioms for $[\sigma]$, and the others are straightforward.

For $k = 0$, we cannot have $P_\sigma\psi \in \Gamma$. And the definition of $\sim_{i,\varphi}$ ensures that $K_i\psi \in \Gamma$ iff $\psi \in \Delta$, for every Δ with $\Gamma \sim_{i,\varphi} \Delta$.

For $k + 1$**:**

Case $P_\sigma\psi$**.** ($\Rightarrow$) Suppose $P_\sigma\psi \in \Gamma$. By the predecessor lemma, since Γ has P-depth > 0, there is a unique predecessor Δ, which by definition is such that $\Gamma R_\sigma \Delta$, $\mathrm{Pre}_\sigma \in \Delta$ and $\psi \in \Delta$. So by the IH, $M_\varphi, \Delta \models \psi$, which means $M_\varphi, \Gamma \models P_\sigma\psi$.

($\Leftarrow$) Suppose $M_\varphi, \Gamma \models P_\sigma\psi$. Then there is some Δ such that $\Gamma R_\sigma \Delta$ and $M_\varphi, \Delta \models \psi$. By the IH, $\psi \in \Delta$, and by the predecessor lemma, Δ is the unique predecessor of Γ. Thus, since $\psi \in \Delta$, $P_\sigma\psi \in \Gamma$, by definition of R_σ.

Case $K_i\psi$. ($\Rightarrow$) Suppose $K_i\psi \in \Gamma$. Take any Δ such that $\Gamma \sim_{i,\varphi} \Delta$. We want to show $\psi \in \Delta$. Let Γ' be the unique σ predecessor of Γ, as defined by the predecessor lemma, and let Δ' be the τ predecessor of Δ. The definition of $\sim_{i,\varphi}$ ensures that $\Gamma' \sim_{i,\varphi} \Delta'$ and $\sigma \sim_i \tau$. Also note that $d_p([\sigma]K_i\psi) = d_p(K_i\psi) - 1$, so $d_p([\sigma]K_i\psi) \in \varphi_k$, by the maximality lemma. So since $K_i\psi \in \Gamma$ and $\text{Pre}_\sigma \in \Gamma$ by assumption, it must be that $P_\sigma[\sigma]K_i\psi \in \Gamma$ due to the Interaction with $[\sigma]$ axiom. This implies $[\sigma]K_i\psi \in \Gamma'$ by construction of predecessors. By the Action-Knowledge axiom

$$[\sigma]K_i\psi \leftrightarrow (\text{Pre}_\sigma \rightarrow \bigwedge\{K_i[\tau]\psi : \sigma \sim_i \tau\})$$

$K_i[\tau]\psi \in \Gamma'$, so $[\tau]\psi \in \Delta'$ by the IH. By consistency, $\psi \in \Delta$.

To describe this in terms of a picture, we want to consider the square given in Fig. 3.5. In order to show that $M, \Gamma \models K_i\psi$, we go through the Action-Knowledge axiom and show that $[\sigma]K_i\psi \in \Gamma'$, which implies that $K_i[\tau]\psi \in \Gamma'$. Since any world Δ which i cannot distinguish from Γ has a predecessor Δ' such that $\Gamma' \sim_{i,\varphi} \Delta'$, we argue that $[\tau]\psi \in \Delta'$, and thus $\psi \in \Delta$.

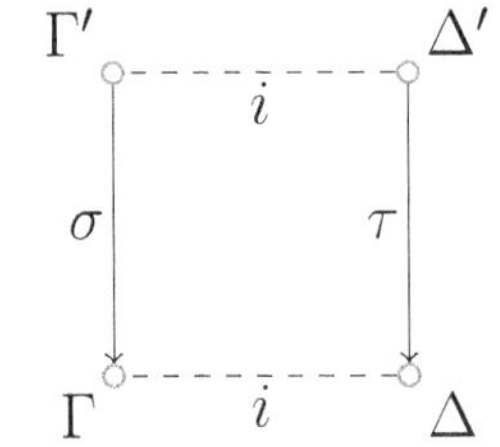

Fig. 3.5 Illustrating the $\Rightarrow$ direction of the K_i case

($\Leftarrow$) Let $M_\varphi, \Gamma \models K_i\psi$. Let Γ' be the unique σ predecessor of Γ. Take any Δ' such that $\Gamma' \sim_{i,\varphi} \Delta'$, and τ such that $\sigma \sim_i \tau$. If $\text{Pre}_\tau \notin \Delta'$, then trivially, $[\tau]\psi \in \Delta'$. Otherwise, take the τ-successor Δ of Δ'. By definition of $\sim_{i,\varphi}$, $\Gamma \sim_{i,\varphi} \Delta$, so $M_\varphi, \Delta \models \psi$ by assumption. By the IH, $\psi \in \Delta$, so $[\tau]\psi \in \Delta'$. This implies that for every Δ' in the $\sim_{i,\varphi}$ equivalence class of Γ', $[\tau]\psi \in \Delta'$. Thus, $\bigwedge\{K_i[\tau]\psi : \sigma \sim_i \tau\} \in \Gamma'$, which implies $[\sigma]K_i\psi \in \Gamma'$. Thus, $K_i\psi \in \Gamma$.

Theorem 3.3 (***Completeness***) *If* $\models \varphi$, *then* $\models \varphi$.

Proof Suppose $\not\models \varphi$. Then $\neg\varphi$ is consistent. Construct the canonical model $M_{\neg\varphi}$. Since $\neg\varphi$ is consistent, there is an MCS Γ in the canonical model with $\neg\varphi \in \Gamma$. Then, by the Truth Lemma, $M_{\neg\varphi}, \Gamma \models \neg\varphi$, which implies $\not\models \varphi$.

Thus, in spite of the fact that there are no reduction axioms, a canonical model can be constructed for any consistent formula, since such formulas can only express a finite amount of information about the past.

3.3 Expressive Power and Variations

Having discussed the formal results relating to DEL+H, the final section of the chapter will discuss the expressive power of the new system, as well as briefly consider variants of the past modal operator introduced in this chapter.

3.3.1 Bisimulation

One thing we ought to remark upon is that we no longer have stabilization under bisimulation. In product update, it is possible to arrive at a state in which subsequent models produced by the update are bisimilar to previous ones. That is, we can have $M \times A$ bisimilar to M, or even more complicated cases of "looping", in which $(M \times A) \times A$ is bisimilar to M, but not to $(M \times A)$. This is treated in detail in Sadzik (2006). However, our temporal modality is sufficiently expressive to distinguish between worlds and their ancestors, so there will never be a case in which $M \times A$ is bisimilar to M. If M cannot be collapsed under bisimulation, then, even after taking updates, every world in subsequent product models can be uniquely defined by a formula in $\mathcal{L}_{\text{DEL}+H}$.

However, we can still use the notion of bisimulation to apply to our new kinds of models, since we could have bisimulation instead between entire trees. Viewed structurally, the P_σ modality is simply a diamond modality, so a product model in our new sense can just be seen as a multimodal frame, for which bisimulation is perfectly well defined.

3.3.2 Common Knowledge and Unsuccessful Updates

One interesting addition in expressive power occurs if we consider the addition of common knowledge to our language. For a group of agents G, let $R(G)^*$ be the reflexive transitive closure of all the $\sim_i$ relations for $i \in G$. Then the semantics for $C_G\varphi$ are as follows:

$$M, w \models C_G\varphi \text{ iff for all } v \text{ s.t. } \langle w, v\rangle \in R(G)^*, M, v \models \varphi.$$

However, the typical problem with common knowledge in dynamic epistemic logic, is that there is no reduction axiom for formulas for the form $[\sigma]C_G\varphi$. In order to provide a reduction axiom, instead of using the ordinary common knowledge operator, we can instead use a relativized common knowledge operator $C_G(\varphi, \psi)$, which expresses that every G-path which consists exclusively of φ worlds ends in a ψ world van Benthem et al. (2005). Then our ordinary common knowledge operator $C_G\varphi$ is definable as a special case of the relativized version, by $C_G(\top, \varphi)$. Let us define $[\![\varphi]\!]$ as

$$[\![\varphi]\!] = \{w \in W : M, w \models \varphi\}$$

and then we have the semantics for $C_G(\varphi, \psi)$:

$$M, w \models C_G(\varphi, \psi) \text{ iff for all } v \text{ s.t. } \langle w, v\rangle \in (R(G) \cap [\![\varphi]\!])^*, M, v \models \psi.$$

A natural language paraphrase of this operator is "If φ were announced, it would be common knowledge among G that ψ was the case before the announcement." And this is in fact a statement about what agents knew in the past, so it seems quite natural to model it using our past modality. Then it is perhaps not surprising that in $\mathcal{L}_{\mathrm{DEL}+H}$ with common knowledge, we can define relativized common knowledge, since we have the following equivalence:

$$C_G(\varphi, \psi) \equiv [!\varphi]C_G P_{!\varphi}\psi.$$

So the past operator captures something quite natural about relativized common knowledge, even though it does not make it redundant. It also captures some of our intuitions about ordinary common knowledge and public announcements. For instance, we might intuitively consider the formula stating that φ becomes common knowledge after it is announced to be true:

$$[!\varphi]C_G\varphi$$

But Moore sentences give us fairly simple counterexamples, such as $\varphi \equiv p \wedge \neg K_i p$, expressing something like "p is true, and i doesn't know it." These represent cases of unsuccessful updates. Clearly the following formula must always be false:

$$[!(p \wedge \neg K_i p)]K_i(p \wedge \neg K_i p)$$

So after the announcement, φ is not common knowledge , since φ's being announced makes it false. However, perhaps our intuition about common knowledge is better captured by the following formula:

$$[!\varphi]C_G P_{!\varphi}\varphi$$

In other words, what *actually* becomes common knowledge is not φ, but that φ was true just before the announcement. So the problem of expressing just what is learned through a public announcement of φ can be dealt with by using the P_σ modality.

3.3.3 Axiom Variants

To obtain one alternative P-operator, we could drop the indexing condition and simply have an un-indexed $P\varphi$, with the following semantics:

$$M, w \models P\varphi \text{ iff } \exists v, \sigma \text{ such that } w = (v, \sigma) \text{ and } M, v \models \varphi.$$

Clearly, $M, w \models P_\sigma\varphi \Rightarrow M, w \models P\varphi$. We could also modify the axioms in order to account for dropping the indices.

Past Looking Axioms (Unindexed)

(Atomic Permanence)	$Pq \leftrightarrow (P\top \wedge q)$
($\neg$-Reduction)	$P\neg\varphi \leftrightarrow (P\top \wedge \neg P\varphi)$
($\wedge$-Reduction)	$P(\varphi \wedge \psi) \leftrightarrow (P\varphi \wedge P\psi)$

In this case, since we do not need to know which action was performed, we only need to know that some action was, so we know that a world was obtained by performing some action when $P\top$ holds there. This is also why we do not need to axiomatize the interaction of this operator with $[\sigma]$, since we do not care if σ was the actual action which led us to any particular world. Similarly, the unique arrows axiom becomes superfluous.

Furthermore, if we chose to drop the indexing condition, we could even add in an iterated past operator $P^*\varphi$, such that

$$M, w \models P^*\varphi \text{ iff } \exists v_1, \ldots, v_n, \sigma_1, \ldots, \sigma_n \text{ such that } R_{\sigma_1}(w, v_1), \ldots, R_{\sigma_n}(v_{n-1}, v_n)$$
$$\text{and } M, v_n \models \varphi \text{ (assuming only finitely many updates)}$$

This simply states that $P^*\varphi$ is true if φ is true at some point in the transitive closure of the backward-pointing arrows. We could include optional axioms for this operator as well. And we could also allow for composition ; of past steps, since the Kleene-$*$ can be seen as a generalization of that.

PDL Style Axioms

(Composition Axiom)	$P_{\sigma;\tau}\varphi \leftrightarrow P_\sigma P_\tau \varphi$
(Kleene-$*$ Axiom)	$P^*\varphi \leftrightarrow \varphi \vee PP^*\varphi$

However, it is worth noting that an un-indexed operator might actually be more complicated than the indexed one – at least in certain cases. If our action models are only allowed to have finitely many possible actions $\sigma_1, \ldots, \sigma_n$, then $P\varphi$ is clearly equivalent to $P_{\sigma_1}\varphi \vee \ldots \vee P_{\sigma_n}\varphi$. However, if we permitted infinitely many events in our action models, then the un-indexed operator would represent a genuine increase in expressive power. For instance, $\neg P\top$ would only be true at a world with no predecessor. But given infinitely many possible actions, the language with indexed past would require an infinitary formula to express the same statement. So dealing with this version of the operator would require more care.

Further discussion of alternative past operators will have to be postponed—however, the introduction of past iteration would hopefully not be as problematic as the introduction of future iteration would be Miller and Moss (April 2005). Action models will have only finite pasts, a fact which could perhaps be exploited in proving completeness. But this remains an open problem.

References

Baltag A, Moss L, Solecki S (1998) The logic of public announcements, common knowledge and private suspicions. In: Proceedings TARK 1998, Morgan Kaufmann Publishers Inc., pp 43–56

van Benthem J, van Eijck J, Kooi B (2005) Common knowledge in update logics. In: Proceedings of the 10th Conference on Theoretical Aspects of Rationality and Knowledge

van Ditmarsch H, van der Hoek W, Kooi B (2007) Dynamic epistemic logic. Synthese library, Springer, New York

Miller J, Moss L (April 2005) The undecidability of iterated modal relativization. Studia Logica 79:373–407

Sadzik T (2006) Exploring the iterated update universe. In: Report PP-2006-26, Institute for Logic, Language, and Compuation, University of Amsterdam

Chapter 4
Exploring the Power of Converse Events

Guillaume Aucher and Andreas Herzig

4.1 Introduction

4.1.1 Aim: Reason About Perception of Events

Accounting for various modes of perception of events is the aim of a family of formal systems called dynamic epistemic logics. They systems were proposed in a series of publications most prominently by Plaza, Baltag, Gerbrandy, van Benthem, van Ditmarsch, van der Hoek, and Kooi Plaza (1989), Gerbrandy and Groeneveld (1997), Gerbrandy (1999), van Benthem (2006), van Ditmarsch (2002); van Ditmarsch et al. (2007b). Dynamic epistemic logics add dynamics to Hintikka's epistemic logic via transformations of its models.

The focus of dynamic epistemic logics is on particular events that are called updates. Updates can be seen as a more general class than the class of announcements made to the agents. The simplest case of updates are public announcements à la Plaza (1989); when the input is propositional such announcements correspond to AGM expansion operations (Alchourrón et al. 1985). Another example are group announcements à la Gerbrandy (1999) and Gerbrandy and Groeneveld (1997). Note that BMS-updates differ from Katsuno-Mendelzon-like updates as studied in the AI literature since these updates always involve a factual change in the situation at stake (Katsuno and Mendelzon 1992).

In Baltag (2000), Baltag et al. (1998); and Baltag and Moss (2004) and elsewhere, Baltag et col. proposed a dynamic epistemic logic that was very influential. We refer to it in this chapter by the term BMS. It has been shown that their account subsumes all other dynamic epistemic logics, justifying our acronym. The semantics of BMS is based on two kinds of models: a static model M^s (called state model by Baltag, s in M^s for $s$tatic) and a (finite) event model A (called epistemic action model by Baltag). M^s models the actual world and the agents' beliefs about it, and is nothing but a good old epistemic model à la Hintikka. A models the actual event taking

G. Aucher (✉)
Faculty of Sciences, Technology and Communication (FSTC), University of Luxembourg, 6 rue Richard Coudenhove – Kalergi L-1359, Luxembourg
e-mail: guillaume.aucher@uni.lu

P. Girard et al. (eds.), *Dynamic Formal Epistemology*, Synthese Library 351,
DOI 10.1007/978-94-007-0074-1_4,

place and the agents' beliefs about it. An agent's beliefs can be incomplete (event a occurred, but agent cannot distinguish occurrence of a from occurrence of a') and even unsound (a occurred, but agent wrongly perceived some a'). M^s and A are then combined by a restricted product construction which defines the situation after the actual event took place, viz. the resulting actual world, and the agents' beliefs about it.

In this chapter, our first aim is to enrich the (dynamic) epistemic language with a modal operator expressing what was true before an event occurr*ed*. Our second aim is to propose a unified language which does not refer in its syntax to an event model as done in the BMS formalism. Indeed, as its name says, this model is a semantic object. So it seems to us inappropriate to introduce it directly into the syntax of the language (although the way it is actually done in the BMS formalism is formally correct).

4.1.2 Semantics of Events: Products vs. Accessibility Relations

Expressing within the BMS formalism what was true before an event a occurr*ed*, i.e. giving semantics to the converse event a^- is not simple partly because the formal definition of what is true after an event a occurs is already rather involved.

On the other hand, in PDL (Harel et al. 2000), the effects of events are interpreted as transition relations on possible worlds, and not as restricted products of models as in BMS. Converse events a^- can then easily be interpreted by inverting the accessibility relation associated to a. The resulting logic is called the tense extension of PDL. To this we then add an epistemic accessibility relation. We call (tensed) Epistemic Dynamic Logic EDL the combination of epistemic logic and PDL with converse.[1]

A semantics in terms of transition relations is more flexible than the BMS product semantics: we have more options concerning the interaction between events and beliefs. In Section 4.2, we will propose an account that captures this relationship more explicitly than the BMS product semantics does by means of constraints on the respective accessibility relations: a no-forgetting and a no-learning constraint, and a constraint of epistemic determinism.

4.1.3 Translating BMS into EDL

To demonstrate the power of our approach we will provide a translation from BMS to EDL. To do so, we will express the structure of an event model A by a nonlogical theory $\Gamma(A)$ of EDL, and prove that any formula φ is valid in BMS if and only if

[1] EDL is related to Segerberg's Doxastic Dynamic Logic DDL (Segerberg 1995, 1999). But research on DDL focusses mainly on its relation with AGM theory of belief revision, and studies particular events of the form $+\varphi$ (expansion by φ), $*\varphi$ (revision by φ), and $-\varphi$ (contraction by φ).

it is a logical consequence of $\Gamma(A)$ in EDL. We will also show that $\Gamma(A)$ actually characterizes the EDL-models which are generated in the "BMS style" by an event model A.

So, unlike BMS, we avoid referring to a semantical structure (i.e. the BMS event model A) in the very definition of the language. Encoding the structure of a BMS event model A by a nonlogical theory $\Gamma(A)$ of EDL is done thanks to converse events. For example $[a]B_j(\langle a^-\rangle\top \vee \langle b^-\rangle\top)$ expresses that agent j perceives the occurrence of a as that of either a or b.

4.1.4 Organization of the Chapter

This chapter is organized as follows. In Section 4.2 we introduce a language of belief, events and converse events. Then we provide a semantics for that language, and define our logic EDL. In Section 4.3 we give BMS's restricted product semantics for the fragment of the language without converse, and define its logic, also called BMS. In Section 4.4 we provide two embeddings of BMS into EDL: a "semantic" one and a "syntactic" one based on a theory $\Gamma(A)$ associated to each event model A (we prove that the consequences of $\Gamma(A)$ in EDL match the BMS-validities). In Section 4.5 we compare our formalism with van Benthem and Pacuit's logic ETL and other related work. Finally, we conclude in Section 4.6.

4.2 EDL: Epistemic Dynamic Logic with Converse

4.2.1 The Language $\mathcal{L}_{\mathsf{EDL}}$ of EDL

In this chapter, Φ is a countable set of propositional symbols, G is a finite set of agent symbols, and E is a *finite* set of event symbols. (Finiteness of E will be crucial for our results, cf. Definition 4.4.).

Definition 4.1 (Language $\mathcal{L}_{\mathsf{EDL}}$) The language $\mathcal{L}_{\mathsf{EDL}}$ is defined as follows

$$\mathcal{L}_{\mathsf{EDL}} : \varphi ::= \top \mid p \mid \neg\varphi \mid \varphi \wedge \varphi \mid B_j\varphi \mid [a]\varphi \mid [a^-]\varphi,$$

where p ranges over Φ, j over G and a over E.

The dual modal operators $\langle a\rangle$ and $\langle a^-\rangle$ are defined as follows: $\langle a\rangle\varphi$ abbreviates $\neg[a]\neg\varphi$; $\langle a^-\rangle\varphi$ abbreviates $\neg[a^-]\neg\varphi$.

We define the language $\mathcal{L}_{\mathsf{BMS}}$ as the sub-language of $\mathcal{L}_{\mathsf{EDL}}$ without converse operators a^- and the language $\mathcal{L}$ as the sub-language of $\mathcal{L}_{\mathsf{EDL}}$ without dynamic operators a^- and a.

The formula $[a]\varphi$ reads "φ will hold after every possible occurrence of event a". $[a^-]\varphi$ reads "φ held before a". So $[a]B_j[a^-]\bot$ is an $\mathcal{L}_{\mathsf{EDL}}$-formula that is not in $\mathcal{L}_{\mathsf{BMS}}$.

4.2.2 Semantics of EDL

When designing models of events and beliefs the central issue is to account for the interplay between these two concepts. In our PDL-based semantics this is done by means of constraints on the respective accessibility relations.

Definition 4.2 (EDL-model, no-forgetting, no-learning, epistemic determinism) An *EDL-model* is a tuple $M = (W, R, \mathcal{R}, V)$ such that

- W is a non-empty set of possible worlds;
- R: $G \to 2^{W \times W}$ assigns an accessibility relation to each agent;
- $\mathcal{R}$: $E \to 2^{W \times W}$ assigns an accessibility relation to each possible event; and
- V: $\Phi \to 2^W$ is a valuation.

We write R_j and $\mathcal{R}_a$ instead of $R(j)$ and $\mathcal{R}(a)$, and define $R_j(w) = \{v \mid wR_jv\}$ and $\mathcal{R}_a^{-1}(v) = \{w \mid w \in \mathcal{R}_a^{-1}(v)\} = \{w \mid v \in \mathcal{R}_a(w)\}$.

Moreover an EDL-model satisfies the constraints of *no-forgetting*, *no-learning* and *epistemic determinism*:

nf If $v' \in (\mathcal{R}_a \circ R_j)(w)$ then there is $b \in E$ such that $v' \in (R_j \circ \mathcal{R}_b)(w)$.
nl If $(\mathcal{R}_a \circ R_j \circ \mathcal{R}_b^{-1})(w) \neq \emptyset$ then $(R_j \circ \mathcal{R}_b)(w) \subseteq (\mathcal{R}_a \circ R_j)(w)$.
ed If $v_1, v_2 \in \mathcal{R}_a(w)$ then $R_j(v_1) = R_j(v_2)$.

The *no-forgetting* principle says that if after an event a agent j considers a world v' possible, then before this event a agent j already considered possible that there was an event b leading to this world (see Fig. 4.1, left). So everything agent j considers possible after the performance of an event stems from what she considered possible before the event. This principle is a generalization of the perfect recall principle (Fagin et al. 1995).

To understand the *no-learning* principle, also known as no miracles (van Benthem and Pacuit 2006), assume that agent j perceives the occurrence of a as that of $b_1, b_2 \ldots$ or b_n. Then, informally, the *no-learning* principle says that *all* such alternatives resulting from occurrence of $b_1, b_2, \ldots, b_n$ in j's alternatives before

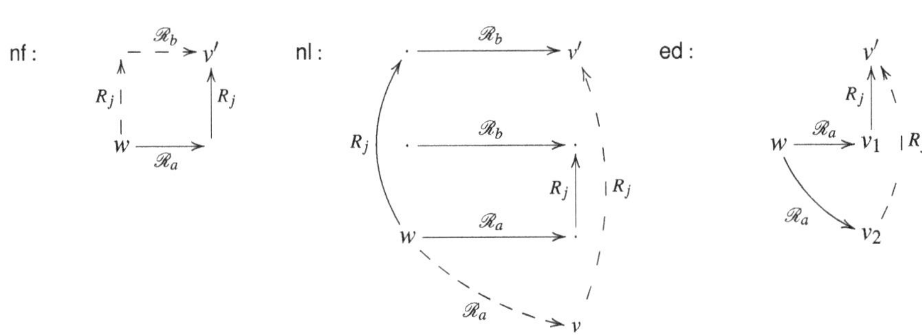

Fig. 4.1 *No-forgetting*, *no-learning* and *epistemic determinism* constraints

a are indeed alternatives after a. In a sense there is no miracles: everything the agent was supposed to consider possible after the event is indeed considered possible after the event (if the latter actually takes place). Formally, assume that agent j perceives b as a possible alternative of a, i.e. $(\mathcal{R}_a \circ R_j \circ \mathcal{R}_b^{-1})(w) \neq \emptyset$. If at w world v' was a possible outcome of event b for j, then v' is possible for j at some $v \in \mathcal{R}_a(w)$ (see Fig. 4.1, middle).

Finally, the *epistemic determinism* principle says that an agent's epistemic state after an event does not depend on the particular nondeterministic outcome. Formally, suppose we have $w\mathcal{R}_a v_1$ and $w\mathcal{R}_a v_2$. Then ed forces that the epistemic states at v_1 and v_2 are identical: $R_j(v_1) = R_j(v_2)$ (see Fig. 4.1, right).

These three constraints delimit the class of events E we consider. Our events are such that the epistemic state of an agent after the occurrence of an event depends only on the previous epistemic state of the agent and on how the event is perceived by the agent, and *not* on which facts hold in the world before or after the event. This feature of our events is formally captured by Proposition 4.1 below: $R_j(w)$ is the epistemic state of the agent before the event and $A_{a,w} = \{b \in E \mid (\mathcal{R}_a \circ R_j \circ \mathcal{R}_b^{-1})(w) \neq \emptyset\}$ is intuitively the set of events that agent j considers as possibly occurring while event a is in fact occurring at world w. For example the event of an agent testing whether φ is the case is not an event of the set of atomic events E. Indeed the epistemic state of this agent after the test (the agent knowing whether φ is true) depends on the actual state of the world (whether φ is true or not). In this example the no-learning constraint is violated. Another example of an event which is not dealt with by our formalism is that of tossing a coin and looking at it. In this example, the epistemic state of the agent after the toss depends on the state of the world after the event, i.e. whether the coin lands heads or tails up. Here the epistemic determinism constraint is violated. On the other hand, both public and private announcements are dealt with by our framework. More generally, any kind of announcement (public, private...) about any kind of information (epistemic, stating that an event just occurred...) is dealt with by our framework. Our events are sometimes called ontic events, feedback-free events or uninformative events (Herzig et al. 2000, de Lima 2007).

Proposition 4.1 *Let* $M = (W, R, \mathcal{R}, V)$ *be a tuple.* M *is an* EDL*-model, i.e.* M *satisfies* nf, nl, ed, *iff for all* $j \in G$, *all* $w \in M$, *all* $a \in E$, *all* $w' \in \mathcal{R}_a(w)$,

$$R_j(w') = \bigcup \{\mathcal{R}_b(v) \mid b \in A_{a,w}, v \in R_j(w)\} \qquad (*)$$

where $A_{a,w} = \{b \in E \mid \mathcal{R}_a \circ R_j \circ \mathcal{R}_b^{-1}(w) \neq \emptyset\}$.

Proof Assume M satisfies nf, nl and ed.

- Let $v' \in R_j(w')$. Then $v' \in (\mathcal{R}_a \circ R_j)(w)$. So by nf there is $b \in E$ and $v \in R_j(w)$ such that $v' \in \mathcal{R}_b(v)$. So $(\mathcal{R}_a \circ R_j \circ \mathcal{R}_b^{-1})(w) \neq \emptyset$ and $b \in A_{a,w}$. So $v' \in \bigcup\{\mathcal{R}_b(v) \mid b \in A_{a,w}, v \in R_j(w)\}$.

- Let $v' \in \bigcup\{\mathcal{R}_b(v) \mid b \in A_{a,w}, v \in R_j(w)\}$. Then there is $b \in E$ such that $v' \in (R_j \circ \mathcal{R}_b)(w)$ and $(\mathcal{R}_a \circ R_j \circ \mathcal{R}_b^{-1})(w) \neq \emptyset$. So by nl, $v' \in (\mathcal{R}_a \circ R_j)(w)$, i.e. there is $w'' \in \mathcal{R}_a(w)$ such that $v' \in R_j(w'')$. Then by ed, $v' \in R_j(w')$.

- Assume M satisfies (*).

  nf Assume that $v' \in (\mathcal{R}_a \circ R_j)(w)$. Then there is $w' \in \mathcal{R}_a(w)$ such that $v' \in R_j(w')$. By (*) there is $b \in A_{a,w}$ and $v \in R_j(w)$ such that $v' \in \mathcal{R}_b(v)$. So there is $b \in E$ such that $v' \in R_j \circ \mathcal{R}_b(w)$.

  nl Assume that $(\mathcal{R}_a \circ R_j \circ \mathcal{R}_b^{-1})(w) \neq \emptyset$ and $v' \in (R_j \circ \mathcal{R}_b)(w)$. Then there is $v \in R_j(w)$ and $b \in A_{a,w}$ such that $v' \in \mathcal{R}_b(v)$. So $v' \in R_j(w')$ for all $w' \in \mathcal{R}_a(w)$, i.e. $v' \in (\mathcal{R}_a \circ R_j)(w)$.

  ed is clearly fulfilled.

Definition 4.3 (Truth conditions for $\mathcal{L}_{\mathsf{EDL}}$) The semantics of $\mathcal{L}_{\mathsf{EDL}}$ is defined inductively as follows. Let M be an EDL-model and $w \in M$.

$$
\begin{array}{ll}
M, w \models \top & \\
M, w \models p & \text{iff } w \in V(p) \\
M, w \models \varphi \wedge \varphi' & \text{iff } M, w \models \varphi \text{ and } M, w \models \varphi' \\
M, w \models B_j\varphi & \text{iff for all } v \in R_j(w), M, v \models \varphi \\
M, w \models [a]\varphi & \text{iff for all } v \in \mathcal{R}_a(w), M, v \models \varphi \\
M, w \models [a^-]\varphi & \text{iff for all } v \in \mathcal{R}_a^{-1}(w), M, v \models \varphi.
\end{array}
$$

Truth of φ in a EDL-model M is written $M \models \varphi$ and is defined as: $M, w \models \varphi$ for every $w \in M$. Let Γ be a set of $\mathcal{L}_{\mathsf{EDL}}$-formulas. Validity of φ in a class of EDL-models $\mathcal{M}$ is written $\mathcal{M} \models \varphi$ and is defined as $M \models \varphi$ for all $M \in \mathcal{M}$. The (global) consequence relation is defined by:

$\Gamma \models_{\mathsf{EDL}} \varphi$ iff for every EDL-model M, if $M \models \psi$ for every $\psi \in \Gamma$ then $M \models \varphi$.

For example we have

$$\{[b]\varphi, \langle a\rangle B_j\langle b^-\rangle\top\} \models_{\mathsf{EDL}} [a]B_j\varphi \quad (*)$$

and

$$\models_{\mathsf{EDL}} (B_j[b]\varphi \wedge \langle a\rangle B_j\langle b^-\rangle\top) \to [a]B_j\varphi \quad (**)$$

Note that in $(*)$, $B_j[b]\varphi$ instead of $[b]\varphi$ is not needed because we use the global notion of logical consequence $\models_{\mathsf{EDL}}$. Now, consider $\varphi = \bot$ in $(**)$: $B_j[b]\bot$ means that perception of event b was unexpected by agent j, while $\langle a\rangle B_j\langle b^-\rangle\top$ means that

j actually perceives a as b. By our no-forgetting constraint it follows that $[a]B_j\bot$, i.e. unexpected events make agents go crazy. In fact, one would like to avoid agents believing inconsistencies: in such situations some sort of belief revision should take place. We do not investigate this further here.

4.2.3 Completeness

Definition 4.4 (Proof system of EDL) The logic EDL is defined by the multi-modal logic K for all the modal operators B_j, $[a]$ and $[a^-]$, plus the axioms schemes Conv_1, Conv_2, NF, NL and ED below:

$$\begin{array}{ll} \mathsf{Conv}_1 & \vdash_{\mathsf{EDL}} \varphi \rightarrow [a]\langle a^-\rangle\varphi \\ \mathsf{Conv}_2 & \vdash_{\mathsf{EDL}} \varphi \rightarrow [a^-]\langle a\rangle\varphi \\ \mathsf{NF} & \vdash_{\mathsf{EDL}} B_j \bigwedge_{a\in E} [a]\varphi \rightarrow \bigwedge_{a\in E} [a]B_j\varphi \\ \mathsf{NL} & \vdash_{\mathsf{EDL}} \langle a\rangle\hat{B}_j\langle b^-\rangle\top \rightarrow ([a]B_j\varphi \rightarrow B_j[b]\varphi) \\ \mathsf{ED} & \vdash_{\mathsf{EDL}} \langle a\rangle B_j\varphi \rightarrow [a]B_j\varphi \end{array}$$

Conv_1 and Conv_2 are the standard conversion axioms of tense logic and converse PDL. NF, NL and ED respectively axiomatize no-forgetting, no-learning and epistemic determinism.

We write $\Gamma \vdash_{\mathsf{EDL}} \varphi$ when φ is provable from the set of formulas Γ in this axiomatics.

One can then show that EDL is strongly complete:

Proposition 4.2 *For every set of $\mathcal{L}_{\mathsf{EDL}}$-formulas Γ and $\mathcal{L}_{\mathsf{EDL}}$-formula φ,*

$$\Gamma \models_{\mathsf{EDL}} \varphi \text{ iff } \Gamma \vdash_{\mathsf{EDL}} \varphi.$$

Proof The proof follows from Sahlqvist's theorem (Sahlqvist 1975): all our axioms NF, NL, ED are of the required form, and match the respective constraints nf, nl, ed.

4.3 BMS: Static Models, Event Models, and Their Products

We here present a star-free version of Baltag's dynamic epistemic logic BMS without the iteration operator $*$ and without common belief (Baltag et al. 1998, Baltag and Moss 2004). We have the same sets of propositional symbols Φ, agent symbols G and event symbols E. We recall that as before, G and E are finite.

4.3.1 Semantics

4.3.1.1 Static Models

are standard epistemic models of the form $M^s = (W, R, V)$, where W is a set of possible worlds, R: $G \to 2^{W \times W}$ assigns an accessibility relation to each agent, and V: $\Phi \to 2^W$ is a valuation.

4.3.1.2 Event Models

are of the form $A = (E, R, Pre)$, where E is a *finite* set of possible events, R: $G \to 2^{E \times E}$ assigns an accessibility relation to each agent, Pre: $E \to \mathcal{L}$ is a precondition function associating epistemic formulas to possible events.

Intuitive interpretation. Informally, $Pre(a)$ is the *pre*condition that a world must fulfill so that the event a can take place in this world. For example $Pre(a) = \top$ means that event a can take place in any world. When we have $R_j(a) = \{b\}$ then the occurrence of a is perceived by agent j as the occurrence of b; when $R_j(a) = \{b_1, b_2\}$ then the occurrence of a is perceived by agent j indistinguishably as the occurrence of b_1 or b_2; etc.

4.3.1.3 Product Construction

Given a static model $M^s = (W, R, V)$ and an event model $A = (E, R, Pre)$, their *product* $M^s \otimes A$ is a static model describing the situation after the event described by A occurred in M^s:

$$M^s \otimes A = (W', R', V')$$

where the new set of possible worlds is $W' = \{(w, a) \mid M^s, w \models Pre(a)\}$, the new valuation is $V'(p) = \{(w, a) \mid w \in V(p)\}$, and the new static accessibility relation is defined by

$$(w_1, a_1)R'_j(w_2, a_2) \text{ iff } w_1 R_j w_2 \text{ and } a_1 R_j a_2.$$

While the truth condition for the epistemic operator is just as in Hintikka's epistemic logic and in EDL, the product construction gives a semantics to the $[a]$ operator which is quite different from that of PDL and EDL. It highlights that BMS is a dynamic extension of epistemic logic, while EDL is an epistemic extension of PDL.

$$M^s, w \models [a]\varphi \text{ iff } M^s, w \models Pre(a) \text{ implies } M^s \otimes A, (w, a) \models \varphi$$

Finally, validity of φ in BMS (noted $\models_{\mathsf{BMS}} \varphi$) is defined as usual as truth in every world of every BMS-model. Note that validity means validity w.r.t. a fixed event model A.

4.3.2 Completeness

Suppose we are given an event model A. The axiomatics of BMS is made up of the principles of the multi-modal logic K for the modal operators B_j and $[a]$, together with the following axioms (Baltag et al. 1998, Baltag and Moss 2004).

(A1) $\vdash_{\mathsf{BMS}} [a]p \leftrightarrow (Pre(a) \rightarrow p)$
(A2) $\vdash_{\mathsf{BMS}} [a]\neg\varphi \leftrightarrow (Pre(a) \rightarrow \neg[a]\varphi)$
(A3) $\vdash_{\mathsf{BMS}} [a]B_j\varphi \leftrightarrow (Pre(a) \rightarrow B_j[b_1]\varphi \wedge \ldots \wedge B_j[b_n]\varphi)$
where $b_1, \ldots, b_n$ is the list of all b such that aR_jb.

We write $\vdash_{\mathsf{BMS}} \varphi$ when φ is provable from these principles. Note that this axiomatization depends on a particular event model A. (We might have written $\vdash^A_{\mathsf{BMS}} \varphi$.)

For example for every event model A where $Pre(a) = \top$, $Pre(b) = p$, and $R_j(a) = \{b\}$ we obtain $\vdash_{\mathsf{BMS}} [a]B_jp$. Indeed, $\vdash_{\mathsf{BMS}} [a]B_jp \leftrightarrow (Pre(a) \rightarrow B_j[b]p)$ and $\vdash_{\mathsf{BMS}} B_j[b]p$ because $\vdash_{\mathsf{BMS}} [b]p$.

4.4 From BMS to EDL

In this section we provide two embeddings of BMS into EDL: a "semantic" one (Section 4.4.1) and a "syntactic" one (Section 4.4.2). This duality will allow us to state a representation theorem in Section 4.4.3 relating these two equivalent characterizations of BMS in EDL.

For the syntactic embedding we will use a particular EDL-theory that encodes syntactically the structure of a given BMS event model A.

Definition 4.5 (Theory of an event model) Let $A = (E, R, Pre)$ be an event model. The *theory of* A, written $\Gamma(A)$, is made up of the following non-logical axioms:

(1) $p \rightarrow [a]p$ and $\neg p \rightarrow [a]\neg p$, for every $a \in E$ and $p \in \Phi$;
(2) $\langle a\rangle\top \leftrightarrow Pre(a)$, for every $a \in E$;
(3) $[a]B_j\big((\langle a_1^-\rangle\top \vee \ldots \vee \langle a_n^-\rangle\top) \wedge ([b_1^-]\perp \wedge \ldots \wedge [b_n^-]\perp)\big)$,
where $a_1, \ldots, a_n$ is the list of all a_i such that $a_i \in R_j(a)$, and $b_1, \ldots, b_n$ is the list of all b_i such that $b_i \notin R_j(a)$;
(4) $\hat{B}_jPre(b) \rightarrow [a]\hat{B}_j\langle b^-\rangle\top$, for every (a, b) such that $b \in R_j(a)$.

Axioms 1 encode the fact that events do not change propositional facts of the world where they occur (see the definition of $V'(p)$ in Section 4.3.1). Axioms 2 encode the fact that an event a can occur in a world iff this world satisfies the precondition of event a (see the definition of W' in Section 4.3.1). Axioms 3 encode the Kripke structure of the event model. Axioms 4 encode the definition of R'_j (see the definition of R'_j in Section 4.3.1).

Example 4.1 Consider that $G = \{i, j\}$ and $\Phi = \{p\}$. In Fig. 4.2 we recall the event models A_1 and A_2 corresponding respectively to the public announcement of φ and

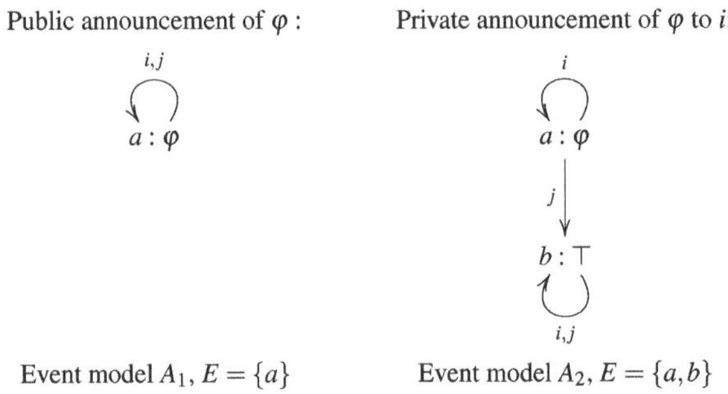

Fig. 4.2 Event models for public announcement and private announcement

the private announcement of φ to A, where $\varphi \in \mathcal{L}$. Here, $Pre(a) = \varphi$ in both models and $Pre(b) = \top$.

Applying Definition 4.5, we obtain that $\Gamma(A_1)$ contains $p \to [a]p$ and $\neg p \to [a]\neg p$ by item (1), $\langle a \rangle \top \leftrightarrow \varphi$ by item (2), $[a]B_i(\langle a^- \rangle \top)$, $[a]B_j(\langle a^- \rangle \top)$, $\hat{B}_i\varphi \to [a]\hat{B}_i\langle a^- \rangle \top$ by item (4), $\hat{B}_j\varphi \to [a]\hat{B}_j\langle a^- \rangle \top$ by item (4).

Besides, $\Gamma(A_2)$ contains $p \to [a]p$ and $\neg p \to [a]\neg p$, $p \to [b]p$ and $\neg p \to [b]\neg p$ by item (1), $\langle a \rangle \top \leftrightarrow \varphi$, $\langle b \rangle \top \leftrightarrow \top$ by item (2), $[a]B_i(\langle a^- \rangle \top \wedge [b^-]\bot)$, $[a]B_j(\langle b^- \rangle \top \wedge [a^-]\bot)$, $[b]B_i(\langle b^- \rangle \top \wedge [a^-]\bot)$, $[b]B_j(\langle b^- \rangle \top \wedge [a^-]\bot)$ by item (3), $\hat{B}_i\varphi \to [a]\hat{B}_i\langle a^- \rangle \top$, $\hat{B}_i\top \to [b]\hat{B}_i\langle b^- \rangle \top$, $\hat{B}_j\top \to [a]\hat{B}_j\langle b^- \rangle \top$, $\hat{B}_j\top \to [b]\hat{B}_j\langle b^- \rangle \top$ by item (4).

4.4.1 A "Semantic" Embedding

We first introduce the notion of forest generated in the BMS style by a static model and an event model (which is just as in Yap's construction (Yap 2006)).

Definition 4.6 Let M^s be a static model and A an event model. We define the tuple $Forest_{\mathsf{EDL}}(M^s, A) = (W, R, \mathcal{R}, V)$ by $W = \bigcup_n W^n$, $V(p) = \bigcup_n V^n(p)$, $\mathcal{R}_a = \bigcup_n \mathcal{R}^n_a$, and $R_j = \bigcup_n R^n_j$, where the tuples $M^n = (W^n, R^n_j, \mathcal{R}^n_a, V^n)$ are defined inductively as follows.[2]

- $M^0 = M^s$
- $M^{n+1} = M^n \otimes_{\mathsf{EDL}} A = (W^{n+1}, R^{n+1}, \mathcal{R}^{n+1}, V^{n+1})$ where
  - $W^{n+1} = W^n \cup \{(w, a) \mid w \in W^n \text{ and } M^s, w \models Pre(a)\}$;
  - $R^{n+1}_j = R^n_j \cup \{((w_1, a_1), (w_2, a_2)) \mid w_1 R^n_j w_2 \text{ and } a_1 R_j a_2\}$;
  - $\mathcal{R}^{n+1}_a = \mathcal{R}^n_a \cup \{(w, (w, a)) \mid w \in W^n\}$;

[2] Note that we use $\otimes_{\mathsf{EDL}}$ to distinguish our product construction here from the BMS product that we write $\otimes_{\mathsf{BMS}}$ from now on to avoid confusion.

- $V^{n+1}(p) = V^n(p) \cup \{(w, a) \mid w \in W^n \text{ and } w \in V^n(p)\}$.

$Forest_{\mathsf{EDL}}(A)$ is defined as the class of all tuples $Forest_{\mathsf{EDL}}(M^s, A)$ where M^s is a static model.

$Forest_{\mathsf{EDL}}(M^s, A)$ is obviously an EDL-model and it is generated by the event model A. So it seems natural that the syntactic encoding $\Gamma(A)$ of this event model be true in $Forest_{\mathsf{EDL}}(M^s, A)$.

Proposition 4.3 *Let M^s be a static model and let A be an event model. Then $Forest_{\mathsf{EDL}}(M^s, A)$ is an EDL-model and $Forest_{\mathsf{EDL}}(M^s, A) \models \Gamma(A)$.*

Proof The proof that $Forest_{\mathsf{EDL}}(M^s, A)$ is an EDL-model is standard. So we only prove the second part of the proposition. Conditions (1) and (2) of Definition 4.5 are clearly fulfilled. As for condition (3), let $w \in W^\infty$, then w' is such that $w\mathcal{R}_a w'$ iff $w' = (w, a)$. Now $(w, a)R_j u$ iff $u = (v, b)$ with $wR_j v$ and $aR_j b$ by definition of $\otimes_{\mathsf{EDL}}$. So for all u such that $(w, a)R_j u$, there are b and v such that $aR_j b$ and $v\mathcal{R}_b u$. This proves that $Forest_{\mathsf{EDL}}(M^s, A), w \models_{\mathsf{EDL}} [a]B_j(\langle a_1^- \rangle\top \vee \ldots \vee \langle a_n^- \rangle\top)$ where $a_1, \ldots, a_n$ is the list of all a_i such that $aR_j a_i$. Finally, concerning condition (4), assume $Forest_{\mathsf{EDL}}(M^s, A), w \models_{\mathsf{EDL}} \hat{B}_j Pre(b)$ and $w\mathcal{R}_a(w, a)$. Then there is v such that $wR_j v$ and $v\mathcal{R}_b(v, b)$. So by definition of $\otimes_{\mathsf{EDL}}$, because $aR_j b$, we have $(w, a)R_j(v, b)$. Hence $Forest_{\mathsf{EDL}}(M^s, A), (w, a) \models_{\mathsf{EDL}} \hat{B}_j\langle b^-\rangle\top$ and finally $Forest_{\mathsf{EDL}}\ (M^s, A), w \models_{\mathsf{EDL}} [a]\hat{B}_j\langle b^-\rangle\top$.

By nature of the EDL setting, $Forest_{\mathsf{EDL}}(M^s, A)$ explicitly represents the iterations of the BMS update product by A *ad infinitum*, starting from the initial static model M^s. Therefore the following proposition is also not surprising.

Proposition 4.4 *Let M^s be a static model and let A be an event model. Then for all $\varphi \in \mathcal{L}_{\mathsf{BMS}}$,*

$$M^s, w \models_{\mathsf{BMS}} \varphi \text{ iff } Forest_{\mathsf{EDL}}(M^s, A), w \models_{\mathsf{EDL}} \varphi.$$

(We added subscripts to $\models$ in order to help the reader to distinguish the two kinds of models.)

Proof We first prove a lemma.

Lemma 4.1 *Let $k \geq 0$. Then $(M^s \otimes_{\mathsf{BMS}} A)^k, (w, a)$ is bisimilar to $M^{k+1}, (w, a)$ (in notation: $(M^s \otimes_{\mathsf{BMS}} A)^k, (w, a) \leftrightarroweq M^{k+1}, (w, a)$), where $(M^s \otimes_{\mathsf{BMS}} A)^k$ is the result of the iteration process applied k times to the static model $M^s \otimes_{\mathsf{BMS}} A$ and the event model A.*

Proof We prove it by induction on k.

$k = 0$: $(M^s \otimes_{\mathsf{BMS}} A)^0 = M^s \otimes_{\mathsf{BMS}} A$, and $M^1 = M^s \otimes_{\mathsf{EDL}} A$. Then by definition of $\otimes_{\mathsf{EDL}}$, we clearly have $(M^s \otimes_{\mathsf{BMS}} A)^0, (w, a) \leftrightarroweq M^1, (w, a)$.

$k+1$: $(M^s \otimes_{\mathsf{BMS}} A)^{k+1} = (M^s \otimes_{\mathsf{BMS}} A)^k \otimes_{\mathsf{EDL}} A$. Now $(M^s \otimes_{\mathsf{BMS}} A)^k, (w,a) \leftrightarrows M^{k+1}, (w,a)$ by induction hypothesis. So $(M^s \otimes_{\mathsf{BMS}} A)^k \otimes_{\mathsf{EDL}} A, (w,a) \leftrightarrows M^{k+1} \otimes_{\mathsf{EDL}} A, (w,a)$ because for any static models M and M', if $M, w \leftrightarrows M', w'$ then $M \otimes_{\mathsf{EDL}} A, w \leftrightarrows M' \otimes_{\mathsf{EDL}} A, w'$.
Then $(M^s \otimes_{\mathsf{BMS}} A)^{k+1}, (w,a) \leftrightarrows M^{k+2}, (w,a)$.

For any formula φ we define the integer $\delta(\varphi)$ as the *maximum number of nested event operator occurrences* as follows:

- $\delta(p) = 0$
- $\delta(\varphi_1 \wedge \varphi_2) = \max(\delta(\varphi_1), \delta(\varphi_2)))$
- $\delta(\neg\varphi) = \delta(B_j\varphi) = \delta(\varphi)$
- $\delta([a]\varphi) = \delta([a^-]\varphi) = \delta(\varphi) + 1$

We set $\mathcal{P}(k)$: "For all $\varphi \in \mathcal{L}_{\mathsf{BMS}}$ such that $\delta(\varphi) = k$, $M^s, w \models_{\mathsf{BMS}} \varphi$ iff $M^k, w \models_{\mathsf{EDL}} \varphi$", where M^s is the static model and M^k is the iteration of the product construction.

We prove $\mathcal{P}(k)$ for all k by induction on k.

$k = 0$: Then φ is epistemic so $\mathcal{P}(0)$ holds by definition of $\otimes_{\mathsf{EDL}}$.
$k+1$: We prove it by induction on φ.

- $\varphi = [a]\varphi'$. We have the following cases:
  $M^s, w \models_{\mathsf{BMS}} [a]\varphi'$
  iff if $M^s, w \models_{\mathsf{BMS}} Pre(a)$ then $M^s \otimes_{\mathsf{BMS}} A, (w,a) \models_{\mathsf{BMS}} \varphi'$
  iff if $M^s, w \models_{\mathsf{BMS}} Pre(a)$ then $(M^s \otimes_{\mathsf{BMS}} A)^k, (w,a) \models \varphi'$ by Induction Hypothesis because $\delta(\varphi') \leq k$,
  iff if $M^s, w \models_{\mathsf{BMS}} Pre(a)$ then $M^{k+1}, (w,a) \models_{\mathsf{EDL}} \varphi'$ by Lemma 4.1
  iff if $M^{k+1}, w \models_{\mathsf{EDL}} Pre(a)$ then $M^{k+1}, (w,a) \models_{\mathsf{EDL}} \varphi'$
  iff $M^{k+1}, w \models_{\mathsf{EDL}} [a]\varphi'$ by definition of $\otimes_{\mathsf{EDL}}$
  iff $M^{k+1}, w \models_{\mathsf{EDL}} \varphi$.
- $\varphi = \varphi_1 \wedge \varphi_2$ works by Induction Hypothesis.
- $\varphi = B_j\varphi'$ works as well.
- $\varphi = p$ is impossible because $k+1 \geq 1$.

Then we can easily prove that for all φ such that $\delta(\varphi) = k$, $M^k, w \models_{\mathsf{EDL}} \varphi$ iff $Forest_{\mathsf{EDL}}(M^s, A), w \models_{\mathsf{EDL}} \varphi$. Then for all k, for all φ such that $\delta(\varphi) = k$, $M^s, w \models_{\mathsf{BMS}} \varphi$ iff $Forest_{\mathsf{EDL}}(M^s, A), w \models_{\mathsf{EDL}} \varphi$, i.e. for all $\varphi \in \mathcal{L}_{\mathsf{BMS}}$, $M^s, w \models_{\mathsf{BMS}} \varphi$ iff $Forest_{\mathsf{EDL}}(M^s, A), w \models_{\mathsf{EDL}} \varphi$.

As a corollary of Proposition 4.4 we get the following "semantic" embedding of BMS into EDL.

Theorem 4.1 *Let A be an event model, and let* $\varphi \in \mathcal{L}_{BMS}$. *Then*

$$\models_{BMS} \varphi \text{ iff } Forest_{EDL}(A) \models_{EDL} \varphi.$$

This theorem illustrates formally the intuition that the fragment of the class of EDL-models that embeds the BMS semantics is the class of EDL-models $Forest_{\mathsf{EDL}}(A)$.

4.4.2 A "Syntactic" Embedding

In this section we prove that $\Gamma(A)$ correctly encodes the event model A from a syntactic point of view, in the sense that for every formula $\varphi \in \mathcal{L}_{\mathsf{BMS}}$,

$$\vdash_{\mathsf{BMS}} \varphi \text{ iff } \Gamma(A) \vdash_{\mathsf{EDL}} \varphi. \quad (***)$$

To do so, we first prove that the axiom of determinism stated in the following proposition is a logical consequence of $\Gamma(A)$ in EDL . This is comforting because the axiom of determinism is indeed valid in BMS .

Proposition 4.5 *Let A be an event model. For every* $\varphi \in \mathcal{L}_{BMS}$ *we have* $\Gamma(A) \models_{EDL} \langle a\rangle\varphi \to [a]\varphi$.

Proof Let $A = (E, R, Pre)$ be a given event model, and let M be an EDL-model such that $M \models \psi$ for every $\psi \in \Gamma(A)$. Assume $w_0\mathcal{R}_a v_0$ and $w_0\mathcal{R}_a u_0$ with $v_0 \neq u_0$. We are going to show that u_0 and v_0 are bisimilar.

Z^e is defined to be an epistemic bisimulation between models M_1 and M_2 if Z^e is a bisimulation between the restriction of these models to epistemic accessibility relations. Let $Z^e := \{(w, w) : w \in W\} \cup \{(v_0, u_0)\}$. We are going to show that Z^e is an epistemic bisimulation. To do so, we need to prove

1. $u_0 \in V(p)$ iff $v_0 \in V(p)$ for all $p \in \Phi$;
2. if $v_0 R_j v'$ then $u_0 R_j v'$;
3. if $u_0 R_j u'$ then $v_0 R_j u'$.

(1) is guaranteed by the first item of Definition 4.5. (2) and (3) are guaranteed by epistemic determinism: ed makes that $R_j(u_0) = R_j(v_0)$.

Now from Z^e, we are going to build up a bisimulation. We proceed as follows.

$Z^0 = Z^e$;

$Z^{n+1} = \{(u_{n+1}, v_{n+1}) \mid u_n\mathcal{R}_a u_{n+1} \text{ and } v_n\mathcal{R}_a v_{n+1} \text{ for some } a \in E \text{ and } u_n Z^n v_n\}$;

$Z = \bigcup_{n\in\mathbb{N}} Z^n$.

We are going to show that Z is a bisimulation.

1. We first show that Z is an epistemic bisimulation: we prove by induction on n that every Z^n is an epistemic bisimulation.
   We have already proved that Z^0 is an epistemic bisimulation. Assume it is true for Z^n and $u_{n+1}Z^{n+1}v_{n+1}$. Then there are u_n, v_n such that $u_n Z^n v_n$, $u_n\mathcal{R}_a u_{n+1}$ and $v_n\mathcal{R}_a v_{n+1}$.

   (a) $u_n \in V(p)$ iff $v_n \in V(p)$ because Z^n is an epistemic bisimulation. So $u_{n+1} \in V(p)$ iff $v_{n+1} \in V(p)$ by Definition 4.5 (1).

(b) Assume $u'_{n+1} \in R_j(u_{n+1})$. Then by nf, there are u'_n and b such that $u'_n \in R_j(u_n)$ and $u'_{n+1} \in \mathcal{R}_b(u'_n)$.

Then there is $v'_n \in W$ such that $v'_n \in R_j(v_n)$ and $v'_n Z^n u'_n$ by induction hypothesis. But $M, u'_n \models Pre(b)$ because $M, u'_n \models \langle b \rangle \top$ and Definition 4.5 (2). Besides for all $\varphi \in \mathcal{L}$, $M, v'_n \models \varphi$ iff $M, u'_n \models \varphi$ because Z^n is an epistemic bisimulation by induction hypothesis. So $M, v'_n \models Pre(b)$ because $Pre(b) \in \mathcal{L}$.

Then there is v'_{n+1} such that $v'_{n+1} \in \mathcal{R}_b(v'_n)$ by Definition 4.5 (2). So $v'_{n+1} \in (R_j \circ \mathcal{R}_b)(v_n)$.

Besides $M, u_n \models \hat{B}_j Pre(b)$, so $M, v_n \models \hat{B}_j Pre(b)$ by induction hypothesis and because $\hat{B}_j Pre(b) \in \mathcal{L}$. So $M, v_n \models [a]\hat{B}_j \langle b^- \rangle \top$ by Definition 4.5 (4).

But $M, v_n \models \langle a \rangle \top$, so $M, v_n \models \langle a \rangle \hat{B}_j \langle b^- \rangle \top$. So $(\mathcal{R}_a \circ R_j \circ \mathcal{R}_b^{-1})(v_n) \neq \emptyset$. So $(R_j \circ \mathcal{R}_b)(v_n) \subseteq (\mathcal{R}_a \circ R_j)(v_n)$ by nl. So there is $v''_{n+1} \in \mathcal{R}_a(v_n)$ such that $v'_{n+1} \in R_j(v''_{n+1})$. Then by ed, $v'_{n+1} \in R_j(v_{n+1})$.

Besides $u'_n Z^n v'_n$ and $u'_{n+1} \in \mathcal{R}_b(u'_n)$, $v'_{n+1} \in \mathcal{R}_b(v'_n)$.

So by definition of Z^{n+1}, $u'_{n+1} Z^{n+1} v'_{n+1}$.

So there is v'_{n+1} such that $v'_{n+1} \in R_j(v_{n+1})$ and $u'_{n+1} Z^{n+1} v'_{n+1}$

(c) The case $v'_{n+1} \in R_j(v_{n+1})$ is similar.

So for all $n \in \mathbb{N}$, Z^n is an epistemic bisimulation. Henceforth Z is also an epistemic bisimulation.

2. Now we are going to show that Z is a full bisimulation. Assume uZv for some $u, v \in W$. Then $uZ^n v$ for some $n \in \mathbb{N}$.

(a) If $u' \in \mathcal{R}_a(u)$ then $M, u \models Pre(a)$ by Definition 4.5 (2). So $M, v \models Pre(a)$ because Z is an epistemic bisimulation and $Pre(a) \in \mathcal{L}^C$.
So there is v' such that $v\mathcal{R}_a v'$. But then $u' Z^{n+1} v'$ by construction of Z^n. So $u' Z v'$.

(b) Similarly we prove that if $v' \in \mathcal{R}_a(v)$ then there is u' such that $u' \in \mathcal{R}_a(u)$ and $u' Z v'$.

Now, we prove the two directions of $(***)$ by means of two propositions.

Proposition 4.6 *Let A be an event model, and let $\psi \in \mathcal{L}_{\mathsf{BMS}}$. If $\not\models_{\mathsf{BMS}} \psi$ then $\Gamma(A) \not\models_{\mathsf{EDL}} \psi$.*

Proof We have to prove that if there is a static model M^s and a w in M^s such that $M^s, w \not\models \psi$ then M^s can be turned into an EDL-model M such that $M \models \Gamma(A)$ and $M, w' \not\models \psi$ for some w' of M. We naturally consider the EDL-model $M = Forest_{\mathsf{EDL}}(M^s, A)$ and $w' = w$.

By Proposition 4.3 we have $Forest_{\mathsf{EDL}}(M^s, A) \models \Gamma(A)$.

By Proposition 4.4 we have $Forest_{\mathsf{EDL}}(M^s, A), w \not\models \psi$.

Proposition 4.7 *Let A be an event model, and let $\psi \in \mathcal{L}_{\mathsf{BMS}}$. If $\models_{\mathsf{BMS}} \psi$ then $\Gamma(A) \models_{\mathsf{EDL}} \psi$.*

Proof We take advantage of the complete axiomatization of BMS-validities given in Baltag et al. (1998) and Baltag and Moss (2004), and show that the BMS-axioms are EDL-valid, and that the BMS-inference rules preserve EDL-validity. As the inference rules of BMS and EDL are identical (i.e. modus ponens and necessitation) it is clear that the BMS-inference rules preserve EDL-theorem hood. It is straightforward to show that every instance of the BMS-axioms not involving dynamic operators is EDL-valid. So what remains is to prove that the BMS schemas

R1 $[a]p \leftrightarrow (Pre(a) \rightarrow p)$
R2 $[a]\neg\varphi \leftrightarrow (Pre(a) \rightarrow \neg[a]\varphi)$
R3 $[a]B_j\varphi \leftrightarrow (Pre(a) \rightarrow B_j[a_1]\varphi \wedge \ldots \wedge B_j[a_n]\varphi)$

where $a_1, \ldots, a_n$ is the list of all a_i such that aR_ja_i, are logical consequences of $\Gamma(A)$ in EDL.

R1 Axiom R1 can be proved by the nonlogical axioms (1) $p \rightarrow [a]p$ and (2) $\langle a\rangle\top \leftrightarrow Pre(a)$ of the theory $\Gamma(A)$ in Definition 4.5.

R2 For the left-to-right direction of R2 we have

$$\Gamma(A) \models_{\mathsf{EDL}} ([a]\neg\varphi \wedge Pre(a) \wedge [a]\varphi) \rightarrow \bot$$

because of the nonlogical axiom (2) $\langle a\rangle\top \leftrightarrow Pre(a)$ of Definition 4.5. For the right-to-left direction, on the one hand we have $\Gamma(A) \models_{\mathsf{EDL}} \neg Pre(a) \rightarrow [a]\bot$ again by the nonlogical axiom (2) of Definition 4.5, and on the other hand $\Gamma(A) \models_{\mathsf{EDL}} \neg[a]\varphi \rightarrow [a]\neg\varphi$ by Proposition 4.5.

R3 For the left-to-right direction of R3, let M be an EDL-model such that $M \models_{\mathsf{EDL}} \Gamma(A)$ and suppose

$$M, w \models_{\mathsf{EDL}} [a]B_j\varphi \wedge Pre(a),$$

and suppose $M, w \models_{\mathsf{EDL}} \neg B_j[b]\varphi$ for some b such that aR_jb. So there must exist worlds w' and v' such that wR_jw', $w'\mathcal{R}_bv'$ and $M, v' \models \neg\varphi$. Therefore $M, w' \models Pre(b)$ by nonlogical axiom 4.5 (2), and $M, w \models_{\mathsf{EDL}} \hat{B}_jPre(b)$. As aR_jb, our nonlogical axiom 4.5 (4) tells us that $M, w \models_{\mathsf{EDL}} \hat{B}_jPre(b) \rightarrow [a]\hat{B}_j\langle b^-\rangle\top$, and hence $M, w \models_{\mathsf{EDL}} [a]\hat{B}_j\langle b^-\rangle\top$. As by hypothesis $M, w \models_{\mathsf{EDL}} Pre(a)$, by nonlogical axiom 4.5 (2) $(\mathcal{R}_a \circ R_j \circ \mathcal{R}_b^{-1})(w) \neq \emptyset$. By the constraint nl on EDL-models we have

$$(R_j \circ \mathcal{R}_b)(w) \subseteq (\mathcal{R}_a \circ R_j)(w),$$

i.e. $v' \in (\mathcal{R}_a \circ R_j)(w)$. As we have supposed that $M, w \models_{\mathsf{EDL}} [a]B_j\varphi$, we must have $M, v' \models_{\mathsf{EDL}} \varphi$, which is contradictory.

For the right-to-left direction of R3, we know that $\Gamma(A) \models_{\mathsf{EDL}} \neg Pre(a) \rightarrow [a]\bot$ again by the nonlogical axiom 4.5 (2), so it remains to prove that

$$\Gamma(A) \models_{\mathsf{EDL}} (B_j[a_1]\varphi \wedge \ldots \wedge B_j[a_n]\varphi) \rightarrow [a]B_j\varphi.(^*)$$

where $a_1, \ldots, a_n$ is the list of all a_i such that aR_ja_i. Suppose $M, w \models_{\mathsf{EDL}} B_j[a_1]\varphi \wedge \ldots \wedge B_j[a_n]\varphi$, and suppose $M, w \models_{\mathsf{EDL}} \neg[a]B_j\varphi$. The latter implies that there are worlds v and v' such that $w\mathcal{R}_a v R_j v'$ and $M, v' \models_{\mathsf{EDL}} \neg\varphi$. By the constraint nf, there is $b \in E$ such that $v' \in R_j \circ \mathcal{R}_b(w)$.

Now, by the nonlogical axiom 4.5 (3) we have

$$[a]B_j\left((\langle a_1^- \rangle \top \vee \ldots \vee \langle a_n^- \rangle \top) \wedge ([b_1^-] \perp \wedge \ldots \wedge [b_n^-] \perp)\right),$$

where $a_1, \ldots, a_n$ is the list of all a_i such that $a_i \in R_j(a)$ and $b_1, \ldots, b_n$ is the list of all b such that $b_i \notin R_j(a)$. Hence $M, v' \models_{\mathsf{EDL}} (\langle a_1^- \rangle \top \vee \ldots \vee \langle a_n^- \rangle \top) \wedge ([b_1^-] \perp \wedge \ldots \wedge [b_n^-] \perp)$. So $b \in R_j(a)$. Then $M, w \models_{\mathsf{EDL}} B_j[b]\varphi$ by (*). So $M, v' \models_{\mathsf{EDL}} \neg\varphi$, which is contradictory.

Putting these two results together we obtain the following "syntactic" embedding of BMS into EDL.

Theorem 4.2 *Let A be an event model, and let $\varphi \in \mathcal{L}_{\mathsf{BMS}}$. Then*

$$\vdash_{\mathsf{BMS}} \varphi \text{ iff } \Gamma(A) \vdash_{\mathsf{EDL}} \varphi$$

Proof Follows easily from Propositions 4.6 and 4.7 by soundness and completeness of BMS and EDL.

This theorem also provides another syntactic characterization of the BMS validities. This syntactic characterization is just made of $\Gamma(A)$ together with the axiomatization of EDL.

4.4.3 A Representation Theorem

Theorems 4.1 and 4.2 give us two characterizations of the BMS logic within EDL. A semantic one: $Forest_{\mathsf{EDL}}(A)$, and a syntactic one: $\Gamma(A)$. From these two results we get easily the following representation theorem.

Theorem 4.3 *Let M be an EDL-model and A be an event model.*

$$M \models \Gamma(A) \textit{ iff } M \textit{ is bisimilar to some } \mathsf{EDL} - \textit{model of } Forest_{\mathsf{EDL}}(A).$$

Proof The right to left direction follows from Proposition 4.3. The left to right direction follows easily from Theorems 4.1 and 4.2.

4.5 Comparison with ETL and Other Related Work

Another approach studying information change over time is Epistemic temporal Logic ETL (Parikh and Ramanujam 2003) (or equivalently interpreted systems Fagin et al. (1995) as shown by Pacuit (2007)). In this section we are going to compare EDL with ETL from the standpoint of Pacuit (2007), van Benthem and Pacuit (2006); and van Benthem et al. (2007) where converse events are introduced as well. We will also study their relationships with the BMS framework and some of its extensions.

4.5.1 Basics of ETL

Let $\mathcal{E}$ be any set. Elements of $\mathcal{E}$ are called *events*, and elements of the set of finite strings $\mathcal{E}^*$ *histories*. For any two sets X and Y, XY is the set of sequences consisting of an object in X followed by one in Y. Given $h \in \mathcal{E}^*$, the *length* of h ($\mathsf{len}(h)$) is the number of events in h. Given $h, h' \in \mathcal{E}^*$, we write $h \preceq h'$ if h is a prefix of h'. Let λ be the empty string. For a set of histories $\mathcal{H} \subseteq \mathcal{E}^*$, $\mathsf{FinPre}_{-\lambda}(\mathcal{H}) = \{h \mid h$ is non-empty and there is $h' \in \mathcal{H}$ such that $h \preceq h'\}$. Given an event $a \in \mathcal{E}$, we write $h \prec_a h'$ if $h' = ha$.

Definition 4.7 Let $\mathcal{E}$ be any set of events. A *protocol* is a set $\mathcal{H} \subseteq \mathcal{E}^*$ with $\mathsf{FinPre}_{-\lambda}(\mathcal{H}) \subseteq \mathcal{H}$. An *ETL- model* is a tuple $(\mathcal{E}, \mathcal{H}, R, V)$ where $\mathcal{E}$ is a finite set of events, $\mathcal{H}$ is a protocol, $R : G \to 2^{\mathcal{H}\times\mathcal{H}}$ assigns an accessibility relation $R(j) = R_j$ to each agent $j \in G$, and $V : \Phi \to 2^{\mathcal{H}}$ is a valuation.

So note that in an ETL-model *events are deterministic* which is not necessarily the case in an EDL-model. The language of ETL is the same as the language $\mathcal{L}_{\mathsf{EDL}}$ of EDL. Truth conditions are defined as usual and we only recall those for the temporal operators.

- $h \models [a]\varphi$ iff $h' = ha \in \mathcal{H}$ and $h' \models \varphi$.
- $h \models [a^-]\varphi$ iff $h = h'a$ for some $h' \in \mathcal{H}$ and $h' \models \varphi$.

ETL-models might satisfy additional constraints listed below.

Definition 4.8 Let $T = (\mathcal{E}, \mathcal{H}, R, V)$ be an ETL-model. T satisfies:

- *Propositional Stability* iff for all $h \in \mathcal{H}, a \in \mathcal{E}, h \models p$ iff $ha \models p$;
- *Perfect Recall* iff for all $h, h'' \in \mathcal{H}, a \in \mathcal{E}$ such that $ha \in \mathcal{H}$ and $h'' \in R_j(ha)$ there is $h' \in R_j(h)$ and $a' \in \mathcal{E}$ such that $h'' = h'a'$;[3]
- *No Miracles* iff for all $h, h' \in \mathcal{H}, a, a' \in \mathcal{E}$ with $ha, h'a' \in \mathcal{H}$, if there are $h'', h''' \in \mathcal{H}$ with $h''a, h'''a' \in \mathcal{H}$ such that $h'''a' \in R_j(h''a)$ and $h' \in R_j(h)$, then $h'a' \in R_j(ha)$;

[3] Note that this definition of perfect recall taken from van Benthem and Liu (2004) is slightly different from the definition of perfect recall in van Benthem and Pacuit (2006) and van Benthem et al. (2007).

- *Weak No Miracles* iff for all $h, h' \in \mathcal{H}, a, a' \in \mathcal{E}$ with $ha, h'a' \in \mathcal{H}$, if there is $h'' \in \mathcal{H}$ with $h''a' \in \mathcal{H}$ such that $h''a' \in R_j(ha)$ and $h' \in R_j(h)$, then $h'a' \in R_j(ha)$;[4]
- *Bisimulation invariance* iff for all epistemically bisimilar $h, h' \in \mathcal{H}$, if $ha \in \mathcal{H}$ then $h'a \in \mathcal{H}$.

Now, given a static model M^s and an event model A, one can naturally define an ETL-model generated in the BMS style, very similarly to the way we defined an EDL-model generated in the BMS style in Definition 4.6.

Definition 4.9 Let $M^s = (W, R, V)$ be a static model and $A = (E, R, Pre)$ be an event model. We define the ETL-model $Forest_{\mathsf{ETL}}(M^s, A) = (\mathcal{E}, \mathcal{H}, R, V)$ as follows.

- $\mathcal{E} = W \cup E$;
- $\mathcal{H} \subseteq WE^*$ and $wa_1 \dots a_n \in \mathcal{H}$ iff $((w, a_1), \dots), a_n) \in W^n$;
- $w'a'_1 \dots a'_n \in R_j(wa_1 \dots a_n)$ iff $((w', a'_1), \dots), a'_n) \in R^n_j(((w, a_1), \dots), a_n))$;
- $wa_1 \dots a_n \in V(p)$ iff $((w, a_1), \dots), a_n) \in V^n(p)$.

The following representation theorem sets some connections between ETL and BMS. It is the counterpart in ETL of our representation Theorem 4.3.

Theorem 4.4 *van Benthem and Liu (2004) An ETL-model T is of the form $Forest_{\mathsf{ETL}}(M^s, A)$ for some static model M^s and some event model A iff T satisfies propositional stability, perfect recall, no miracles and bisimulation invariance.*

However, the right to left direction of this theorem does not hold in general if we use the standard BMS framework (Baltag et al. 1998, Baltag and Moss 2004) used in this chapter (in particular if we assume that T is infinite). Indeed to prove this theorem, the preconditions of the event model A might involve infinite conjunctions and disjunctions of epistemic formulae and not a single epistemic formula as in our chapter and in Baltag et al. (1998) and Baltag and Moss (2004). We are going to need this assumption in the 2nd item of Definition 4.5.

4.5.2 ETL and EDL

To compare EDL and ETL we need a notion of "equivalence" between EDL-models and ETL-models. It is captured here formally by the notion of DT-bisimulation defined as follows.

Definition 4.10 Let $M = (W, R, \mathcal{R}, V)$ be an EDL-model and $T = (\mathcal{E}, \mathcal{H}, R, V)$ be an ETL-model. Let Z be a relation between W and $\mathcal{H}$. We define the property of Z being a DT-bisimulation in $w \in W$ and $h \in \mathcal{H}$, noted $Z : M, w \leftrightarrows_{\mathsf{DT}} T, h$, as follows:

---

[4] This notion of *weak* no miracles is only introduced in our chapter.

- If wZh then for all $p \in \Phi$, $w \in V(p)$ iff $h \in V(p)$.
- If wZh and $w' \in R_j(w)$ then there exists $h' \in R_j(h)$ such that $w'Zh'$.
- If wZh and $h' \in R_j(h)$ then there exists $w' \in R_j(w)$ such that $w'Zh'$.
- If wZh and $w' \in \mathcal{R}_a(w)$ then there exists $h' \in \mathcal{H}$ such that $h \prec_a h'$ and $w'Zh'$.
- If wZh and $h' \in \mathcal{H}$ is such that $h \prec_a h'$ then there exists $w' \in \mathcal{R}_a(w)$ such that $w'Zh'$.

We say that M, w and T, h are DT-bisimilar, noted $M, w \leftrightarrows_{\mathsf{DT}} T, h$ iff there is a relation Z such that $Z : M, w \leftrightarrows_{\mathsf{DT}} T, h$.

Naturally, two "equivalent" models satisfy the same formulas:

Proposition 4.8 *Let M be an EDL-model and T be an ETL-model, $w \in M$ and $h \in T$. If $M, w \leftrightarrows_{\mathsf{DT}} T, h$ then for all $\varphi \in \mathcal{L}_{\mathsf{EDL}}$, $M, w \models \varphi$ iff $T, h \models \varphi$.*

We can now express formally that $Forest_{\mathsf{EDL}}$ and $Forest_{\mathsf{ETL}}$ are "equivalent" constructions.

Proposition 4.9 *Let M^s be a static model and A be an event model.*

$$Forest_{\mathsf{EDL}}(M^s, A) \leftrightarrows_{DT} Forest_{\mathsf{ETL}}(M^s, A)$$

Proof Follows easily from the definition of $Forest_{\mathsf{ETL}}(M^s, A)$.

This ends our mathematical preliminaries for the comparison of EDL and ETL. Now, natural questions to ask are: given an ETL-model, can we find an "equivalent" EDL-model? And vice versa: given an EDL-model, can we find an "equivalent" ETL-model? The answers to both questions are negative: first because an ETL-model does not necessarily satisfy the *no-forgetting* and *no-learning* principles; second, because an EDL-model does not necessarily satisfy the *determinism* principle, that is to say $\mathcal{R}_a$ and $\mathcal{R}_a^{-1}$ are partial functions for all events a. Nevertheless we have the following proposition.

Proposition 4.10 *Any ETL-model satisfying perfect recall and weak no miracles is DT-bisimilar to an EDL-model satisfying determinism, and vice versa.*

Proof We just give the corresponding ETL- and EDL-models. The proof that they satisfy perfect recall, weak no miracles and determinism is routine.

Let $M = (W, R, \mathcal{R}, V)$ be an EDL-model satisfying determinism, and let $w \in W$. We define the corresponding ETL-model $T = (\mathcal{E}, \mathcal{H}, R, V)$ as follows.

- $\mathcal{E} = E$;
- $\mathcal{H} = \bigcup_n \mathcal{H}_n$ where
  - $\mathcal{H}_0 = \{v \mid v \in \left(\bigcup_{j \in G} R_j\right)^* (w)\}$
  - $\mathcal{H}_n = \{wa_1 \ldots a_n \mid a_1, \ldots, a_n \in \mathcal{E} \text{ and } (\mathcal{R}_{a_1} \circ \ldots \circ \mathcal{R}_{a_n})(w) \neq \emptyset\}$ for $n \geq 1$;
- $wa_1 \ldots a_n \in R_j(wb_1 \ldots b_m)$ iff $n = m$ and $v_n \in R_j(u_m)$, where $v_n = (\mathcal{R}_{a_1} \circ \ldots \circ \mathcal{R}_{a_n})(w)$ and $u_m = (\mathcal{R}_{b_1} \circ \ldots \circ \mathcal{R}_{b_m})(w)$;

- $V(p) = \{wa_1 \dots a_n \in \mathcal{H} \mid (\mathcal{R}_{a_1} \circ \dots \circ \mathcal{R}_{a_n})(w) \in V(p)\}$, for all $p \in \Phi$.

We write $v_n = (R_{a_1} \circ \dots \circ R_{a_n})(w)$ instead of $\{v_n\} = (R_{a_1} \circ \dots \circ R_{a_n})(w)$. This makes sense because M satisfies determinism.

Let $T = (\mathcal{E}, \mathcal{H}, R, V)$ be an ETL-model. We define the corresponding EDL-model $M = (W, R', \mathcal{R}, V')$ as follows. $W = \mathcal{H}$; $R' = R$; $h' \in \mathcal{R}_a(h)$ iff $h \prec_a h'$; and $V' = V$.

In fact, note that perfect recall is the ETL-version of our no-forgetting principle and weak no miracles is the ETL-version of our no learning principle.

Now we are going to compare the relationships that ETL and EDL entertain with BMS. On the one hand, an EDL-model M validates the BMS logic if $M \models \Gamma(A)$ for some event model A. On the other hand, an ETL-model validates the BMS logic if it satisfies propositional stability, perfect recall, no miracles and bisimulation invariance (according to Theorem 4.4). The following proposition relates these two conditions.

Proposition 4.11 *Let M be an EDL-model and let A be an event model. If $M \models \Gamma(A)$ then M is DT-bisimilar to an ETL-model satisfying propositional stability, perfect recall, no miracles and bisimulation invariance.*

Proof Let $M = (W, R, \mathcal{R}, V)$ be an EDL-model and $w \in W$. We define the ETL-model $T = (\mathcal{E}, \mathcal{H}, R', V')$ as in the proof of Proposition 4.10. The definition makes sense because if $M \models \Gamma(A)$ then M satisfies determinism by Proposition 4.5. We now have to check that T satisfies propositional stability, perfect recall, no miracles and bisimulation invariance.

Propositional stability holds because of the first item of Definition 4.5. Perfect recall holds by Proposition 4.10. We now check that no miracles and bisimulation invariance hold.

No Miracles Let $h, h' \in \mathcal{H}, a, b \in \mathcal{E}$ with $ha, h'b \in \mathcal{H}$. Assume there are $h'', h''' \in \mathcal{H}$ with $h''a, h'''b \in \mathcal{H}$ such that $h'''b \in R_j(h''a)$ and $h' \in R_j(h)$. Then $h = wa_1 \dots a_n$, $h' = wa'_1 \dots a'_n$, $h'' = wa''_1 \dots a''_m$, $h''' = wa'''_1 \dots a'''_m$. Let $u = (\mathcal{R}_{a_1} \circ \dots \circ \mathcal{R}_{a_n})(w) \in W$, $u' = (\mathcal{R}_{a'_1} \circ \dots \circ \mathcal{R}_{a'_n})(w) \in W$, $u'' = (\mathcal{R}_{a''_1} \circ \dots \circ \mathcal{R}_{a''_m})(w) \in W$ and $u''' = (\mathcal{R}_{a'''_1} \circ \dots \circ \mathcal{R}_{a'''_m})(w) \in W$.

Then we have $u' \in R_j(u)$ (1), and $u''' \in (\mathcal{R}_a \circ R_j \circ \mathcal{R})b^{-1}(u'')$ (2). We have to show that $h'b \in R_j(ha)$, i.e. $u' \in (\mathcal{R}_a \circ R_j \circ \mathcal{R}_b^{-1})(u)$. By (2) we have that $(\mathcal{R}_a \circ R_j \circ \mathcal{R}_b^{-1})(u'') \neq \emptyset$. Therefore $M, u'' \models [a]\hat{B}_j\langle b^-\rangle\top$. So $b \in R_j(a)$ by Definition 4.5 (3). Now, $h'b \in \mathcal{H}$, so $\mathcal{R}_b(u') \neq \emptyset$. Therefore $M, u' \models Pre(b)$ and $M, u \models \hat{B}_j Pre(b)$. So $M, u \models [a]\hat{B}_j\langle b^-\rangle\top$ by Definition 4.5(4). But $ha \in \mathcal{H}$. So $\mathcal{R}_a(u) \neq \emptyset$. Therefore $M, u \models \langle a\rangle\hat{B}_j\langle b^-\rangle\top$, i.e. $\mathcal{R}_a \circ R_j \circ \mathcal{R}_b^{-1} \neq \emptyset$.

So by the no-learning constraint $R_j \circ \mathcal{R}_b(u) \subseteq \mathcal{R}_a \circ R_j(u)$ (3). But $h'b \in \mathcal{H}$. So $\mathcal{R}_b(u') \neq \emptyset$. Therefore there is $v \in \mathcal{R}_b(u')$, so $v \in R_j \circ \mathcal{R}_b(u)$. So $v \in \mathcal{R}_a \circ R_j(u)$ by (3). Finally $u' \in \mathcal{R}_a \circ R_j \circ \mathcal{R}_b^{-1}(u)$.

Bisimulation invariance Let $h, h' \in \mathcal{H}$ which are epistemically bisimilar such that $ha \in \mathcal{H}$. Then we have $h = wa_1 \dots a_n$ and $h' = w'a'_1 \dots a'_n$. Let $u = (\mathcal{R}_{a_1} \circ \dots \circ \mathcal{R}_{a_n})(w) \in W$ and $u' = (\mathcal{R}_{a'_1} \circ \dots \circ \mathcal{R}_{a'_n})(w') \in W$. Then $\mathcal{R}_a(u) \neq \emptyset$, so $M, u \models Pre(a)$. Therefore $M, u' \models Pre(a)$ because u and u' are epistemically bisimilar. So $M, u' \models \langle a \rangle \top$ by Definition 4.5 (3), i.e. $\mathcal{R}_a(u') \neq \emptyset$. So $h'a \in \mathcal{H}$.

Note that the converse of Proposition 4.11 does not hold in general for the same reason that the right to left direction of Theorem 4.4 does not hold in general if we adopt the standard BMS framework.

4.5.3 Other Related Work

Still in the ETL paradigm the authors in van Ditmarsch et al. (2007a) show how to translate a BMS formula satisfied in a static model into an ETL formula satisfied in an interpreted system. So their approach is less general than ours because it only deals with the model checking problem. Starting from the BMS formalism, Yap (2006) and Sack (2008, 2010) introduce a "yesterday" temporal modal operator to the BMS language expressing what was true before the last event; Sack gets a complete characterization. To prove completeness (Sack 2010) also introduces a separate component expressing that an event just occurred but this is not a converse *modal* operator like ours. However he does introduce a converse modal operator for public announcement logic but does not provide a completeness proof for it Sack (2008).

Another approach embedding the BMS formalism into a formalism that also deals with events and beliefs on the same formal level is proposed by van Eijck (2004) and van Benthem et al. (2006). They map the BMS formalism to (epistemic) propositional dynamic logic (refining a similar result for *automata* propositional dynamic logic (van Benthem and Kooi 2004)). However they do not resort to converse events and translate directly event models into a transformation on PDL programs.

In a previous publication of ours (Aucher and Herzig 2007), the constraint of no-forgetting and condition (3) on $\Gamma(A)$ of Definition 4.5 on EDL-models of Definition 4.2 were replaced by the following ones

nf' if $v(\mathcal{R}_a \circ R_j \circ \mathcal{R}_b^{-1})v'$ then vR_jv';
(3)' $\vdash_{\mathsf{BMS}} [a]B_j\varphi \leftrightarrow (Pre(a) \rightarrow B_j[a_1]\varphi \wedge \dots \wedge B_j[a_n]\varphi)$
where $a_1, \dots, a_n$ is the list of all b such that $b \in R_j(a)$.

Neither do EDL models satisfy nf', nor the other way round. Hence the version of EDL in Aucher and Herzig (2007) cannot be compared with our present version. If we moreover assume that event models are serial then we obtain the same results as here. Here we do not need this last assumption and our condition (3) describes more accurately than (3)' the structure of event models. Our constraint nf is also a better generalization of the principle of perfect recall than nf'.

4.6 Conclusion

We have presented an epistemic dynamic logic EDL whose semantics differs from the BMS semantics. We have shown that BMS can be embedded into EDL. This result allows to conclude that EDL is an interesting alternative to Baltag et al.'s logic, that allows to talk about agents' perception of events just in the same way as BMS does. However, EDL is more expressive than BMS because it allows to talk about past events. This is of interest for example in order to model the Sum and Product puzzle, in the formulation of which the sentence "I knew you didn't know" occurs (van Ditmarsch et al. 2007c). Another of its advantages is that EDL allows for incomplete beliefs about the event taking place and can still draw inferences from this incomplete description of the event, while in BMS the event model has to specify everything. So in a sense EDL seems more versatile than BMS to describe events.

On the other hand, the power of event models (actually called action models in BMS) is not completely exploited in the BMS approach. Indeed, the philosophy of the BMS approach is to represent events in the same way as situations are represented in epistemic logic by means of static models. But unlike a static model, an event model does not have a genuine valuation to describe possible events.An obvious extension of the BMS formalism would be to add a valuation to event models in order to describe possible events more precisely. Then we could define epistemic languages for event models completely identical to the various epistemic languages we already defined for static models, except that the propositional letters of these languages would describe possible events instead of possible worlds. This would allow to express things about events that are *currently* taking place, and not only to express things before or after the occurrence of events as in EDL. This would also allow to update/revise events by other events which is a phenomenon that often occurs in everyday life. It is not possible to model such phenomena in EDL because the accessibility relations for events are set once and for all. This new approach is explored in Aucher (2009, 2010).

References

Alchourrón C, Gärdenfors P, Makinson D (1985) On the logic of theory change: Partial meet contraction and revision functions. Journal of Symbolic Logic 50:510–530

Aucher G (2009) BMS revisited. In: Proceedings of Theoretical Aspects of Rationality and Knowledge (TARK 2009), pp 24–33

Aucher G (2010) Characterizing updates in dynamic epistemic logic. In: Proceedings of the twelfth International Conference on the Principles of Knowledge Representation and Reasoning (KR 2010), Toronto, Canada, to appear

Aucher G, Herzig A (2007) From DEL to EDL: exploring the power of converse events. In: Mellouli K (ed) Proceedings of the 9th European Conference on Symbolic and Quantitative Approaches to Reasoning with Uncertainty (ECSQARU 2007), Springer Verlag, LNCS, vol 4724, pp 199–209

Baltag A (2000) A logic of epistemic actions. Tech. Rep., CWI, http://www.cwi.nl/abaltag/papers.html

Baltag A, Moss LS (2004) Logics for epistemic programs. Synthese 139(2):165–224

Baltag A, Moss LS, Solecki S (1998) The logic of public announcements, common knowledge, and private suspicions. In: Proceedings of the TARK'98, Morgan Kaufmann, pp 43–56

van Benthem J (2006) One is a lonely number: on the logic of communication. In: Chatzidakis Z, Koepke P, Pohlers W (eds) Logic Colloquium'02, ASL & A.K. Peters, Wellesley MA, pp 96–129, Tech. Rep. PP-2003-07, ILLC, Amsterdam (2002)

van Benthem J, Kooi B (2004) Reduction axioms for epistemic actions. In: Schmidt R, Pratt-Hartmann I, Reynolds M, Wansing H (eds) AiML-2004: Advances in Modal Logic, University of Manchester, number UMCS-04-9-1 in Tech. Rep. Series, pp 197–211

van Benthem J, Liu F (2004) Diversity of agents in games. Philosophia Scientiae 8(2):163–178

van Benthem J, Pacuit E (2006) The tree of knowledge in action: Towards a common perspective. In: Governatori G, Hodkinson I, Venema Y (eds) Advances in modal logic Volume 6, King's College Press, London, pp 87–106

van Benthem J, van Eijck J, Kooi B (2006) Logics of communication and change. Information and Computation 204(11):1620–1662

van Benthem J, Gerbrandy J, Pacuit E (2007) Merging frameworks for interaction: DEL and ETL. In: Samet D (ed) Theoretical aspect of rationality and knowledge (TARK XI), Brussels, pp 72–82

van Ditmarsch H, Ruan J, van der Hoek W (2007a) Model checking dynamic epistemics in branching time. In: Formal approaches to multi-agent systems 2007 (FAMAS 2007), Durham UK

van Ditmarsch HP (2002) Descriptions of game actions. Journal of Logic, Language and Information (JoLLI) 11:349–365

van Ditmarsch HP, van der Hoek W, Kooi B (2007b) Dynamic epistemic logic. Synthese library, Springer, New York

van Ditmarsch HP, Ruan J, Verbrugge R (2007c) Sum and product in dynamic epistemic logic. Journal of Logic and Computation 18(4):563–588

van Eijck J (2004) Reducing dynamic epistemic logic to PDL by program transformation. Tech. Rep. SEN-E0423, CWI

Fagin R, Halpern JY, Moses Y, Vardi MY (1995) Reasoning about knowledge. MIT Press, Cambridge

Gerbrandy J (1999) Bisimulations on planet kripke. PhD thesis, University of Amsterdam

Gerbrandy J, Groeneveld W (1997) Reasoning about information change. Journal of Logic, Language and Information 6(2):147–169

Harel D, Kozen D, Tiuryn J (2000) Dynamic logic. MIT Press, Cambridge

Herzig A, Lang J, Longin D, Polacsek T (2000) A logic for planning under partial observability. In: Proceedings of the National (US) Conference on Artificial Intelligence (AAAI'2000), Austin, Texas, pp 768–773

Katsuno H, Mendelzon AO (1992) On the difference between updating a knowledge base and revising it. In: Gärdenfors P (ed) Belief revision, Cambridge University Press, pp 183–203 (preliminary version in Allen, J.A., Fikes, R., and Sandewall, E., eds., Principles of Knowledge Representation and Reasoning: Proceedings of the 2nd International Conference, pages 387–394. Morgan Kaufmann Publishers, 1991)

de Lima T (2007) Optimal methods for reasoning about actions and plans in multi-agent systems. PhD thesis, Université de Toulouse, Toulouse

Pacuit E (2007) Some comments on history based structures. Journal of Applied Logic 5(4):613–624

Parikh R, Ramanujam R (2003) A knowledge based semantics of messages. Journal of Logic, Language and Information 12(4):453–467

Plaza JA (1989) Logics of public communications. In: Ras ZW (ed) Proceedings of the Fourth International Symposium on Methodologies for Intelligent Systems (ISMIS 1989), North-Holland

Sack J (2008) Temporal languages for epistemic programs. Journal of Logic, Language and Information 17(2):183–216

Sack J (2010) Logic for update products and steps into the past. Annals of Pure and Applied Logic 161(12):1431–1461

Sahlqvist H (1975) Completeness and correspondence in the first and second order semantics for modal logics. In: Kanger S (ed) Proceedings of the 3rd Scandinavian Logic Symposium 1973, North Holland, no. 82 in Studies in Logic

Segerberg K (1995) Belief revision from the point of view of doxastic logic. Bulletin of the IGPL 3:534–553

Segerberg K (1999) Two traditions in the logic of belief: bringing them together. In: Ohlbach HJ, Reyle U (eds) Logic, language and reasoning: essays in honour of Dov Gabbay, Trends in Logic, vol 5, Kluwer Academic Publishers, Dordrecht, pp 135–147

Yap A (2006) Product update and looking backward. Prepublications PP-2006-39, ILLC

Chapter 5
Modal Logic for Qualitative Dynamics

Darko Sarenac

> *This qualitative study, once completed, will be of the greatest utility to the numerical calculation of the function. Furthermore, this qualitative study will be in itself, of primary interest. Many important questions in Analysis and Mechanics in fact reduce to just this.*
>
> Henri Poincaré

5.1 Introduction

The goal of the present study is to introduce a general formalism in which different dynamic modal logics can be compared and categorized. Our analysis will conform to the relatively standard analysis of dynamical systems that dates back at least as far as Henri Poincaré's work on the three-body problem at the end of the 19th century. The modern incarnation of this study, the theory of complex systems, that these days includes chaotic and nonlinear dynamical systems as well as their better behaved cousins, linear systems is among the hottest scientific pursuits. The top prizes in this pursuit including the understanding of multicellular organisms, ecosystems, social species such as ants, bees, and primates (humans), social and economic complexes. As the name indicates, the theory of *complex* systems, the study of such objects is a difficult endeavor. We as a community of scientists, logicians, and philosophers need all the help that we can get, and as many diverse points of view as we can muster. Our systematic approach to dynamic modal logics has as one of its aims bringing logical approaches one small step closer to the research community that studies complex dynamics. Once we can see various logical systems as instantiations of the same dynamic set of phenomena, we can not only compare the logics among themselves, but we can also participate in an active exchange of results between now numerous fields that study various aspects of complex dynamics: from mathematics, physics, chemistry, and biology, all the way up to computer science, economics, sociology and anthropology. We will use *Iterative Function Systems (IFS)*, a concept familiar from various approaches to dynamics, as the underlying system in which,

D. Sarenac (✉)
Department of Philosophy, Colorado State University, Fort Collins, CO 80523-1781, USA
e-mail: darko@colostate.edu

P. Girard et al. (eds.), *Dynamic Formal Epistemology*, Synthese Library 351,
DOI 10.1007/978-94-007-0074-1_5,

we will claim, a number of interesting modal logics can be fruitfully interpreted.[1] Furthermore, modal systems interpreted in IFS, we will argue, mostly instantiate what is known in the dynamical system community as *local* perspective. We will for comparison present a *global* modal system and argue that a class of such global systems is important to the modal study of dynamics, not only for its computational advantages, but also for the readily available and fully transparent view of underlying dynamics.

We will use somewhat freely concepts borrowed from the theory of dynamical systems, mostly from its mathematical aspects. Strogatz (2001) and Thompson and Stewart (1993) provide an excellent introduction to the field of complex dynamics. We will mostly use basic concepts and explain the ones we use, but if further reference is needed, the two books should provide sufficient background. On the modal logic side, we assume that the reader is familiar with standard modal logic, and at least to some extent, Dynamic Epistemic Logic. No material that could have not been gained through a careful study of Blackburn et al. (2001) will be needed.

5.1.1 Modal View of Dynamics

Phase spaces, formal models of change in dynamical systems, cry out for an interpretation in the language of modal logic. If there were no other reasons, and there are plenty that are philosophically and mathematically fundamental, one would set out to interpret phase spaces modally just because of the view widespread in the philosophical literature that such dynamic models are somehow essentially superior to the more standard possible world models.[2] In our view, the philosopher's conviction notwithstanding, the *phase space* is really only slightly peculiar when viewed as a frame of modal logic. It is peculiar only in that it combines Kripke and topological modal semantics in a single frame.[3] In the standard phase spaces, we have a topological space and a single change tracking function f, both of which are readily amenable to various modal interpretations. The topology will interpret a variety of topological modalities, while the function f, as a special case of a binary relation will interpret a variety of Kripke style relational modalities. Our main goal

---

[1]The conception of an Iterative Function System is somewhat more liberal than that of the computer scientist Michael Barnsley who popularized the usage in Barnsley (1988). We decided to stick with this name for we found it the most evocative of the role that such systems play in modal logic.

[2]For instance, Manuel De Landa in his 2002 book *Intensive Science and Virtual Philosophy* makes such a claim at the end of Chapter 1. He sees dynamical systems, phase spaces, and vector spaces as deeply metaphysically distinct from and preferable to the possible world approach to formal metaphysics.

[3]This combination of the two modal semantics has been introduced quite independently from any dynamic considerations. For instance in various combinations of modal logics such a products or fusion models, it has become commonplace to have both Alexandroff and metric topologies alongside each other. The former is just a transitive, reflexive Kripke frame. Results about their interaction on a single frame are also fairly common these days.

here is to explore modal languages that have been used for interpreting the function f and the various topological properties of the underlying metric space modally. Our main contribution is the introduction of a class of novel qualitative dynamical modalities that in our opinion have an essential role to play in the modal approach to the theory of the dynamical systems. As we hope the presentation will make clear, our language readily generalizes to a number of more specific kinds of function systems that extend or generalize IFS, the general setting for modal thinking about dynamics we introduce next.

5.2 Iterated Function Systems and Some General Notes on Dynamical Systems

We need a general mathematical description of a *Complex Dynamical System* that will enable us both to systematize the taxonomy of the existing *Dynamical Modal Logics* and motivate introduction of new classes of dynamic logics. The mathematical structure that we propose below, IFS, in our view strikes a healthy balance between including as large and as feasible a class of logical systems that claim to be dynamic, and respect for the historical usage of the term "dynamical" in mathematical physics where the term originated.

Let X be some metric topological space, and let $\mathcal{F} = \{f_1, \ldots, f_n\}$, $f_i : X \to X$ for $i \in \{1, \ldots n\}$ be a set of functions on X.

Definition 5.1 (IFS) We call $\mathfrak{X} = (X, \mathcal{F})$ an *iterated function system*, or *IFS* for short.

In the simplest case, $\mathcal{F}$ is a singleton function. In such a case, we write $\mathfrak{X} = (X, f)$.

Example 5.1 For a simple but interesting example, take for instance the closed interval $[0, 1] \subseteq \mathbb{R}$ as the underlying metric space, and the *Tent Map* as the time function f. Tent map is defined as follows:

$$f(x) = rx, \ if \ x < \frac{1}{2}, \ and \tag{5.1}$$

$$f(x) = r(1-x) \ otherwise \tag{5.2}$$

For different values of $r \in (0, 2]$ the behavior of f varies wildly. For instance, at $r = 0.75$, the point 0 acts as an attractor. If we think of time in the IFS as a repeated application of the function f, then over time for any $x \in [0, 1]$, the *orbit* $x, f(x), f(f(x)), f(f(f(x))), \ldots$ will converge towards 0. Take for instance $x = 0.6$. The sequence is then 0.6, 0.3, 0.225, 0.16875, 0.1265625, 0.095 and so on. The sequence clearly approaches 0. The choice of 0.6 was arbitrary. We could have started anywhere in the interval as the point 0 acts as an attractor in the system.[4]

[4] See below for a detailed discussion of attractors.

Note here that we can describe the global behavior of our IFS by simply saying that all orbits, wherever they start, tend towards 0, a pretty concise summary of the dynamics.

If we however set $r = 2$, the behavior becomes much more complex. Now behavior of the IFS becomes *chaotic* in the technical sense. Roughly, taking an average orbit, it will over time come close to *any* point in the interval. The orbit is thus bouncing back and forth around the interval. *Average* is an important term here, as many orbits, in fact countably many of them, will behave in an orderly fashion. For instance if we start with $x = \frac{2}{3}$, the orbit infinitely repeats $\frac{2}{3}$ as $2\left(1 - \frac{2}{3}\right) = \frac{2}{3}$.

VISUALIZING AN IFS

A good intuitive feel for why IFS provides a good model of dynamical systems is provided by *Conway's game of life.* Although the underlying class of spaces Conway uses is different – his spaces are discrete and often finite – the idea behind the formalism of the game is the same. One has a space and a set of change functions. Varying the spaces, the initial setup in the space, and the change functions often produces beautiful, intriguing, and evocative dynamic patterns. The examples of programs illustrating the game of life and related dynamic setups are ubiquitous on the Internet.

The internet encyclopedia `Wikipedia` entree on Conway's game of life contains several striking examples. A nice animation of the dynamics of hexagonal variant of Conway's game that brings the dynamics to life, can be seen at:
`http://en.wikipedia.org/wiki/File:Oscillator.gif.`

5.2.1 Time and Space as Dynamical Control Variables

5.2.1.1 Kinds of Time

Let $\mathfrak{X} = (X, f)$ be an IFS. We can categorize classes of IFS systems according to the properties of f. Thus for example an IFS can be continuous, open, homeomorphic, interior, affine, linear, etc. as reflected by the properties of f. For instance, one would use linear transformations when one is interested in preserving lines but not, say, angles. If one is not interested in preserving lines, but only general shape properties of an object such as connectedness, one would use homeomorphisms.

A further important distinction is between *deterministic* and *stochastic* IFS. The terminology is borrowed from the theory of dynamical systems and differential equations.[5] The two concepts are meant to capture the difference between the kind of change where the next state of the system is rigidly determined, versus the kind of

[5] We mention the distinction between differential and difference equations throughout the chapter. The difference is, in short, that differential equation model time as a continuum–time is modeled

change that can result in one of finitely many states, each with some likelihood. For instance, every time you press the same key on your computer's word processing program, the same symbol appears on the screen. This would be a clear case of deterministic dynamics. In contrast, every time you toss a dice, one of six different options happens, each one in this case with the probability $\frac{1}{6}$. This is a case of a stochastic process. Bellow are formal renderings of this difference.

Definition 5.2 An IFS $\mathfrak{X} = (X, f)$ is said to be *deterministic* if for all $x, y \in X$, if $x \neq y$ then $f(x) \neq f(y)$.

In formal speak, f is nonconvergent. As we have seen in the example of the keyboard, a deterministic IFS models systems that do not change the underlying "atoms" of the space. Spatial relationships between *atoms* change overtime, but the parts remain the same. If you for instance drive your car around town, the matter that the car is made of remains (largely) the same, but what changes are relationships among different parts. Wheels turn, engine parts move, etc. Or for a different example, if you are molding clay, the piece of clay with its molecules remains the same, but the positions of different molecules change over time. Thus the idea of a deterministic IFS is based on a physical intuition according to which the "number" of basic parts, atoms, molecules, remains the same while their topological and geometric arrangements become more or less intricate over time. Furthermore, for any arrangement X at time k, the function f determines the exact arrangement at any subsequent time $k + l$. There is no indeterminism in the system.

The stochastic model retains this basic intuition that the number of spatial parts remains the same, but allows for nondeterministic change. We cannot tell beforehand which of the several available states the current state will be transferred into. Thus, dice toss cannot be modeled deterministically as we can't tell which of the faces the dice will land on. One is to imagine the change function f specifying that $f(x)$ is 1 with the probability $\frac{1}{6}$, 2 with the probability $\frac{1}{6}$, etc.

Formally, let P be a standard probability measure. For a sentence A, we write $P(A) = r$, where $0 \leq r \leq 1$.

Definition 5.3 An IFS $\mathfrak{X} = (X, f)$ is a *stochastic* IFS iff

(i) for all $x \in X, \exists y_1, \ldots, y_n$, $P(f(x) = y_1) = q_1, \ldots, P(f(x) = y_n) = q_n$, where $0 < q_i \leq 1$ for $i \in \{1, \ldots, n\}$;
(ii) $q_1 + \ldots + q_n \leq 1$;
(iii) for all $z \notin \{y_1, \ldots, y_n\}$, $P(f(x) = z) = 0$.

Rather than assigning a fixed successor state y to any state x, the stochastic IFS assigns a finite set of "possible" successors $y_1, \ldots y_n$ to each state x. Each successor is reached with a nonzero probability and the combined probability that one of the successors is reached is 1. The three clauses simply ensure that the transitions are

on a real line $\mathbb{R}$, while the difference equation model discrete time–time is modeled as the natural numbers $\mathbb{N}$.

well behaved probabilistically. In a stochastic IFS, f is best thought of as a relation rather than function and the vertices $(x, f(x))$ are weighted. Each x has at least one relatum y, no more than finitely many relata, and the sum of the weights of the vertices is 1. Thus a stochastic IFS is really modal logical frame (X, R) with the serial, finitely branching, weighted relation R. If the weights on the edges are ignored, the standard modal logical framework arises. We can, then use the standard modal language and explore the dynamics in this way. An interesting question is:

In this study, we will mostly be dealing with deterministic IFS's with a single iterative function f and an underlying metric space. An occasional remark may be made about other kinds of spaces, but their systematic study is largely left untouched here. We think that the stochastic IFS's are definitely worth studying for their own modal logical merits. For instance, even the question of what would count as a suitable modal language for expressing interesting properties of stochastic IFS's is nontrivial.

Question 5.1 *What is the [set of] modal language[s] suitable for exploring the frames with the weighted relation?*

Even the idea of weighted Kripke frame is worth some independent attention. Weighted graphs have gotten a fair amount of attention from mathematicians and computer scientists, but not studied in great detail from the logical perspective. Part 1 of this volume contains a number of important contribution to the study of time in dynamic logic, but also in our thinking about time in dynamical systems in general. Part 3 contains some important contributions on the stochastic notion of time and the its role in dynamic thinking.

5.2.1.2 Some Spatial Variations

For some time now, the wider dynamical systems community has recognized that the structure of space matters.[6] The examples where underlying space influences the diversity of dynamical behaviors in the model abound. For instance, in the dynamical models of evolution, paying attention to the intricacies of spatial patterns has proven useful. A longstanding puzzle of how a species can evolve reproductive self-control has recently been solved using spatial tools. Essentially the researchers realized that a sufficiently high level of spatial segregation among the members of the same species enabled a subgroup that was only moderately eager in their reproductive practice to survive while the larger reproductively overly zealous group went extinct as the result of the environmental over-exploitation. In researchers' own words, using spatial techniques in the model enabled them to show: "how spatially distributed populations avoid overexploiting resources due to the local extinction of over-exploitative variants..."[7]

---

[6] The importance of space in dynamics was first brought to my attention by J. van Benthem and A. Baltag.

[7] See work of Bar Yam at the *New England Institute for Complex System* for the research on the importance of spatial patterns in understanding of evolution. In particular their recent *Nature*

Another striking example from evolutionary biology goes in the opposite direction: rather then disconnecting subgroups, the setting insists on connectedness. Random acts of kindness on their own have been shown insufficient for the evolution of altruism. If, however, an adequate number of such acts are peppered in the space with an appropriate structure, that is, a space that is well-connected, altruism becomes an evolutionarily stable strategy. The structure of space helps us be nicer to each other!

Returning to our exploration of properties of IFS, there are several important kinds of spaces that do not fit our description of the IFS built on a metric topology, but which can be easily accommodated if necessary. First, the requirement that the topology be metric is rather strong and it is often relaxed in the actual study of dynamical systems. There is a large variety of non metric spaces, such as network spaces, lattices, and various Kripke spaces, that are of great significance in the study of complex systems, both for their simplicity and their structural symmetry. Second, there is no reason to limit oneself to just one topology per space. Some of the most successful dynamical logics consist of more than one topology. A familiar example is the Dynamic Epistemic Logic which represents epistemic possibilities for agents as (S4) or (S5) structures which are essentially transitive reflexive Kripke structures and hence fall in the class of Alexandroff spaces. We discuss DEL as an IFS below in detail. Third, there is the singleton topology. This choice of topology is essentially a signal that spatial properties can be ignored. They add no interesting dynamics to the system. Finally, as a negative observation, it is important to note that commonly when one deals with high density grids or lattices in computational representations of dynamics, one often does so simply as a reasonable approximation of some standard metric space and not out of a strong theoretical commitment.

To sum up, occasionally it is well worth relaxing the IFS requirement on spatial properties of the dynamical systems, especially to include graph-like structures and multiple interacting topologies. Graph-like structures often provide one with interesting insight into dynamical systems even when the structure of the graph falls short of meeting the standard of a metric topology or even the low standard of a general topology. Furthermore, as we are sometimes interested in multiple layers of conceptual structure, representing several layers of properties by allowing for several topologies in one system often proves fruitful.

5.2.2 Time, Change, and Dynamics

Intuitively, we think of iterative function systems as temporally evolving systems. Formally, they are essentially systems of *difference equations*. For any given state x in our underlying topological space X and any starting time t, the deterministic IFS gives us a unique *future* state of the system. Thus we can think of f as determining

paper Goodnight et al. (2008) describes the research we mention here. The group also studies the importance of spatial patterns in a variety of other kinds of dynamical systems. Examples range from negotiation of tasks, to large scale design projects, such as design of cars or airplanes.

the unique step by step evolution of the dynamical system $\mathfrak{X}$. For any topological, geometric, or a perfectly randomly assembled object $O \subseteq X$, f gives us the unique trajectory of O in time. In particular it tells us what O "looks like" after n steps for any n. The possible changes of O thus depend on the kind of function that f is. If f is for instance a homeomorphism, and O is say a doughnut–to use the well worn topological example – then $f^n(O)$ is some topological equivalent of the original doughnut O. It could, say, be a cup with a single-holed handle, to continue the overused example, but it could not be a bottle (without a handle). Different maps would enforce stronger or weaker relation between O and $f^n(O)$. If f is an arbitrary map, then we could not predict any particular property of $f^n(O)$ including persistence in time, and if f is a rigid transformation, then we could say a fair bit about O's geometrical features after n stages of applying f assuming that O was reasonably geometrically coherent to begin with. We are mostly interested in well behaved maps that are at the very minimum continuous, but one could sensibly deal with a much wider range of functions. The particular application determines the strength of the function f and the properties preserved by f over time. It is a quite curious observation that in thinking about the world dynamically, the features of time are in some sense determined by our particular goals in exploring the given system. If we are modeling a spatial system, but we are only interested in its topological properties, then our time will be represented by some "topological"function. Time will be a continuous function or a homeomorphism. The more spatial details we are interested in seeing-exploring-predicting, the stronger the function f representing time. Putting things in terms of change rather than time, the stronger the function f, the less change it allows over some fixed period of time. Thus a function f that is a rigid transformation will never allow a ball to be transformed into a box without loss of identity. Thus, the change allowed in a class of objects will determine the way the properties of time apply to that class of objects. The claim here need not be taken overly metaphysically. Simply, the IFS model in which f is a strong geometrical function will be a rather lousy model for a temporal evolution of Play-Doh on a desk of a kindergartener. Such an IFS will hopefully be a pretty good model for changes that your car undergoes during a normal week of operation. Hopefully no nonrigid transformations, squishing or mangling are taking place. There is a lot more we can say about the time/change function f of an IFS. In some sense, continuity is an external property of time function f. Continuity tells you how time interacts with space, but time has inherent features too. For instance, time can be discrete, dense, continuous, or even finite.

5.2.3 CFS: Time as Continuum

IFS is the preferred view of dynamical systems in computer science. Natural representation of time in a computer is discrete; one could of course simulate continuous quantities of time via some discrete approximation, but why not just be honest and realize that the dynamics in a digital computer is always discrete. It is after all designed to be so; hence the designation *digital* rather than *analogue* computer. The main contrast class comes from the physical and mathematical sciences where dynamics and time are most often viewed as a continuum. We can call this approach

CFS approach, standing for *continuous-time function systems*. The two approaches are very closely related and to a large degree complementary. In our view, it bears fruit to explore and understand them side by side. CFS too, may be defined over a metric space X as follows. Let $\mathcal{F} = \{f_1, \ldots, f_n\}$, $f_i : \mathbb{R} \to X$ for $i \in \{1, \ldots n\}$ be a set of differentiable continuous functions onto X, then:

Definition 5.4 (CFS) We call $\mathfrak{X} = (X, \mathcal{F})$ a *continuous-time function system–CFS* for short.

Thus, CFS is essentially a system of differential equations. The main difference between an IFS and a CFS is in their respective notions of time. In an IFS or a difference system, time is discrete. It makes sense to talk about the initial time, and then a sequence of discrete times that follow the initial moment. The functions f_i are said to order the set X temporally, and for $r < q$, $[f_i(r), f_i(q)]$ is a closed temporal interval. If we have more than one function in $\mathcal{F}$, then the notion of time is nondeterministic.

Example 5.2 In IFS it makes sense to say:

(i) In the fifth stage of the evolution of the dynamics of X, the system was stable. In the sixth stage an event B happened and it destabilized the system X in the stage immediately following that one.

In CFS, the time is continuous and the notion of "next moment" does not make sense. Instead of (i), we could say:

(ii) the dynamics of X was stable for a while, but it then destabilized shortly following an event e at $f(r)$.

It is worth noting that there are standard ways of "translating" the difference equation systems to differential systems, but as we know from say temporal logics, the continuous nature of time introduces further interesting formal complications. Thus the move from discrete or even dense time to the continuous time is certainly not trivial. What is often gained, however, is a certain amount of smoothness in the formalism itself. More about the difference between discrete and continuous dynamic logics below.

Example 5.3 For an additional example of the contrast between IFS and CFS consider the difference between the changes in the amount of money in your bank account against the changes in your body weight (expressed in kilograms). Your body weight is a paradigmatic physical quantity that is changing continuously in time. If today you weigh 99 kg and yesterday you weighed 97 kg, a wild change indeed, you have transitioned smoothly from 97 to 99 kg. Another way to put this is that if at time t_1 your weight was 97 kg and at t_2 it was 99 kg, then for any weight w between 97 and 99 kg there was some time t', $t_1 \leq t' \leq t_2$, and your weight was w at t'. Put yet another way, there was no sudden jumps in weight over time, but you rather transitioned smoothly from 97 to 99 kg.[8]

---

[8] In calculus the existence of such smooth transitions is supported by the "Intermediate Value Theorem".

The amount of money in your bank account, in contrast, consists of a discrete series of "jumps". As withdrawals alternate with deposits, the total amount in your account jumps from an old total to a new total. The change is sudden and momentary, and no smooth transition takes place. The latter process can be modeled quite appropriately as a sequence of discrete moments and the totals at any such moment.

The former process seems to be more amenable to modeling smoothly in continuous time. One could certainly model one's weight changes in discrete time and measure, say, daily. In fact, practicality may require one to do so. One would be hard pressed to model bank transactions continuously, though even such a wild formal twist may be useful in financial applications. The decision of whether to go with continuous or discrete time is best left to a case by case approach and particularities of the application at hand.

As significant as the differences between IFS and CFS seem at first, it turns out the two are much closer formally than it first appears, although in logic too, with the exception of temporal logic, the discrete conception of time is the better understood one. There currently remains a fair amount of work to be done on better appreciating the continuous time systems. In the remainder of the chapter, we will treat the two together dealing with significant differences as they arise. In contrast to the standard quantitative approach, our approach here is of a global kind. The best known local approach to dynamics, DTL, which we look at next, also uses discreet time.

5.2.4 Dynamic Topological Logic, DTL

Dynamic Topological Logics as *modal logics* have first been looked at by two Russian-N. American teams, Artemov, Davoren and Nerode at Cornell (Artemov et al. 1997) and Kremer and Mints at Stanford (Kremer and Mints 1997).[9] The two approaches are formally largely identical. They are both based on the simplest IFS, that is, the time considered is discrete, and both use a fairly natural modal language with two temporal and one spatial modality.[10] The main difference – Artemov et al. allow for multiple functions f_i whereas Kremer and Mints treat an IFS with a single

[9] The Russian connection may not be entirely coincidental. The Russians have been known to advance the mathematics of dynamical systems in the 1940s and 1950s when not much interest in such systems existed outside the Soviet Union, at least not in mathematics. It seems that the Western interest in the mathematics of complex systems originates in the late 1960s and early 1970s as a result of various formal and other scientific advances. Lorenz's "discovery" of the butterfly effect in the weather systems, Maynard Smith's work in mathematical biology, Mandelbrot's work on fractals, and somewhat later on work by Feigenbaum and others in physics of chaos. Though this oft repeated view of history seems overly simplified, there is a definite spike of interest in mathematics of dynamical systems across the disciplinary boundaries in that period. A decently philosophically informed historical summary can be found in say Peter Smith's book *Explaining Chaos* Smith (1998).

[10] Plus their existential duals.

function f-turns out not to be of a great formal consequence. We will follow Kremer and Mints's presentation in calling the logic and the system DTL, for Dynamic Topological Logic. The idea for the system is natural. One starts with the simplest IFS, $\mathfrak{X} = (X, f)$. The topological space X of the IFS interprets spatial properties, while the function f determines the temporal behavior of the system. The language chosen to express the properties of X will determine what spatial properties can be expressed, while the language chosen to express the temporal properties will interpret f.

In DTL, one largely concentrates on points $x \in X$ and sets of such points, but one also gains some of the global dynamical perspective by looking at the orbits of a point x. The orbit o_x is a function $o_x : \mathbb{N} \rightarrow X$ where $o_x(0) = x$ and for all $n > 0, o_x(n) = f^n(x)$, that is, the result of n applications of f to x. Another way to view o_x is as a countable sequence $\{x, f(x), f^2(x), f^3(x), \ldots\}$. Essentially, one is interested in general tendencies and behaviors of orbits. Notice that the notion of an orbit extends quite naturally to the continuous time. One simply takes a path along f, that starts with x. In a deterministic CFS such path is clearly unique. o_x is still a function, just that this time the domain is the positive real numbers rather than naturals.

5.2.5 Modalities and Their Semantics

The three modalities in the language of DTL are $\Box$-the standard topological interior modality of McKinsey and Tarski (1944), $\bigcirc$-the temporal next moment modality, and $*$-the temporal henceforth modality. Without going into to much formal detail, here are the semantic renderings of the three modalities:

$\Box\varphi$ is true at some point $x \in X$ if there is an open neighborhood U of x, and $\forall y \in U$, φ is true at y,

$\bigcirc\varphi$ is true at x if φ is true at $f(x)$,

$*\varphi$ is true at x if for all $n \in \mathbb{N}$, $f^n(x)$ makes φ true, where $f^n(x)$ is the result of n successive applications of f to x.

The modalities $\Diamond$-the topological closure operator; and F-"sometime in the future" are also used and of interest, though they are definable as $\Diamond := \neg\Box\neg$ and $F := \neg * \neg$ and thus not needed as primitive.

One could further strengthen the language by adding expressive power to the spatial component of the language. It is well known that for instance topological connectedness cannot be expressed in the language of $\Box/\Diamond$. Furthermore, one could add additional operators to express some metric information contained in the IFS models. The language, however, is already extremely powerful in the sense expressed by measuring computational complexity of the satisfaction problem (SAT) for the resulting logics.

5.2.6 Some Computational Properties of DTL and Its Fragments

It has been argued that the main advantage of modal formalisms over, say, their first and second-order counterparts lies in their relatively low computational complexity.

It is often noted that a modal logic and a first-order logic over the same class of domains commonly have radically different computational properties. Modal logics, it is said, allow for some quantification while keeping the logic in question not only decidable, but also of a very low complexity.[11] If this indeed is the main advantage of Modal Logic over other related counterparts, then DTL does not fare too well. As it has been shown in (Konev et al. 2006a, b), DTL becomes undecidable and even recursively nonaxiomatizable under some very weak assumptions about the time function f. Konev et al. prove a series of high complexity results concerning DTL over some rather natural classes of models. They do so by connecting DTL with products of modal logics over transitive frames which are known to be of high complexity. Putting these results in perspective, we now know that the full DTL interpreted over the class of arbitrary topological spaces with time interpreted as a homeomorphism is at a par in computational complexity with the full first-order logic. In fact, most of the interesting fragments of the full language of DTL are also undecidable and thus the best one can hope for are sufficiently tractable axiomatizations. Put strongly, DTL provides us with very little computational reason to use modal languages rather than, say, full first-order languages or even full second-order languages in reasoning about dynamical systems. Non-axiomatizable modal logics are formally interesting in their own right, of course, and their study has provided us with invaluable insights into reasoning about dynamical systems, but one wonders if there is any way to recover some of the nice computational features of modal logic while keeping with the sprit of DTL's modal view of dynamical systems. We think so, but a change in perspective is needed. One needs to shift from the usual local perspective of modal logic to a more global perspective. This shift in approach is the hallmark of the dynamical system behavior, as we will argue, and the main interest in and the main force behind the complex system thinking is derived from the availability of the global perspective.

5.2.7 Poincare and Topology of Dynamical Systems

As Poincare has observed in his study of the three body problem, for any given dimension and from a specific global perspective, there are only a relatively small number of kinds of dynamical behaviors that are worth distinguishing. That is, if one looks at the global tendencies of a changing system rather than its local behavior, one can isolate a certain number of points to which the local motion is attracted. He called these points attractors.

With this observation, Poincare essentially fathered the study of complex dynamical systems. In his researches, he was responding to the contest called by King Oscar II of Sweden to finally solve the problem of calculating the interactions of a set of three heavenly bodies based on their mutual gravitational influence. The

[11] For the computational perspective on modal logic see for instance Blackburn et al. (2001). For a general introduction on computational complexity see Papadimitriou (1994).

two-body variant of the problem was solved by Newton himself. The three-body and the n-body generalization proved a bit of an embarrassment to the mathematical community. The problem went unsolved for about 300 years! Even for the great mind of Poincare, the three-body problem turned out to be a hard nut to crack. Having worked on an analytic solution for several years, Poincare concluded that the conceptual apparatus needed to understand such systems was hopelessly tricky. He gave up, but not before he did enough work to win King Oscar's prize[12] and had discovered the topology of attractors of a phase space and classified all possible attractors in one dimensional space, i.e., the real line. He did not, however, solve the three body problem. That had to wait for another dozen or so years, 1912 to be precise. An excellent dissection of various aspects of the three-body problem, from both historical and mathematical perspectives, can be found in the wonderfully subtle discussion of Diacu (1996).

Poincare's results concerning *attractors* implied that one can say a great deal about the dynamics of a particular complex system with a weak conceptual apparatus that lacks the capacity for detailed local descriptions. Thus, instead of the complete phase space in all of its detailed glory, to obtain a reasonably complete dynamical picture, one needs only the information about a small number of distinguished points and their relationships with their neighbors. Equipped with such information one can predict how any particular run of the system will evolve without calculating the details of the trajectory. Here is a particularly simple example that nicely illustrates the global topological perspective introduced by Poincare.

5.2.7.1 Dynamics of a One-Dimensional System

We consider the following simple single function IFS that we will call RS.

Example 5.4 ($IFS_{RS} = (\mathbb{R}^*, x^2)$) The space is the set of real numbers $\mathbb{R}$ together with $+\infty$, $-\infty$. We call this set $\mathbb{R}^*$. The sole "change" function f is x^2. We stipulate that x^2 behaves as a fixed point on $+\infty$, and sends $-\infty$ to $+\infty$, that is, $f(+\infty)^2 = +\infty$ and $f(-\infty)^2 = +\infty$. Thus for any point $r \in \mathbb{R}^*$, will in the next moment move to $f(r)$, i.e., r^2.

How will objects in this space behave in the long term? For instance, if the initial condition is 345.65, what will the path look like over many applications of the change function f? What if the initial condition places us at 0.5? Will they converge toward the same point? Poincare has provided us with a general answer to this type of question. Essentially, instead of computing the trajectory of the function x^2 starting with our distinguished points and comparing those trajectories, a tedious task indeed, we can look at the global topology of the system and answer more or less immediately where the trajectories are headed. Poincare's method tells us that

[12] No one lesser than Karl Weierstrass advised the King that Poincare's contribution was substantial enough. This made the fact that there was an actual mistake in the original proposal so much more embarrassing. Poincare later fixed the mistake arriving at the topological accomplishments that we describe here. See Diacu (1996) for historical details.

there are three distinguished points in this IFS, and those three alone, while ignoring all the uncountably many others, tell us much of what we care to know about this simple dynamics. For instance, we know that a path starting at 345.65 will rapidly tend towards $+\infty$. This follows since 345.65 is a positive number and an increasing infinite sequences of squares of positive numbers will have $+\infty$ as their limits. In fact all orbits with initial states in the set that we will call the *basin of attraction*, $(1, +\infty] \cup (-\infty, -1)$, will converge to $+\infty$.

There is a curious bit of dynamics in the initial behavior of the orbits that start in $(-\infty, -1)$. Before initiating their steady march towards $+\infty$, they first "jump" into the positive numbers. Except for that initial leap, the two sets of numbers have the same dynamics. This jump, however, is sufficient to ensure that the system described here is not deterministic. It is not stochastic either, but rather it is *over determined.* Every orbit except the one starting with 0 has two distinct starting points. For instance, the two orbits, $(-2, 4, 16, \ldots)$ and $2, 4, 16, \ldots)$ are identical but for their starting position. If we think of our IFS as a physical dynamics, the situation is curios indeed. Two rather distant events have exactly the same causal consequences. The problem in this particular IFS is systematic. Every sequence of events that does not begin at the origin has two possible beginnings. Perhaps a theistic accommodationist would find this scenario plausible, but for the rest of us it really shows why we are interested mostly in deterministic and stochastic IFS's. In the example below to which we apply modal languages, we will ensure that this curious causal behavior is ruled out.

The three distinguished points are $0, 1, +\infty$. Each one is a fixed points of f,[13] but only two, 0 and $+\infty$, are stable fixed points or attractors. If we were to perturb the initial position away from 0 or $+\infty$ by some small margin ε, with enough time, the trajectory of the new starting point would come arbitrarily close to the attractor. The size of ε is crucial here. To see this, take for instance the attractor 0. If we perturb the starting position by more than say 1, the new wayward trajectory would tend towards $+\infty$. Any stable attractor has a non negligible *basin of attraction* surrounding it. The basin of attraction of 0 is the open interval $(-1, 1)$. It is called the basin of attraction as every orbit with its starting position inside the basin of attraction of 0 will have 0 as its limit.[14]

$+\infty$ behaves in a way similar to 0. It is a stable attractor. As we said earlier, its basin of attraction is $(1, +\infty] \cup (-\infty, -1)$. Any trajectory with the initial point in this basin will have $+\infty$ as its limit. 1 is also a fixed point, that is, $f(1) = 1$, but its basin of attraction is just a singleton consisting of the point itself. What that means is that any perturbation of the trajectory that begins with the point 1, however small, sends the new trajectory drifting away towards a different attractor.

[13] A point x is a fixed point of a function f, if $f(x) = x$.

[14] This is the standard notion of a limit. Here is a quick informal reminder: Let x be in the basin of attraction of y. Then, for any distance δ however small, there is a natural number n, and every $f^m(x)$ for $m > n$, the distance between $f^m(x)$ and y is smaller than δ. That is, the distance between $f^m(x)$ and y gets smaller and smaller as m increases. Recall that $f^m(x)$ is a shorthand for $f(\ldots(f(x)))$, with f applied m times. So $f^3(x)$ is $fff(x)$.

Points that exhibit such unstable behavior are called *repellors*; the flow of the nearby trajectories is diverted away from them.

In one dimensional space with f continuous, there are three kinds of fixed points.[15] Attractors: they attract neighboring orbits from both left and right; repellors: they repel neighboring orbits from both sides, bipolar fixed points: they attract orbits on the right and repel those on the left, or vice versa. This is in an important topological sense a complete taxonomy of dynamical behaviors in one dimension. All other points in one dimension can be labelled "transients". The kinds of flow one gets in the limited topological arrangement of one dimension are of course limited; the kinds of flow in two dimensions will get more complicated, and further dimensionality will add additional complexity of attractors. In each case, however, there is a limit to the eco-diversity of attractors. We can use the taxonomy of attractors of a space and their interrelations to capture the logic of the space in a modal logical setting.

Before we move one, it is worth noting that the IFS = $(\mathbb{R}^*, x^2)$ is at the same time a CFS, ignoring the slight glitch that the time is defined over $\mathbb{R}^*$ rather than $\mathbb{R}$. The function x^2 is certainly continuous and differentiable, we just need to change our outlook on time. Essentially, the event 2-time units after x would be $f(x+2)$ rather than $ff(x)$ and the orbit that starts at x assuming that $f(r) = x$ and that $r \neq x$ is $[f(r), f(q))$ with q being the least $q' > r$ such that $f(q)$ is a fixed point. The IFS $(\mathbb{R}^*, x^3)$ we introduce below can similarly be transformed into a CFS.

5.3 A Case Study: IFS=$(\mathbb{R}^*, x^3)$ via Some Qualitative Modal Languages

We now closely examine a deterministic IFS space $RC = (\mathbb{R}^*, x^3)$ and consider some possible modal interpretations over this simple dynamic space.[16] The space is a lot like the IFS $RS = (\mathbb{R}^*, x^2)$ that we examined above, except that it exhibits some further symmetries. The main differences are:

1. The distinguished points of RC are now five fixed points: $0, -1, 1, -\infty, +\infty$ (compared to the three fixed points in RS).
2. Three of the five points, $0, +\infty, -\infty$ are stable attractors, and two, $-1, 1$, are repellors.

The long term dynamic behavior of any object in RC can be approximated from the two facts above as we can readily infer the basins of attraction for the five fixed points. For completeness of presentation, we list the basins of each point. We label the basin of attraction of a point x, B_x. Then,

---

[15] This does not hold for general maps. A good example is the Tent Map that we defined in Section 2. It is capable of much more complex behavior, including chaos and hence *strange attractors.*

[16] RC stands for the real line with the cubing function. This is in contrast to the earlier RS, the same underlying topology with the squaring function capturing change.

$B_{-\infty} = [-\infty, -1),\ B_{-1} = [-1],\ B_0 = (-1, 1),\ B_1 = [1],\ B_{+\infty} = (1, +\infty]$.[17]

5.3.1 RC and the Local Language of DTL

The first example of a local modal approach to RC is provided by DTL. What makes this approach local is that it does not explicitly account for the attractor level global topological information of the system, at least no mention of such information is made explicitly in the language. If present at all, such global information emerges bottom up from the local detailed description of the model. The topological modalities $\Box$ and $\Diamond$ are interpreted in the standard metric topology over $\mathbb{R}^*$ with the appropriate adjustments to accommodate $+\infty$ and $-\infty$. Further, let $[\varphi]$ be the set of points that make φ true. Let $\propto A$ stand for "the largest open subset of A".[18]

$$[\Box\varphi] = \propto [\varphi].$$

Let $\star A$ be the smallest closed set that contains A.[19]

$$[\Diamond\varphi] = \star[\varphi].$$

For any

$$r \in [\varphi],\ \sqrt[3]{r} \in [\bigcirc\varphi].$$

This follows from the standard definition

$$[\bigcirc\varphi] = f^{-1}([\varphi]).$$

Finally, the interpretation of $*$ as an infinite conjunction leads to the definition:

$$[*\varphi] = \bigcap_{n \geq 0} f^{-n}([\varphi]).$$

As we know from the complexity results mentioned earlier, the unrestricted language of DTL is quite potent, and since x^3 is a continuous function, the DTL_{RC} axioms build on the axioms for DTL over $\mathbb{R}$ with an arbitrary continuous function. This axiomatization is unknown, but the logic is known for instance to be stronger

---

[17] We write $[1]$ for the closed singleton $\{1\}$. Also, it a bit unconventional to think of repellors as having a basin of attraction, but in our view treating their basin as a singleton helps systematize the set of distinguished points, and it facilitates their definition as the fixed points with the singleton basin of attraction.

[18] It is an easy topological observation that this function is (i) well-defined, (ii) the open subset is unique.

[19] Again, some fiddling with complements and the definition of a closed set shows this set to exist and to be unique.

than DTL over the real plane $\mathbb{R}^2$ with an arbitrary continuous function (the latter logic is known).[20] Finding DTL_{RC} would barely amount to much more than a curiosity, and at any rate, our goal in this chapter is philosophical rather than formal. Based on general principles at least, the question of axiomatization seems to be substantially less difficult than the corresponding question for DTL over the reals with unrestricted continuous functions. Answering this question may be an easy, approachable case study in applying modal techniques to a particular dynamical system. We will not concern ourselves here with the following questions, but they are well worth formulating:

Question 5.2 *What is the logic of DTL_{RC}? Is it decidable? Finitely axiomatizable?*

We remark though, that given the simplicity of the function f, and the amount of structure that RC model has, it would not be too surprising if this logic was computationally better behaved that the general DTL over continuous functions.

What does interest us is the issue of the expressive power of this language. DTL has just enough of expressive power that enables it to express some rather structured tilling problems [see (Konev et al. 2006a, b)], but the approach in such construction assumes that the function f is a homeomorphism. That assumption makes the models of DTL satisfy strong grid interaction properties known as *Church-Rosser* and *Commutativity*, which in turn make the language and logic behave like a class of highly complex modal product languages. Weakening the assumption from homeomorphism to a continuous function simply bars this avenue for assessing the complexity of DTL, but the question of whether this weakening of the assumption actually makes the logic less complex or even perhaps decidable is still – at least as of the time of writing this chapter – open. In the case of highly specialized underlying model, however, even if the general case turns out to be undecidable or not even recursively axiomatizable, this special case may turn out to be of low complexity. So what can we say in this complex language? It is often claimed that what you loose on the complexity side of things, you gain in expressivity. Though no real guarantees exist here. You may have an unfortunately designed language that simply expresses all the wrong formal properties. The proofs of high complexity goes through, but no interesting new properties become definable. In our case, the question is what interesting properties of dynamical systems are expressible? Mints and Kremer show that this propositional language can express some interesting topological theorems [see Kremer and Mints (2007), Section 3].

In the systems based on $\mathbb{R}$, we would minimally like to be able to express that a point is fixed, a repellor, and attractor respectively. Fixed points are defined as points that validate $p \rightarrow \bigcirc p$ or equivalently $p \rightarrow *p$. It turns out, however, that when one tries to extend this reasoning further, even with the assumption that the function f is continuous and the rather strong assumption that the space is $\mathbb{R}$, one runs into problems trying to define even the obvious global properties like being an attractor or repellor. In a longer paper, we would define a notion of bisimulation and actually

[20] See Kremer and Mints (2007) for up to date account of the state of DTL.

prove that the two properties are undefinable, but here we just observe that the two are not definable in any obvious way.[21]

There are several ways to extend the DTL language while preserving the local perspective. Most extensions pertain to the spatial fragment, but there are also interesting temporal variants. Among the spatial extensions, adding a universal modality that enables one to express *topological connectivity* and adding some amount of expressivity over the metric properties count as the most obvious. The main temporal proposal would base the logic on an CFS, a continuous time based dynamical system. This system would presumably have only one modality as the next operator, $\bigcirc$ does not make sense here. Related is an exploration of $\Box, *$ fragments of the logic. The $\Box, \bigcirc$ fragment has been studied extensively, but little or no attention is paid to the other obvious option, $\Box, *$.

5.3.2 Qualitative Modal Operators

5.3.2.1 A Simple Global Language for RC

Whatever the actual complexity of DTL_{RC} turns out to be, the fact that DTL plays an important role in the spectrum of modal dynamic logics is undeniable, as is the fact that it is a language that is detail oriented and perhaps best suited to the applications where a great deal of precision is needed and where one readily sacrifices computational efficiency for the extra added detail. The language introduced in this section is on the opposite side of the spectrum. The high level topological global language is best suited for understanding the rough global topological structure of the dynamics embodied in RC.

The language is so weak that it does not need the full detail of the RC's IFS. Instead, we obtain a finite model by filtering through most of the points out of RC. We preserve all the fixed points, as well as the barest outlines of their basins of attraction. Everything else is disposed of. We call the procedure of shrinking the size of the IFS, *attractor filtering*.

The *attractor filtering* of an IFS X is procedure designed to collapse large often uncountable sets of points that form the IFS into more manageable, in fact often finite set. The goal is to preserve as much of the global dynamics of the IFS as possible while eliminating all the extraneous detail. We call the new collapsed frame

---

[21] Tamar Lando found a curious class of models that defeat all reasonable attempts at defining these two properties. One of the models consists of a countable sequence of points approaching 0 from the right, each of the points is an attractor for some sequence of the form $f(x)$, $f^2(x)$, $f^3(x)$, but no such sequence approaches 0 itself. The model seems to be indistinguishable in the language of DTL from either the model that has 0 as an attractor, or conversely a model that has 0 as a repellor. So curiously, although DTL has explosive complexity, it makes it difficult to define even the simplest of global dynamic properties.

AF_X. $AF_X = (Y, R)$ is a pair consisting of the underlying set Y and a set of relations R. We define AF_X in two stages. First we define the set Y.

Definition 5.5 Given an IFS X, we obtain the set Y_X by *attractor filtering* of X.

(i) For every fixed point r in X, we add y_r to Y_X.
(ii) For each fixed point r, we add up to two new points to Y_X. If there is an orbit o_x for $x < r$, and r is the limit of o_x, we add a point o_L to Y_X. Similarly, if there is an orbit o_x for $r < x$, with r as limit, we add a point o_R to Y_X.
(iii) No other points are added to Y_X.

Thus, for RC, the attractor filtering, Y_{RC}, consists of the following nine points. For simplicity of exposition we let i abbreviate $-\infty$ and j abbreviate $+\infty$.

$$y_i,\ o_R^i,\ y_{-1},\ o_L^0,\ y_0,\ o_R^0,\ y_1,\ o_L^j,\ y_j$$

Next, we need to choose a relation set appropriate to the dynamics we wish to capture. This is admittedly a harder task. In the simple model RC we can get away with adding just two relation, R_A and R_R to capture the simple dynamics. To understand what these relations do, we need to look at set Y_X. In some sense, Y_X gets its significance by mixing points and orbits. For example, o_L^0 is clearly a representative of a class of orbits that approach 0 from the left. Now, what about y_0? Is it an orbit consisting of all 0s or just a point? We in fact don't need to answer this question. It is in a sense akin to the relation between light and wave/particle dichotomy. Thus, we treat y_0 as both a representative of an orbit, and a point that attracts other orbits. The relation R_A holds between two points z, w if in the original model the sequence z is attracted to w. Thus, since any sequence that starts in $(-1, 0)$ is attracted to 0, we say that o_L^0 which represents all such sequences is related to y_0, $R_A(y_0, o_L^0)$. Similarly, $R_A(y_0, o_R^0)$, $R_A(y_i, o_R^i)$, $R_A(y_j, o_L^j)$, and slightly less obviously, $R_A(y_i, y_i)$, $R_A(y_{-1}, y_{-1})$, $R_A(y_0, y_0)$, $R_A(y_1, y_1)$, and $R_A(y_j, y_j)$.

The second relation, R_R, records the pairs of a sequence and a point where the sequence drifts away from the point. Thus since all orbits starting in either $[-\infty, -1)$ or $(-1, 0)$ are repelled away from -1, both $R_R(y_{-1}, o_L^0)$ and $R_R(y_{-1}, o_R^i)$ hold. Furthermore, $R_R(y_1, o_R^0)$, and $R_R(y_1, o_L^j)$ also hold. Bellow is the graphical representation of the attracting and repelling in the model. The arrow with the tip on the lower side represents attracting, and the relation R_A is the inverse of the arrow in the sense that $R_A xy$ iff $y \rightharpoondown x$. Similarly the arrow with the upper tip represents repelling, but this time the relation is as it is, not an inverse: $R_R xy$ iff $x \rightharpoonup y$.

$$y_i \leftharpoondown o_R^i \leftharpoonup y_{-1} \rightharpoonup o_L^0 \rightharpoondown y_0 \leftharpoondown o_R^0 \leftharpoonup y_1 \rightharpoonup o_L^j \rightharpoondown y_j$$

As far as the language goes, we predictably add the pair of modalities $\Box_A$ and $\Box_R$. To enable us to talk readily about the dynamical system globally and express claims like "all fixed points are attractors," and other related propositions about the overall behavior of the system, we add the global modality U. Semantically $\Box_A$ and $\Box_R$ are standard modalities interpreted via their corresponding relations R_A and R_R:

$\Box_A\varphi$ is true at a point x, if all points y such that R_Axy make φ true;

$\Box_R\varphi$ is true at a point x, if all points y such that R_Rxy make φ true.

The modality U is not dependent on a relation,

$U\varphi$ is true at x if every point y makes φ true.

Each modality has an associated existential variant: $\Diamond_A$, $\Diamond_R$, and E, all defined in the obvious way.

Notice that repellors and attractors are now defined as a simple matter of accounting. For instance, saying that a point has three R_A successors in one-dimensional model will ensure that it is an attractor. For end points, two attractors and, for the left end point, no relata along either relation to the left. Similar story goes for the right end point, and attractor in general is then a disjunction. Repellors are defined simply by saying that an attractor has a repellor relatum.

This just is the simplest one dimensional example of a dynamical system and the taxonomy of attractors is relatively simple. As we mention briefly above, there really are only three different kinds of fixed points, and the dynamical systems are the ones that can be assembled by combining such attractors on a line, not much to report really. It is still however somewhat surprising that,

PROPEvery one-dimensional IFS with a finite number of fixed points is decidable in the three-modal language of AF. The claim follows from the fact that the filtered model will contain only finitely many points. Extending to the general case for all one-dimensional spaces would likely not be more difficult.

5.3.3 Modal Languages for Higher Dimensional Dynamical Systems

This however does not make the approach trivial. As the number of dimensions increases, the complexity of the behaviors that attractors are capable of increases rapidly. Already in two dimension, we have *saddle nodes*, nodes which attract orbits along one dimension of approach and repel orbits along the other dimension. Furthermore, in two dimensions, the behavior is further complicated as the result of the fact that we now not only have fixed points, which endlessly repeat a single point, but we now have *periodic orbits* that circle periodically along some finite sequence of points. Furthermore, such periodic orbits themselves can be repellors or attractors. One can additionally have circles on the plane that serve as attractors or repellors, and they can be both at once, say repelling in the interior of the circle and attracting on the exterior. Then there are *quasi periodic* behaviors where periodic behavior is not quite achieved, but points periodically remain close enough to the original points of period. For example, an orbit of quasi period 3 may have the sequence 3.4, 6.7, 54.2, 3.6, 6.5, 54.9, 3.2.6.6, 54.5, ... Thus, although the orbit does not return to the exact starting point after three iterations, it remains close enough to the starting point every three iterations, and close enough to the second point in the sequence on 2nd, 5th, 8th,... period. Another way to see this, is by noticing that though not exactly periodic, if one truncates enough decimal places, one ends up with a periodic sequence: $3, 6, 54, 3, 5, 54, 3, 6, 54, \ldots$

Such more complex behaviors would clearly require a more sophisticated modal language to capture interesting phenomena in their global behavior, and in fact as pure repellors and attractors are relatively rare, our language would not be of much use, but a language in its spirit, where the IFS is filtered resulting in a small set of distinguished points. The remaining points are then related via a set of relations most of which essentially have to do with repelling and attracting orbits and the points, or in this case sets of points, that do the attracting and repelling.

There are even more complex behaviors with such obscure labels as *riddled basins*, *fractal boundaries*. Riddled basins are areas around a point that contain intertwined attracted and repelled regions, wheres fractal boundaries are boundaries of a basin of attraction that have fractal properties such as nondifferentiablility and fractal dimension. It seems like an interesting challenge to devise a small set of modal operators capable of expressing various such topological properties of dynamical systems and perhaps introduce a modal logical classification of kinds of dynamics.[22]

So far we have seen the most local class of modal logics of dynamical systems, DTL, and the most global one, based on the attractor filtering of IFS. There are a variety of options in the middle. One can for instance weaken the language of DTL and explore the properties of such weakened DTL system. Completeness and decidability of some such fragments have been explored by Kremer, Mints and others. It would also be interesting to see what are some of expressive features of such languages. Further, one can turn the complexity argument on its head and argue that since DTL has computational properties of such high complexity, why not try strengthen the language to add extra expressive power with the only restriction that the computational properties of the extensions be no worse than those of the systems they started with. So one starts with the class of axiomatizable DTL models, perhaps any extension that preserves axiomatizability is a fair game. There is a variety of desirable spatial properties that one may think of adding. From the simplest like the universal modalities U and E, to the more exotic extension towards stronger quantification or metric and various geometric languages.

Then on the temporal side, one may wish to see what would happened if the IFS were to be replaced by CFS, or continuum based time. It goes without much argument that the high complexity of time in DTL and its interaction with space give result in the high complexity of the system. The fragment of the language that involves $\Box$ and $\bigcirc$ only has been looked at extensively. Curiously, however, the fragment that only involves $\Box$ and $*$ has not plaid a major role in DTL community. Some properties of such combination are already known from the current author's

---

[22] A great source on dynamical systems in general, and variety of global behavior is Strogatz (2001).

work on the products of topological modal logics, and some extensions of that work by Kremer.[23]

Question 5.3 (Some research questions) *Adding the universal modality U to DTL. Does the universal modality increase the complexity of DTL over the class of homeomorphisms (continuous functions)? Does it simplify or further complicate the axiomatizations, when applicable? Zacharyaschev et al. have proposed some interesting modal metric languages. Same concerns as above for such language extensions.*

What is the general logic for the $\Box/*$ fragment of DTL over the class of homeomorphisms (continuous functions)? What are computational properties of this fragment?

Let $\bigoplus$ be a single temporal modality interpreted over the class of CFS with a single time function f. What is the most plausible semantic interpretation for $\bigoplus$? For instance, is it "all moments hereafter" or "all moments in an interval starting with the current point" or something entirely different? Let's call such logic DCTL for Dynamic Continuous-time Temporal Logic. What is the logic of DCTL over the class of homeomorphisms (continuous functions)? What is the complexity?

We are moving on now from the more extreme ends of the spectrum, a fine-grained local perspective of DTL, and global pattern based perspective of AF, towards logics that combine aspects of both global pattern based thinking and the honest labour of detailed approach.

5.3.4 Dynamic Epistemic Logic and the IFS Perspective

If one were to simply poll the number of researchers in the field, the Dynamic Epistemic Logics (DEL) form by far the most important and most studied class of dynamic logics. Not only has it yielded some of the most interesting technical questions in the field, but it has also sprung the most richly diverse class of enhancements and offshoots – logic of action, probabilistic dynamic logic (see Chapter 2), the belief variant, to name just a few. Furthermore, it is probably the best philosophically motivated logic in the dynamic realm. What can the IFS perspective say about this class of logics? Curiously, although there does not seem to be a clear connection between the IFS and DEL, we have started thinking about the IFS perspective while looking at some metalogical problems in Public Announcement Logic in DEL paradigm. DEL is usually presented as a system that starts with a standard modal epistemic logic. Normally one is already in a multimodal setting, that is, there are n agents interacting epistemically. We will concentrate here on

[23] The logic of $*$ in the most standard case just is $S4$, and so the DTL model that only involves $\Box$ and $*$ is a product of an Alexandroff topology and an arbitrary topology. The complexity and various axiomatizations are known for a number of classes of topologies such as Alexandroff and Alexandroff, Alexandroff and Metric, Alexandroff and singleton $\mathbb{Q}$, etc.

the Public Announcement subclass of DEL. In such models, change in the system is then introduced in form of an externally made public announcement, and the subsequent update that the announcement forces upon the knowledge of the agents in the model. Agent knowledge, as in the standard multi-agent epistemic logic, is represented as a Kripke frame that is either transitive and reflexive (S4) or those two plus symmetric (S5). The multi-agent system is formed by having relations $R_1, \ldots, R_n$ over a single universe, with each relation representing another agent. An announcement then changes the model, and the change in the model comes to represent the change in what the agents know. How different or similar is this DEL set-up from the IFS approach? Here are the main points of similarity/difference:

5.3.5 DEL vs. IFS

Space: 1. By taking all sets $U_x = \{y \mid Rxy\}$ as a base of a topology, one can show that (S4), and (S5) frames are readily seen as topologies. Thus, like in IFS, the base space is a topological space.

2. The topology induced by (S4)-frames is called *Alexandroff* topology. (S5)-frames induce *Almost Discrete* topology over their underlying space.[24] Unlike the topological element of IFS, these topologies are not necessarily metrizable, and they are certainly not the standard Euclidean metric topologies that we have been using in the examples.[25]

3. IFS was defined as having single topology over its underlying space. The multi-agent DEL has multiple topologies.

Time: 1. We insisted that IFS has at least one function representing change, possibly many different ones. Thus, whatever the notion of change happens to be in DEL, there is no reason why it should not be representable in an IFS.

2. On the standard view, the change in DEL literally removes a subset of the DEL model. The change functions f in IFS leaves the underlying space intact.

3. DEL has a designated point, the real world. Let x be the designated world. The announcements restrict the class of functions to the ones where for every n, there is a y, s.t. $f^n(y) = x$. That is, the real world has to survive each update; it has to be in the range of every updating function. This is another way of saying that the announcements have to be truthful.

How important are these differences? Let us see what an IFS_{DEL} looks like before we set out to compare them.

---

[24] A topological space is *Almost Discrete* if every open set is closed.

[25] To see that that, for instance, Almost Discrete Topology is not metrizable, notice that its basic open sets induce a partition of the space exactly as R does in (S5)-frames. The elements of the same partition will not be metrically distinguishable. Any metric would have $d(x, x) = 0$, but for any other y in the same partition as x, $d(x, y) > 0$ by the definition of a metric. In a finite partition, which can not be ruled out, this would make singleton x open which it is not by the fact that it is part of the partition.

Definition 5.6 (IFS_{DEL}) As before we begin with some set of points X. We now, however, need to designate a distinguished point $g \in X$.[26]

i) Let $\mathcal{O} = \mathbb{O}_1, \ldots, \mathbb{O}_l$ be a set of Alexandroff [Almost Discrete] topologies over X.
ii) Let $\mathbb{F} = f_1, \ldots, f_k$ be a set of functions $f_i : X \to X$ that satisfy the following restriction: for each i and all $n \in \mathbb{N}$, $f^n(g)$ is in the range of f^n

Then, $IFN_{DEL} = (\mathcal{O}, \mathbb{F})$.

As a first observation concerning the difference between DEL and IFS, note that we can simply treat DEL and PAL as an argument for expanding the concept of IFS to allow for multi-topology. We already allow multiple time/change functions f_i, so unless there is an independent reason to discriminate against topological multiplicity, it seems like a reasonable accommodation. Moreover, on the issue of metricity of the topologies, we can go either way. We can either argue that requiring the topologies of DEL to be metric makes DEL more like the standard dynamical systems, and, hence, may lead to transfer of useful results from the theory of dynamical systems to DEL. One would be especially tempted to take this topological route with respect to DEL, if one could show that none of the usual theorems about properties of DEL are altered significantly by the new restriction that the spaces be metric rather than the usual (S4) and (S5). One could, conversely, argue that the notion of space in IFS needs to be liberalized, not just for the sake of DEL, but also for the sake of other spaces that we mention together with the definition of IFS above. Although we do have some partiality towards the topological perspective, it is not strong enough to push us to either side of this dilemma. So, take your pick.

On the temporal side, we are, at least initially, interested in restricting functions in $\mathbb{F}$ to the class of update functions. The updates are in the language of multi-agent epistemic logic and thus we will have functions corresponding to atoms, the booleans, epistemic modalities, and perhaps some of the group modalities such as group relative common knowledge and Universal Knowledge modalities. We can now set the required restrictions on our update functions. For instance, standardly, for a propositional variable p, the function f_p, that is, the update based on the announcement that p holds (letting $[p]$ stand for the set of points making p true),

$ran(f_p) = [p]$[27] is the golden standard of Public Announcement Logic. After an announcement of p, p becomes true everywhere in the updated model. It also becomes common knowledge among all the agents at each remaining point. The IFS approach strongly suggests, however, a variety of other ways to update an atom. The obvious weakening would be $ran(f_p) \subseteq [p]$. This weakening would go well with some global restriction on all f_is, say, that all functions have their ranges be an open set. There are other plausible update restrictions. Why, for instance, do we

[26] As in Kripke, g is chosen for *Gaia*, mother earth, or real world.
[27] $ran(f)$ stands for the range of f and $dom(f)$ stands for its domain.

want to insist that no $\neg p$ points survive the p-update? We could instead insist, say, that the probability of $\neg p$ after the update is 0, but that alone by no means entails that no $\neg p$ states survive. For instance, we could insist that $ran(f_p) \cap [\neg p]$ has measure 0. That means that the probability of $\neg p$ given $ran(f_p)$ would still be 0, but there could be as many as a countable infinity of $\neg p$ points there, provided that they are sparsely distributed, that is, the set of such points has no density. One can even have uncountably many point if the remaining set of $\neg p$ points is say the classical Cantor set over [0, 1] contrasted with the rest of the interval [0.1]. The options are literally countless.

There is no good reason to stop here. We can rule out updates that leave more than 1, 3, 5, 10, 25% of the $\neg p$ points. As a matter of fact, if I could have my in-class public announcements be followed by 80% of my students, I would count it a success. Whether such restrictions would ultimately work out in the setting of PAL, depends on a lot more than just defining f_p, but the IFS perspective at least puts a plethora of options to a PAL/DEL researcher to explore.

As it ought to be quite familiar to a DEL action community, further restrictions can be devised in interactions among update functions for various formulas. For instance, it seems plausible that $ran(f_{\Box_i p}) \subseteq ran(f_p)$ should hold if $\Box_i$ interprets topological interior operator. As a homework exercise, what would for instance the restriction $ran(f_{\Box_i p}) = ran(f_p)$ tell us about our set $\mathbb{F}$ if we also insisted that each update give us the biggest possible range?

In fact, as we know from the dual language of DEL that allows for the talk of actions, the update functions for various sentences of the epistemic language have to cohere in a certain sense, and it is also important that they sequentially compose in a plausible manner. Thus one needs a set of principles about pairwise combining of functions that produce such a coherent picture. For instance, it would be foolish to require that $ran(f_p) = [p]$ and that $ran(f_{\Box_i p}) = ran(f_p)$, at least if one does not desire to interpret all propositional variables as open sets. In fact, if a pair of restrictions like this are executable, then $[p]$ is an open set. But even if $[p]$ were an open set to begin with, there is no guarantee that some intervening set of updates has not destroyed the openness. Thus we arrive to the archetypal DEL question: can the update f_φ be carried out at the present time?

What this question suggests is that our set $\mathbb{F}$ of IFS_{DEL} is underspecified. We need to decide what happens when a f_φ does not meet its prescribed restrictions. For instance one announces that $\Box_i p$ and the pair of restrictions $ran(f_p) = [p]$ and $ran(f_{\Box_i p}) = ran(f_p)$ both hold. Do we just say that the preconditions for this announcement are not met if p is not an open set? Or do we perhaps update and throw out one of the restrictions? The standard PAL way is not to update. This is akin to observing that f_φ has crashed after 0 applications. Similarly, if φ does not hold at the designated point g, f_φ crashes immediately, but this could change with a lucky sequence of updates. Thus both, there may be updates that you can make now, or perhaps even make for some finite number of times, which then cease to be updateable. Conversely, there could be an update that requires some finite other set of updates before it can be made. This gives us a classification of kinds of update functions f_φ:

There are f_φ that can:

(i) never be executed in a model M. ($\varphi = p \wedge \neg p$)
(ii) not be executed now, but there is a sequence of updates $\langle f_1, \ldots, f_n \rangle$ of size n, that when executed enables us to execute f_φ. ($\varphi = \Box_i p$ that is false now at g, but becomes true with some updates.)
(iii) be executed once, and never again. Twice and never again, thrice and never again, ... (Once, Moore's formula.)
(iv) [be executed some finite number of times, then again several times but not before another sequence has been executed, ...] (not sure about these)
(v) always be executed infinitely many times, alone, or in combination with other sequences, whatever the circumstances. (Propositional variables true at g are like this.)

We can now turn the questions about classes of formulas such as for instance the well-known one about formulas that are preserved after updates, into questions about relation between updates. For instance, we now ask which update functions never change applicability of other update functions? Which update functions are self-undermining? What exact features make them so? Which functions undermine other functions? Which ones? For instance, a $f_{\neg p}$ update undermines all existential p updates that even if they were executable before this one was made, are not executable afterwards. Thus, one can build a tree of possible executions to try to understand various dependencies among update functions.

5.3.5.1 DEL, Global or Local Logic?

DEL has a curious status with regards to the question of at what level it reasons about its dynamics. On the face of it, it is a local system as it only looks at points and the relations between them, but when one asks the question of where the agents are in the model, one realizes that agents are nowhere to be seen. They certainly are not associated with any given point, as any point except g can disappear without anyone skipping a beat. They are not in the connections either, at least not in any simple sense. Rather, agents and their epistemic properties *emerge* from the structure of the model. Now if the emergence is the hallmark of the global perspective, then DEL model subtly exhibits the global approach.

5.4 Conclusion

We hope to have demonstrated that IFS provides not only a good global system for comparing various systems of Dynamic Modal Logic , but also that thinking about dynamics in this way raises a variety of new and interesting questions. The hope for the future is to look at some of the proposed avenues in greater detail.

Acknowledgments The author wishes to thank Patrick Girard for his enduring patience, indefatigable support and friendship over the years. He is the kind of friend who was there in good times but who did not shy to lend his support and advice when the sailing was not so smooth. It is not an exaggeration to say that without Patrick's support and encouragement during several critical years the author may not have been in the position to be writing this chapter now. In those tough times, Patrick exhibited the kind caring humanity that I will forever appreciate and never forget. On a more directly related level, numerous conversations with him inspired, challenged, and oftentimes redirected and restructured the author's thinking about the topics of both the dynamical systems and their relation to logic. More specifically, without Patrick, the author would not have found the strength and direction to materialize the present chapter in time and form that this high quality volume required. He provided an endless stream of invaluable comments and suggestions which were at the same time critical and direct, and kind and encouraging. It is a great privilege to count him among friends.

The author also wishes to thank Tamar Lando for valuable comments and corrections on an earlier draft.

References

Artemov SN, Davoren J, Nerode A (1997) Modal logics and topological semantics for hybrid systems. Tech. Rep., Cornell University

Barnsley M (1988) Fractals Everywhere. Academic Press Inc., New York

Blackburn P, de Rijke M, Venema Y (2001) Modal logic. Cambridge University Press, Cambridge

Diacu F (1996) The solution of the n-body problem. The Mathematical Intelligencer 18:66–70

Goodnight C, Rauch E, Sayama H, de Aguiar M, Baranger M, Bar-Yam Y (2008) Evolution in spatial predator-prey models and the "prudent predato": The inadequacy of steady-state organism fitness and the concept of individual and group selection. Complexity 13(5):23–44

Konev B, Kontchakov R, Tishovsky D, Wolter F, Zakharyaschev M (2006a) On dynamic topological and metric logics. Studia Logica 84(1):129–160

Konev B, Kontchakov R, Wolter F, Zakharyaschev M (2006b) Dynamic topological logics over spaces with continuous functions. Tech. Rep., available via http://www.csc.liv.ac.uk/frank/

Kremer P, Mints G (1997) Dynamic topological logic. Bulletin of Symbolic Logic 3:371–372

Kremer P, Mints G (2007) Dynamic topological logic. In: M Aiello JvB I Pratt-Hartmann (ed) Handbook of spatial logic, Springer, New York pp 565–606

McKinsey JCC, Tarski A (1944) The algebra of topology. Annals of Mathematics 45:141–191

Papadimitriou C (1994) Computational complexity. Addison-Wesley, New York

Smith P (1998) Explaining Chaos. Cambridge University Press, New York

Strogatz SH (2001) Nonlinear dynamics and Chaos: With applications to physics, biology, chemistry, and engineering (Studies in Nonlinearity). Perseus Books Group, Reading, MA

Thompson J, Stewart H (1993) A tutorial glossary of geometrical dynamics. International Journal of Bifurcation and Chaos 3(2):223–239

Chapter 6
Knowing One's Limits: An Analysis in Centered Dynamic Epistemic Logic

Denis Bonnay and Paul Égré

6.1 Dynamic Logic and Epistemic Paradoxes

Dynamic epistemic logic has been used to explain away various epistemic paradoxes. Van Benthem (2004) showed how the difference between successful and unsuccessful epistemic updates can account for the Fitch paradox. Gillies (2001) proposed a similar approach to Moore's paradox, and Gerbrandy (2007) recently examined the Surprise Examination paradox in the light of dynamic epistemic logic. In all three cases, the paradoxes can be seen to originate in an equivocation between what one may learn or *realize*, and what one may *actually* know. In the most typical case, the case of Moore's paradox, the agent is assumed to learn that a certain fact holds, of which she was not aware. The fact makes crucial reference to the very ignorance of the agent, so that realizing it results in the fact holding no more.

As an example, suppose that it is raining and the agent does not know it. The agent is told so. She realizes that it is raining and that she does not know it. But this changes her epistemic state, in such a way that it is no longer true that she does not know that it is raining. Thus, realizing that $p \wedge \neg Kp$ does not yield knowledge of $p \wedge \neg Kp$. The notion of *epistemic update* used in Dynamic Epistemic Logic can be used to account for that situation. In Dynamic Epistemic Logic, whenever an agent is informed about some *atomic* fact (true proposition), she thereby comes to know that it is true. Such updates on the agent's epistemic state are called *successful*, when updating by some proposition leads to the knowledge of that proposition. Not all updates with true propositions need be successful, however, in particular when *non-atomic* propositions are involved. The three paradoxes we mentioned all involve unsuccessful updates in that sense, namely updates by non-atomic true propositions that, when announced or revealed to the agent, are no longer true *after* they have been announced. As argued by van Benthem, the logical analysis of these paradoxes

D. Bonnay (✉)
Département d'Etudes Cognitives de l'ENS, IREPH/IHPST, 29 rue d'Ulm, 75005 Paris, France
e-mail: denis.bonnay@ens.fr

P. Girard et al. (eds.), *Dynamic Formal Epistemology*, Synthese Library 351,
DOI 10.1007/978-94-007-0074-1_6,

in Dynamic Logic suggests that their paradoxical flavor primarily stems from the illusion that all updates should be successful.

In Bonnay and Égré (2009), we introduced a non-standard semantics for epistemic logic, Centered Semantics, as well as a further generalization called Token Semantics, in order to account for another epistemic paradox, originally due to T. Williamson, and akin to the Surprise Examination paradox. From a model-theoretic point of view, the new semantics was designed to make compatible a notion of inexact knowledge, based upon non-transitive and non-euclidian relations of epistemic accessibility, and the principles of positive and negative introspection, whose validity is equivalent to transitivity and euclideanness of the accessibility relation. At the conceptual level, we intended the semantics to dispel what we saw as an excessive tension between epistemic principles that are plausible when viewed separately, but conflicting when brought together. Thus, Williamson showed that margin for error principles for knowledge (principles of the form: I know that p provided p is true in all contexts sufficiently similar to the actual one), plus knowledge of these principles, are not compatible with positive introspection. Williamson considers this a *reductio* of the introspection principles. Centered semantics, on the other hand, has the remarkable feature that it can validate the principle of margin for error without thereby validating knowledge of the principle.[1] We argued that this was the way to go: "knowledge of the margins" results in a change of the margins, so that in general, it is not safe to assume that an agent knows that her knowledge obeys a fixed margin. In our perspective, Williamson's paradox should be taken as a *reductio* of the knowability of margin for error principles, rather than as a case against introspection.

However, our formal account of the epistemic scenario underlying the paradox remained *static*. We merely showed how to resist knowledge of the margin for error principle while accepting the principle as a valid principle constraining the semantics for knowledge. Nevertheless, the conceptual argument presented in favor of this strategy was essentially *dynamic*. In Bonnay and Égré (2009), we point out that reflection on the limitation of one's knowledge makes it possible to improve on that knowledge. Such reflection is obviously a good thing, but it makes dubious the assumption that knowledge about one's limits is unmodified through the reflection process. In Bonnay and Égré (2009), however, we did not present a worked out elaboration of this dynamic intuition. In this respect, our proposal to construe Williamson's paradox as a *reductio* of knowledge of the margin for error principle remained incomplete.

Given the structural similarities between Williamson's paradox and the Surprise Examination paradox, which are fully explicit in Williamson (2000), and given the broader similarities between our conceptual analysis of Williamson's paradox and previous accounts of epistemic paradoxes based on Dynamic Epistemic Logic, it is worth considering whether a complete account of Williamson's paradox can be

[1]This is a particular case of a general failure of the rule of necessitation over models, see Bonnay and Égré (2009) for details.

provided by merging Dynamic Epistemic Logic and Centered Semantics. This is precisely what this chapter achieves: we propose a Centered Dynamic Epistemic Logic, for which the standard axiomatization remains sound and complete, and we show how it can be used to account for the dynamics of reflection on one's margins.

An important fact about this merge is that we could not have done without centered semantics in the first place: an account based solely upon Dynamic Epistemic Logic and Kripke semantics would not preserve the introspection principles over models of inexact knowledge. Furthermore, the merging creates values for shareholders on both sides. As explained, short of introducing epistemic updates, the reflection process on one's epistemic limitations remained unaccounted for in Centered Semantics. Conversely, the issue of what happens when one reflects upon one's epistemic limitations constitutes an original field of application for Dynamic Epistemic Logic. The Fitch paradox or knowability paradox, which can be seen as derivative from Moore's paradox, looms quite large in discussions on the limits of knowledge. Prima facie, limitations imposed by Moorean sentences on our knowledge can be quarantined. However, failure to distinguish between perceptual alternatives is a very pervasive phenomenon. And so is to some extent the reflection on our limitations: in everyday life, we constantly try to maximize our knowledge by taking into account what we realize that we do not know.

6.2 Centered Semantics with an Update Operator

In this section we present a logic that we call Centered Dynamic Epistemic Logic (CDEL for short). The language of CDEL is the same as the language of DEL, namely an epistemic language with a dynamic update operator. The semantics differs from that of DEL in two respects. Regarding the static knowledge operator, the underlying semantics is Centered Semantics (CS) instead of standard Kripke semantics. Because of that, the semantics of the update operator requires some minor adjustments. In what follows, we first present CS for the basic epistemic language. In the second part, we define CDEL, the dynamic version of CS, for the epistemic language with an update operator. In all that follows, the language $\mathcal{L}$ of static epistemic logic is defined by $\varphi := p|\neg\varphi|(\varphi \wedge \varphi)|K\varphi$, where K is the static knowledge operator. The language $\mathcal{DL}$ of dynamic epistemic logic is the extension of $\mathcal{L}$ defined by: $\varphi := p|\neg\varphi|(\varphi \wedge \varphi)|K\varphi|\ [\varphi]\varphi$, where $[\ ]$ is the update operator. The notation $\langle\varphi\rangle$, which we shall use below, is an abbreviation for $\neg[\varphi]\neg$, namely for the dual of the update operator.[2]

[2]See van Ditmarsch et al. (2007), whose notational conventions are taken up here. More precisely, the language $\mathcal{DL}$ corresponds to the language $\mathcal{L}_{K[]}$ of Public Announcement Logic defined in van Ditmarsch et al. (2007), p. 73, for the case of a single agent.

6.2.1 Centered Semantics

Centered Semantics is a two-dimensional semantics in which epistemic alternatives are relativized to the actual world: when an epistemic operator is evaluated relative to a world that is possible for all we know, the accessible worlds are taken to be those worlds that are already accessible from the actual world. We recall here the definition of truth in CS. It is a two stage definition. Truth is first defined with respect to couples of worlds, and truth at a single world is then defined by diagonalizing. Here are the precise definitions:

Definition 6.1 Truth for couples of worlds:

(i) $\mathcal{M}, (w, w') \models_{CS} p$ iff $w' \in V(p)$.
(ii) $\mathcal{M}, (w, w') \models_{CS} \neg\varphi$ iff $\mathcal{M}, (w, w') \not\models_{CS} \varphi$.
(iii) $\mathcal{M}, (w, w') \models_{CS} (\varphi \wedge \psi)$ iff $\mathcal{M}, (w, w') \models_{CS} \varphi$ and $\mathcal{M}, (w, w') \models_{CS} \psi$.
(iv) $\mathcal{M}, (w, w') \models_{CS} K\varphi$ iff for all w'' such that wRw'', $\mathcal{M}, (w, w'') \models_{CS} \varphi$.

Definition 6.2 $\mathcal{M}, w \models_{CS} \varphi$ iff $\mathcal{M}, (w, w) \models_{CS} \varphi$

Clause (iv) of the definition accounts for the "centered" feature of the semantics. Looking on a picture, we can easily see how clause (iv) works. If we evaluate a

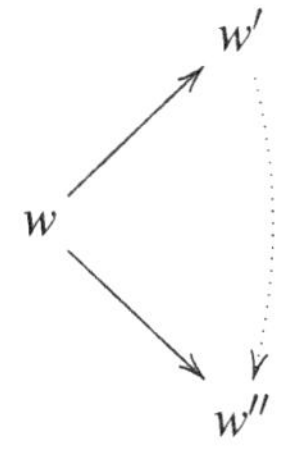

formula at a world w, and if the evaluation process takes us to a world w' accessible from w, we then take as worlds w'' accessible from w' the worlds that are in fact accessible from w. Let us call "evaluation world" the world at which a whole formula is being evaluated, as opposed to other worlds visited during the evaluation process. Semantic evaluation is *centered* on the evaluation world in the following sense: at every step in the evaluation process, the worlds that are accessible are the worlds accessible from the evaluation world. In particular, clause iv) entails that for every w and w': $\mathcal{M}, (w, w') \models_{CS} K\varphi$ iff $\mathcal{M}, (w, w) \models_{CS} K\varphi$ iff $\mathcal{M}, w \models_{CS} K\varphi$.

K45 is sound and complete with respect to CS on the class of all frames, and **S5** is sound and complete with respect to CS on the class of reflexive frames (see Bonnay and Égré (2009)). Centering makes positive and negative introspection automatically satisfied over arbitrary structures. Thus **K45** axioms and theorems turn out to be valid even over non-transitive and non-euclidian frames, just as **S5** axioms and theorems turn out to be valid even on reflexive frames where the accessibility relation is not an equivalence relation.

6.2.2 Centered Dynamic Epistemic Logic

From a static viewpoint, the rationale for using CS rather than standard Kripke semantics is to be able to preserve the introspection principles over structures of inexact knowledge, where the agent's information is not sharply distributed into equivalence classes of worlds (we refer to Bonnay and Égré (2009) for ampler discussion). Taking a dynamic perspective, what we seek now is to provide an analysis of epistemic updates within the framework of CS. The first question on our agenda is of course: how shall we define epistemic updates in this setting?

Let us briefly recall how things work in standard DEL. As explained above, a new dynamic action modality is added to basic epistemic logic. If φ and ψ are formulas, so is now $[\varphi]\psi$. The subformula $[\varphi]$ means *updating* with the information that φ, and $[\varphi]\psi$ is true if ψ is true after updating with φ. For our purposes, in the case of a single agent, $[\varphi]$ will mean that the agent *realizes* that φ is the case. In particular, $[\varphi]K\psi$ states that the agent will know that ψ after realizing φ, namely after receiving the hard information that φ. $[\varphi]$ itself is interpreted by an operation on models. Basically, all worlds which are not φ are cut off, and the standard clause reads:

$$(^*)\ \mathcal{M}, w \models [\varphi]\psi \text{ iff, if } \mathcal{M}, w \models \varphi, \text{ then } \mathcal{M}|\varphi, w \models \psi$$

where $\mathcal{M}|\varphi$ is the restriction of $\mathcal{M}$ to φ worlds. The condition that $\mathcal{M}, w \models \varphi$ guarantees that we are talking about a truthful piece of information.

So what about updates in centered semantics? The natural clause to consider would be:

$$\mathcal{M}, (w, w') \models_{\text{CS}} [\varphi]\psi \text{ iff, if } \mathcal{M}, (w, w') \models_{\text{CS}} \varphi, \text{ then } \mathcal{M}|\varphi, (w, w') \models_{\text{CS}} \psi$$

But one question arises: what do we mean exactly by $\mathcal{M}|\varphi$? The question arises because truth has been relativized in CS, so that various options are available. In this particular context, we can restrict $\mathcal{M}$ either to worlds w'' such that $\mathcal{M}, (w'', w'') \models \varphi$ or to worlds w'' such that $\mathcal{M}, (w, w'') \models \varphi$. The second option is the more natural: accessibility remains relativized to the evaluation world w. The adequacy of this option will show up in subsequent theorems about updates in Centered Semantics. By contrast, choosing the first option would permit "cheating" through epistemic updates, namely worlds that are not direct alternatives to the actual world could become relevant during the evaluation.

To make this clear, we shall therefore write the clause for epistemic updates as clause (v):

$$(\text{v})\ \mathcal{M}, (w, w') \models_{\text{CS}} [\varphi]\psi \text{iff, if } \mathcal{M}, w \models_{\text{CS}} \varphi, \text{ then } \mathcal{M}|\varphi_w, (w, w') \models_{\text{CS}} \psi$$

where $\mathcal{M}|\varphi_w$ is $\mathcal{M}$ restricted to worlds w'' such that $\mathcal{M}, (w, w'') \models_{\text{CS}} \varphi$.

By definition, the semantics DEL for $\mathcal{DL}$ is the basic Kripke semantics augmented with clause (*) for the update operator. We define CDEL for $\mathcal{DL}$ as the basic Centered Semantics (clauses (i)–(iv)) augmented with clause (v) for the update operator. When no assumptions are made on the epistemic accessibility relation, DEL is axiomatized by the logic **K** plus the following recursion axioms RA (see van Ditmarsch et al. (2007)):

$$
\begin{aligned}
&[\varphi]p && \leftrightarrow \varphi \rightarrow p \\
&[\varphi]\neg\psi && \leftrightarrow \varphi \rightarrow \neg[\varphi]\psi \\
&[\varphi](\psi \wedge \chi) && \leftrightarrow [\varphi]\psi \wedge [\varphi]\chi \\
&[\varphi]K\psi && \leftrightarrow \varphi \rightarrow K(\varphi \rightarrow [\varphi]\psi)
\end{aligned}
$$

In the static case, we saw that CS is axiomatized by **K45** over the class of all frames. Likewise, in the dynamic case, CDEL is axiomatized by **K45** and the recursion axioms characteristic of DEL over the class of all frames:[3]

Theorem 6.1 K45 *(resp.* **S5***) plus the recursion axioms is sound and complete with respect to Centered Semantics with updates on the class of all frames (resp. of all reflexive frames).*

What the theorem shows is that CDEL, just like CS, is conservative in terms of axioms, though innovative in terms of models. This is a desirable feature, since our aim in the next section is to use the centered version of dynamic logic to analyze Williamson's paradox in a way that still combines introspection and a notion of inexact knowledge, but taking updates into account.

6.3 The Margin of Error Paradox

What we call a paradox is not intended as such by Williamson, but rather as an argument against the principle of positive introspection, and as part of a more general philosophical argument against the so-called *luminosity* of mental states, namely the idea that there might be propositions φ that are automatically known, namely such that $\varphi \rightarrow K\varphi$ (thus when φ is of the form $K\psi$, positive introspection becomes a particular instance of luminosity). Yet what Williamson's argument establishes is that a number of independently plausible premises lead to contradiction. The form of the argument is also closely related to that of a sorites argument. Because of these two features, the argument in its general form has sometimes been called the *luminosity* paradox (see Leitgeb (2002), Égré (2008)). To give it a name here, we shall call it the *Margin of Error* paradox, because the notion of a margin of error used by Williamson is distinctive of its formulation.

6.3.1 The Paradox

The form of the argument is the following: Mr Magoo is a myopic character who observes a tree at some distance. Magoo is certain that the tree is less than k meters

[3]The proof is given in the Appendix.

high (say less than $k = 20$). Magoo's knowledge about sizes is constrained by a *margin for error principle*, whereby for Magoo to know that the tree is less than k meters high, the tree has to be less than $k - \eta$ meters high, for a particular η. For instance, suppose $\eta = 1$: the principle says that for Magoo to know that the tree is less than 20 m high, the tree cannot measure 19 m, for from where he is Magoo cannot reliably discriminate between sizes that differ by just 1 meter. By assumption, Magoo is also taken to be *aware of this limitation* on his knowledge, to be *positively introspective*, and to *know the consequences* of what he knows. For the argument to go, finally, η must be positive, but can be arbitrarily small.

The structure of the argument is the following. Let '$(s < k)$' be an atomic proposition standing for "The tree is less than k meters high". Similarly, '$(s \geq k)$' is an atomic proposition standing for "The tree is at least k meters high". Starting from assumption (1), (4) follows via (2) and (3) on the basis of the assumed principles for Magoo's knowledge:

(1) $K(s < k)$
(2) $K(s \geq (k - \eta) \rightarrow \neg K(s < k))$
(3) $KK(s < \eta)$
(4) $K(s < (k - \eta))$

(1) is the assumption that Mr Magoo knows the tree to be less that k meters high. If the conditions of Mr Magoo's woodsy observation are sufficiently good and if k is taken to be sufficiently large, (1) is quite uncontroversial. (2) expresses Mr Magoo's knowledge about his own margin for error, namely that it is at least η. (3) follows from (1) by positive introspection. (4) follows from (2) and (3) by closure of knowledge under logical consequence. Here η can be taken to be arbitrarily small. By repeating the argument i times, one may reach the conclusion that Mr Magoo knows the size of the tree to be less than $k - i \cdot \eta$. So long as (2) is taken to hold without restriction for all relevant values of k, one can repeat the argument enough times to reach an absurd conclusion with respect to the actual size of the tree (namely that the tree is of size 0).

The principle of closure under logical consequence is not objectionable in this context, at least as an idealization holding for rational agents – who might have to evaluate tree sizes just as forest wardens do. So there are only two options left: either the principle of positive introspection is to be rejected, thereby blocking the inference from (1) to (3), or (2) is to be rejected, suggesting that one cannot improve on (1) by reflecting on one's limitations. Stressing the fact that η – the estimated margin – can be taken to be arbitrarily small, Williamson construes the paradox as a *reductio* of positive introspection. By contrast, the gist of our dynamic analysis will be to point (2) as the real culprit for the paradox.

The problem is that rejecting (2) seems counterintuitive at first glance. To begin with, the principle of margin for error itself seems quite reasonable. I certainly cannot know by vision alone that a tree that is $k - \eta$ meters tall is less than k meters tall when a difference of η meters in size is too small to be detected by my eye. And it is certainly possible to find a value for η – say 0.01 meters – such that no such difference meets the eye. More generally, our perceptual knowledge is certainly bounded by the limitations of our perception, namely by what we cannot perceptually discriminate.

If the principle of margin for error holds, it seems equally reasonable to grant that we can become aware of this fact. I can certainly realize that my eyesight is far from perfect. Reflecting on this limitation, I can realize that my visual knowledge obeys a margin for error principle. I know for sure that if a tree is 11.99 m tall, I cannot know for sure that it is less than 12 m tall. Hence, if (I know that) I know that the tree is less than 12 m tall, I thereby know that it is also less than 11.99 inches tall, and so starts the sorites.

6.3.2 Knowing and Realizing

The paradoxicality of Williamson's argument originates from the fact that the reasoning seems perfectly valid and the premises sound. Or... could we have been misled? In Dokic and Égré (2009), Dokic and Égré argue that margins for error come in different varieties. My initial knowledge that the tree is less than 12 m tall is purely visual. My acquired knowledge that it is less than 11.99 m tall is not so. If this knowledge is to obey the same margin for error principle as my initial visual knowledge, and if I can know it does, we are in trouble. But why should the knowledge I gained, which is based partly on perception, partly on rational reflection and on drawing inferences, be subject to exactly the same limitations as my initial knowledge? On the contrary, it seems that whatever limitations my visual knowledge was subject to, my reflective knowledge is not subject to *exactly those*, since my reflective knowledge *improves on* my visual knowledge.

The strategy in Dokic and Égré (2009) was to carefully distinguish between kinds of knowledge according to their sources. Indexing knowledge operators accordingly blocks iterations of the reasoning after the first step. One problem with this strategy, however, is that it sprays "plain knowledge" into several varieties of knowledge, and it remains silent regarding the principles governing the generic notion of knowledge, irrespective of its source. Indeed, should this general knowledge also obey a margin for error principle, then a revenge form of the paradox would be lurking around.

We wish to preserve the intuition put forward in Dokic and Égré (2009) that the paradox can be explained by observing that there is a somewhat hidden but crucial assumption that the estimated margin can be kept fixed. However, we also want to make this intuition compatible with the original analysis in terms of a single notion of general knowledge involved throughout the argument. The story we want to tell is essentially dynamic. There is some visual knowledge to begin with. Reflection comes in and results in improved (mixed) knowledge. So there is knowledge at the beginning, and there is knowledge at the end. Yet does the transition itself, namely reflection upon one's limits, qualify as a piece of knowledge? Williamson thinks it does, and thinks that it is safe to assume that there is a fixed margin η such that we can always assume that Mr Magoo's visual margin for error can be known by himself to be at least η. By "always", we mean that (2) is considered to remain true even as Mr Magoo's knowledge of the size of the tree has improved after, say, the first round of reasoning and reflecting on his margin for error.

An alternative analysis treats Mr Magoo's reflection on his limitations as *realizing* something, rather than *knowing* it. At this point, the analogy with Moorean scenarios is telling. *Realizing* that it rains and that I do not know it does not count as *knowing* that it rains and that I don't know it. Indeed, it cannot count as knowledge. The mere fact that one realizes that one does not know something which is true changes the relevant epistemic facts. The same holds of margins of error. Mr Magoo's realizing that his knowledge concerning the tree size is limited changes the relevant epistemic facts. Indeed, he is able to gain new information on the basis of his reflection on his limitations.

6.3.3 Reanalyzing the Paradox with Epistemic Updates

Following these intuitions, the best tool to model Mr Magoo's scenario is epistemic updates, which model from a dynamic perspective what happens when an epistemic agent realizes that something is the case. Using updates is more adequate than using the K operator of epistemic logic, which models from a static perspective what happens when an agent knows that something is the case.

Let us abbreviate "$(s \geq (k - \eta)) \rightarrow \neg K(s < k)$" by $ME(k, \eta)$. $ME(k, \eta)$ states that Mr Magoo's margin is of at least η when it comes to estimating sizes around k. $\mathcal{M}, w \models_{CS} [\varphi]\psi$ does not exactly say that ψ holds in the model one gets from $\mathcal{M}, w$ by updating with φ, because $[\varphi]\psi$ is trivially true when φ is false at w.[4] It says that ψ will be true *if the update is successful*. To express that ψ will hold after the successful update by φ, we need to use the dual of the update operator, namely $\langle\varphi\rangle\psi$. For $\mathcal{M}, w \models \langle\varphi\rangle\psi$ provided $\mathcal{M}, w \models \varphi$ and $\mathcal{M}|\varphi, w \models \psi$.[5] Here then is the new formalization we suggest for the argument:

(1') $K(s < k)$
(2') $ME(k, \eta)$
(3') $K(s < k) \rightarrow [ME(k, \eta)]K(s < (k - \eta))$
(4') $\langle ME(k, \eta)\rangle K(s < (k - \eta))$

On this account, Mr Magoo starts with some knowledge about the size of the tree being less than k (1'). It is a fact that his margin for error is at least η when it comes to estimating heights around k (2'). If Mr Magoo knows something to be of size at least k, and he realizes that his margin for error is at least η when it comes to estimating heights around k, then he will come to know that the size of the tree has

[4]The same is true of course for $\mathcal{M}, w \models [\varphi]\psi$.

[5]See van Ditmarsch et al. (2007), prop. 4.14 p. 78: the formula $\langle\varphi\rangle\psi$ is thereby equivalent to $\varphi \wedge [\varphi]\psi$ and to $\varphi \wedge \langle\varphi\rangle\psi$.

to be less than $k - \eta$. Hence after reflecting on his margin, Mr Magoo does know the size of the tree to be less than $k - \eta$ (4').

We shall now briefly compare with the earlier formalization by Williamson. (1') is the same as (1), the first premise in Williamson's argument. (2) was knowledge of a basic margin for error principle, namely that Mr Magoo's margin is of at least η. (2') states the basic margin of error principle, the epistemic use of which is deferred to (3'). (3') describes the reflection process itself and says that it results in a gain of η in terms of knowledge of heights. Getting (4') as a conclusion means that (4) is true in the situation we get *after* an epistemic update on $ME(k, \eta)$ starting from a situation in which (1) is true. This is a significant difference with the previous static account according to which (4) is true in the same epistemic situation as the one in which premise (1) is taken to be true. Note that (4') admits of a natural temporal reading.[6] It says that when epistemic facts are changed according to what it means to realize that the margin was at least η, *then* it becomes known that the size of the tree is less than $k - \eta$.

What about the soundness of the argument?[7] The argument is certainly valid: (4') follows from (1'), (2') and (3') by propositional logic alone. Just as before, (1) can certainly be assumed to be true in some situation, and (2) will be true as well if we agree that knowledge, or at least perceptual knowledge, obeys a margin for error principle. As a consequence, the question whether the argument is sound boils down to the question whether (3'), that is $K(s < k)) \rightarrow [ME(k, \eta)]K(s < (k - \eta))$ is true (in general or in the particular structures modeling the scenarios under scrutiny).

$ME(k, \eta)$ is equivalent to $K(s < k) \rightarrow s < (k - \eta)$, so (3') is of the form $Kp \rightarrow [Kp \rightarrow q]Kq$. The formula says that if I know p, and if I realize that knowing p implies q, then I know q as well. Should this be fine? Realizing that knowing p implies q will help only if I can use my knowledge that p, that is only if I know that I know p. And then to conclude that q, I need to apply *modus ponens*. So the general principle of which (3') is an instance seems to be acceptable under two closure assumptions about knowledge, namely that knowledge is closed under logical consequence and that knowledge is introspective. This is no surprise since these two principles were used in Williamson's derivation of the paradox but did not appear explicitly in (1')–(4').

Note that the principle is stated for atoms only. It is crucial that it does not apply to any arbitrary formula. Our informal discussion took for granted that the truth of q is preserved under realizing $Kp \rightarrow q$. This is fine because q describes a non-epistemic fact. But if q were replaced with a formula ψ describing an epistemic fact, realizing that ψ is implied by Kp might result in ψ ceasing to be true. As a consequence, it would be absurd to claim that $K\varphi \rightarrow [K\varphi \rightarrow \psi]K\psi$ is valid in general. To see this at a glance, take any tautology $\top$ for φ, we get $K\top \rightarrow [K\top \rightarrow \psi]K\psi$ which is equivalent to $[\psi]K\psi$. What we get is thus a success principle for ψ: realizing that ψ results in knowing that ψ. As we made clear in the first section,

[6] See van Benthem et al. (2009) on merging epistemic updates and temporal logic.

[7] By a *sound* argument, we mean a valid argument whose premises are true.

success principles cannot be assumed to be valid no matter what. Take for ψ the Moorean sentence $p \wedge \neg Kp$. Then $[p \wedge \neg Kp]K(p \wedge \neg Kp)$ is false whenever $p \wedge \neg Kp$ is true to start with, because $K(p \wedge \neg Kp)$ is contradictory.

With this restriction clearly in mind, we think that $Kp \to [Kp \to q]Kq$ is on the face of it a quite reasonable assumption, at least for idealized rational agents. It says that I can actually improve on my knowledge by realizing that some inferential connections hold between my knowing of certain things, that is Kp, and some other things, namely q. Denying such a principle would certainly deprive epistemic updates of some of their interest, since it would severely limit our ability to gain knowledge through updates. Eventually, one's attitude towards this principle might depend on its relationships with other principles, and the fact that positive introspection and closure under logical consequence came up in our informal discussion is bound to suggest various pros and cons. We shall not propose here an independent defense of positive introspection and closure under logical consequence. In some contexts, it might indeed be more realistic to assume that they do not hold. But it seems to us that it should be nevertheless always consistent to assume that they both hold. It would be quite surprising if it were the case that the very idea of idealized rational agents, who are able to know what they know and draw all the consequences of what they know, was intrinsically inconsistent. Therefore, a priori we favor an analysis of the Margin for Error paradox which would show that the paradox can be explained and these principles maintained. This is exactly what we shall offer in the last section of this chapter.

Finally, let us state precisely the connection between our schema and the principles used in Williamson's formalization. Given two schematic formulas A and B in the language of Dynamic Epistemic Logic, we say that *B follows from A modulo closure under logical consequence and the recursion axioms for updates* (notation: $A \vdash_{Cl,RA} B$) if and only if we can get any instance of B from instances of A using only propositional logic, closure of knowledge under logical consequence and the recursion axioms for updates. The following fact holds:[8]

Fact 2 $Kp \to KKp \vdash_{Cl,RA} Kp \to [Kp \to q]Kq$

Thus, our formal rendering in terms of epistemic updates might seem to confirm Williamson's idea that positive introspection is to blame for the paradox. The only option, if we want to resist (4'), is to deny (3'), and denying (3') logically implies denying introspection (if closure under logical consequence is granted, and we agree with Williamson that it should be granted). However, we do not have to deny (4'): in most cases, it might indeed be perfectly reasonable to accept (4'). The problem shows up when the argument is iterated. Our take is that it should be at least coherent

---

[8]See the proof in part B of the Appendix. One might have hoped to get as well the converse $Kp \to [Kp \to q]Kq \vdash_{Cl,RA} Kp \to KKp$. As pointed out to us by Olivier Roy, this is however not the case since one can find a Kripke-structure validating (every instance of) $Kp \to [Kp \to q]Kq$ but invalidating $Kp \to KKp$. We conjecture that it is possible to get a full equivalence by suitably liberalizing the schema $Kp \to [Kp \to q]Kq$ so as to allow not only atoms but also any successful formula.

to assume that the agent knows what he knows, and we wish to explain *why* iterating the argument is problematic in this context. We shall see in the next section that this is precisely the point where using epistemic updates makes a difference.

6.3.4 CDEL does it better

Up to now, we have only discussed the validity of the argument, in its two different forms, and the intuitive plausibility of the premises. Before discussing iterations of the argument, it is well worth looking at the exact truth conditions of the sentences involved. The point we want to make is that our formalization using epistemic updates makes sense only if the semantics for update operators is given by Centered Semantics instead of the standard semantics. DEL does not make the right predictions on the intended models, and this will be our reason for using later on CDEL rather than DEL in order to provide a model-theoretic analysis of what is going when the argument is iterated.

Let us consider a Kripke model $\mathcal{M}_d = \langle W, R_d, V\rangle$, where W is a space of worlds, each of which is indexed by a real number in $\mathbb{R}^+$. Each world w_r is to be thought of as a world at which Mr Magoo's tree is r meters tall. The valuation V is defined accordingly by letting $w_r \in (s \leq k)$ iff $r \leq k$. Mr Magoo's eyesight is characterized by his ability to tell the difference between any two objects whose sizes differ by at least d meters, where d is a real number greater than zero. d is the margin for error which determines Mr Magoo's visual knowledge, and it can be used to define the accessibility relation R_d encoding Mr Magoo's knowledge by setting $w_r R_d w_{r'}$ iff $|r - r'| \leq d$. R_d is symmetric and reflexive, but it is not transitive. Following (Williamson 1992), we shall call models like $\mathcal{M}_d$ *margin models*.[9] Margin models are certainly the natural models to use if one thinks of knowledge as being determined by a margin for error principle, so we take them to be the intended models for Mr Magoo's scenarios.

Centered semantics and Kripke semantics make different predictions when margin models are used.

Fact 3 *The following propositions hold:*

(i) $\mathcal{M}_d \vDash_{CS} ME(k, \eta)$, *for all* $k \in \mathbb{R}^+$ *and* $0 < \eta \leq d$.
(ii) $\mathcal{M}_d \nvDash_{CS} ME(k, \eta)$, *for all* $k \in \mathbb{R}^+$ *and* $\eta > d$.
(iii) $\mathcal{M}_d \vDash_{CS} K(s < k) \rightarrow [ME(k, \eta)]K(s < (k - \eta))$, *for all* $k, \eta \in \mathbb{R}^+$.
(iv) $\mathcal{M}_d \vDash ME(k, \eta)$, *for all* $k \in \mathbb{R}^+$ *and* $0 < \eta \leq d$.
(v) $\mathcal{M}_d \nvDash ME(k, \eta)$, *for all* $k \in \mathbb{R}^+$ *and* $\eta > d$.
(vi) $\mathcal{M}_d \nvDash K(s < k) \rightarrow [ME(k, \eta)]K(s < (k - \eta))$ *for all* k *and* $\eta \in \mathbb{R}^+$.

[9] In (Williamson 1992), Williamson considers an arbitrary space of worlds equipped with a metric, and rephrases the semantics so as to appeal directly to the parameter d with no detour through a defined accessibility relation.

Proof Proofs are left to the reader and we shall just make two remarks. First, regarding the proof of (iii), it is sufficient to analyze the effect of updates like $[ME(k, \eta)]$ on a model $\mathcal{M}_d, w_r$ where $\mathcal{M}_d, w_r \models_{CS} K(s < k)$. What does the set $\{r' \in \mathbb{R}^+ \ / \ (w_r, w_{r'}) \models_{CS} ME(k, \eta)\}$ look like? Since $\mathcal{M}_d, w_r \models_{CS} K(s < k)$, we have $(w_r, w_{r'}) \models_{CS} K(s < k)$ as well. So we are going to take away all the worlds at which $s < (k - \eta)$ is not true. Thus, $\{r' \in \mathbb{R}^+ \ / \ (w_r, w_{r'}) \models_{CS} ME(k, \eta)\} = [0, k - \eta[$, and now $\mathcal{M}_d \backslash [k - \eta, +\infty[, w_r \models_{CS} s < k - \eta$, since all the worlds where the tree is $k - \eta$ tall or taller have been cut off. (vi) follows from Fact 4 below.

$ME(k, \eta)$ is a statement about an approximation of Mr Magoo's margin. It should hold exactly when the approximation is correct, that is whenever η is indeed smaller than the d such that Mr Magoo cannot distinguish between objects whose size differ by no more than d. So (i), (ii) and (iv) and (v) are welcome properties which are shared by Centered Semantics and standard Kripke semantics. (iii) says that premise (3') holds in margin models when evaluated according to Centered Semantics, and (vi) that it fails to hold (quite generally) when evaluated according to standard Kripke Semantics. This is no surprise since, as we said, (3') follows from positive introspection in normal modal logics and the gist of Centered Semantics is to enforce introspection even on non-transitive models such as margin models.

This difference can be traced to a more general contrast between CDEL and DEL. Let us say that a semantics $\models_S$ for a language with update operators is *validity-insensitive* if the following hold: let φ be a formula which is valid on a model $\mathcal{A}$, let ψ be an arbitrary formula and w be any world in $\mathcal{A}$, then $\mathcal{A}, w \models_S [\varphi]\psi$ iff $\mathcal{A}, w \models_S \psi$. A semantics will be said to be *validity-sensitive* if it is not validity-insensitive.

Fact 4 *The following propositions hold:*

- $\models$ *is validity-insensitive*
- $\models_{CS}$ *is validity-sensitive*

Proof $\models$ is validity-insensitive because if φ is valid on $\mathcal{A}$, $\mathcal{A}|\varphi$ is the same as $\mathcal{A}$. This is not true of $\models_{CS}$ and the fact that $\models_{CS}$ is validity-sensitive can be checked on a suitably chosen margin model. For example, set $d = 2$, take w_{11} as the actual world. We have $\mathcal{M}_d, w_{11} \not\models_{CS} K(s < 12.5)$ but $\mathcal{M}_d, w_{11} \models_{CS} [ME(13.5, 1)]K(s < 12.5)$.

Fact 4 brings forward a significant difference between epistemic updates in Centered Semantics and epistemic updates in Kripke semantics. An account of Mr Magoo's scenario in Dynamic Epistemic Logic must capture the intuition that Mr Magoo learns something new when he realizes that his perceptual knowledge obeys a given margin for error. But on margin models, any correct approximation from below of the actual perceptual margin is true everywhere in the model. Therefore, *on these models*, standard Kripke semantics makes the counter-intuitive prediction that realizing that one's knowledge is bounded by a certain margin of error has no epistemic consequence at all. By contrast, as shown in the proof, Centered Semantics correctly predicts that the kind of learning ascribed to Mr Magoo in our scenario does occur, even though the margin for error principle is valid on the considered model. In this respect, Centered Semantics gives a more adequate picture of learning than ordinary Kripke semantics does with regard to margin models.[10]

A complete comparison of the merits of each semantics based on its predictions on margin models should include a discussion of yet another difference.

Fact 5 *The following propositions hold:*

(vii) $\mathcal{M}_d \not\models_{CS} KME(k, \eta)$*, for all* $k \in \mathbb{R}^+$ *and* $0 < \eta \leq d$.
(viii) $\mathcal{M}_d \models KME(k, \eta)$*, for all* $k \in \mathbb{R}^+$ *and* $0 < \eta \leq d$.

This has been discussed to some length in Bonnay and Égré (2009), and we refer the interested reader to our earlier paper. Note that failure of $KME(k, \eta)$ is not as strange as it might seem at first sight, if $ME(k, \eta)$ is something for us to realize, and not something for us to know, as it happens with Moorean sentences. (vii) says that premise (2) is false.[11] We shall not press this point here, but shall rather argue that (1)–(4) are simply not the best way to formalize what is going on.

6.4 Keeping on Reflecting

6.4.1 Once Versus More Than Once

In the previous section, we have offered an alternative to Williamson's analysis of Mr Magoo's inferential story. But what is the point of reanalyzing the argument, if in both cases one reaches basically the same conclusion, namely that Mr Magoo's knowledge improves, under basically the same assumption, namely that introspection holds? As we have suggested, the essential difference shows up when it comes to iterating the argument – recall that there is nothing intrinsically paradoxical with Mr Magoo's reasoning at the first step and that the paradox comes up when the reasoning is repeated an arbitrary number of times.

---

[10]The difference with respect to epistemic updates mirrors a similar difference with respect to the knowledge operator. In Kripke semantics, if $\mathcal{A} \models \varphi$ then $\mathcal{A} \models K\varphi$ (namely the rule of necessitation is valid over models, or *model-valid*, see Bonnay and Égré (2009)). This is not true in Centered Semantics, which is why learning can occur. When assessing the superiority of Centered Semantics, one should nonetheless keep in mind that it is always possible to change the underlying models. What Mr Magoo learns can be described in classical Kripkean terms on a model in which $ME(k, \eta)$ is not true everywhere. The unravelling of $\mathcal{M}_d$, as described in Bonnay and Égré (2009), would yield such a model. The fact remains true that the non-transitive model $\mathcal{M}_d$ validating $ME(k, \eta)$ is arguably the most intuitive and simple formal rendering of Mr Magoo's predicament.

[11]In Bonnay and Égré (2009), we welcomed failure of KME on margin models as a way to resist Williamson's argument, but we failed to provide a complete alternative logical analysis of Mr Magoo's scenario.

If we assume (2) to be true in its general form, that is $K\forall x((s \geq (x - \eta) \rightarrow \neg K(s < x))$,[12] the truth of (1) and (2) leads us to the truth of (4), via introspection. (4) can then replace (1) as a premise, (2) gets instantiated with $k - \eta$ instead of k and we can finally derive the truth of $K(s < k - 2\eta)$ by introspection again. If the initial argument is sound, every iteration of it is sound. After i iterations, for a large enough i, we reach the paradoxical conclusion that $K(s < k - i \cdot \eta)$. This is to us one of the main reasons to reject the formalization by (1)–(4). Intuitively, it is perfectly fine for Mr Magoo to reflect at least once on his limitations. The problem – which is still in need of a precise characterization – comes up when we somehow assume that Mr Magoo can go on like that forever. Therefore any formalization which has it that *if the argument is sound once it is forever sound*[13] seems to us to be misguided.

What happens if the reasoning is iterated along the lines of (1')–(4')? Let us have a look at the following continuation:

(4') $\langle ME(k, \eta)\rangle K(s < (k - \eta))$ (as before, the conclusion of the first argument is the first premise of the second)
(5') $\langle ME(k, \eta)\rangle ME(k - \eta, \eta)$
(6') $\langle ME(k, \eta)\rangle K(s < (k - \eta)) \rightarrow [ME(k - \eta, \eta)]K(s < (k - 2\eta))$
(7') $\langle ME(k, \eta)\rangle\langle ME(k - \eta, \eta)\rangle K(s < (k - 2\eta))$

(5') states that the principle of margin for error with parameters $k - \eta$ and η holds in the new epistemic state obtained trough the previous update. (6') is the second step of Mr Magoo's reflexive process. This time $s < (k - \eta)$ is what is initially known and $s < (k - 2\eta)$ is the further information that might be inferred by reflection. (7') gives the conclusion that after *two* steps of reflection, Mr Magoo comes to know that $s < (k - 2\eta)$.

Is the continuation of the argument sound? Validity is preserved under the scope of a $\langle\ \rangle$ operator. Since (4'), (5') and (6') are exactly analogous to (1'), (2') and (3'), the argument from (4')–(6') to (7') must be valid. But, of course, soundness is not necessarily preserved. (1')-(3') guarantee that (4') is true, since the argument was sound. If introspection holds, (6') will be true as well. But the case of (5'), that is $\langle ME(k, \eta)\rangle ME(k - \eta, \eta)$, is more involved. The epistemic state obtained after successfully updating with $ME(k, \eta)$ is different from the initial epistemic state. The fact that $ME(k, \eta)$ holds in the initial state, even in the generalized form $\forall x\ ME(x, \eta)$, does not ensure that $ME(k - \eta, \eta)$ holds. $\langle\forall x\ ME(x, \eta)\rangle\forall x\ ME(x, \eta)$ means that updating with ME is successful.[14] If it were the case, we would have that *if the argument is sound once it is forever sound.* But this is not so:

---

[12] $K\forall x((s \geq (x - \eta) \rightarrow \neg K(s < x))$ is to be construed as equivalent to $\bigwedge_{i \in \mathbb{R}^+} K((s \geq (i - \eta) \rightarrow \neg K(s < i))$. It seems that nothing important hinges on how the details of this quantification over possible heights are spelled out.

[13] This notion is further elaborated in the next subsection, under the name of "iterative soundness".

[14] See van Ditmarsch et al. (2007) on successful and unsuccessful updates.

Fact 6 *There are margin models $\mathcal{M}_d$ and estimates η such that $\mathcal{M}_d \not\models_{CS} \langle \forall x\, ME(x, \eta)\rangle \forall x\, ME(x, \eta)$.*

Proof Updating with $\forall x\, ME(x, \eta)$ shrinks the margin by η. But then if $\eta > d - \eta$, $\forall x\, ME(x, \eta)$ will be false in the updated model.

In order to illustrate this fact, consider a discrete margin model $\mathcal{M}$ such that $W = \mathcal{N}$, and $d = 1$. Let p_i be the proposition true exactly at index i, and for simplicity, let us write $\mathcal{M}, i \models i$ instead of $\mathcal{M}, i \models p_i$. The model satisfies all margin principles of the form $K\neg(i + 1) \rightarrow \neg i$, both relative to Kripke semantics and to Centered Semantics. In particular, $\mathcal{M}, 17 \models_{CS} K\neg 19 \rightarrow \neg 18$. Unlike with Kripke semantics, $\mathcal{M}, 17 \models_{CS} \langle K\neg 19 \rightarrow \neg 18\rangle K\neg 18$. However, $\mathcal{M}, 17 \not\models_{CS} \langle K\neg 19 \rightarrow \neg 18\rangle K\neg 18 \rightarrow \neg 17$. The reason is that $\mathcal{M}|(K\neg 19 \rightarrow \neg 18)_{17}$ does not contain the pair $(17, 18)$, hence $\mathcal{M}|(K\neg 19 \rightarrow \neg 18)_{17}, 17 \models_{CS} K\neg 18$, but clearly $\mathcal{M}|(K\neg 19 \rightarrow \neg 18)_{17}, 17 \not\models_{CS} \neg 17$. For instance, after realizing that if I know the size of the tree is not 19, it is not 18 either, it is no longer true that if I now know the size not to be of 18, it should not be of 17.

Fact 6 makes it clear that, even though adequate margin principles are valid on margin models, they may become false when the agent realizes that they hold. Again, the intuition, which is fully accounted for in CDEL, is that realizing that the margin is at least η amounts to diminishing the margin, which might end up being less than η. Therefore, in contrast to the formalization in terms of (1)–(4), the suggested formalization of Mr Magoo's reasoning in Dynamic Logic makes a substantial difference between going just *once* through the reasoning and repeating it a certain number of times. It could happen that a true conclusion is reached from true premises by running the argument for the first time, whereas a false conclusion is reached later on. It would mean that one of the extra-premises needed to get that conclusion is false, and this is compatible with all the initial premises being true.

We regard this diagnosis of the paradox as fairly plausible. Mr Magoo can certainly reflect on his perceptual limitations and thus acquire knowledge. It would make little sense for us to deny that. But when this reflective process is captured in terms of knowledge about one's absolute margin for error – this is premise (2) – we get the unwelcome consequence that recognizing Mr Magoo's one shot reasoning as correct commits us to accepting each further repetition of this process as equally correct. However, if we can accept the truth of the premises of the argument *and* its validity without being committed to arbitrarily many iterations of it, we no longer have to reject any of the general principles making the reasoning valid. This way introspection can be safe from blame.

Moreover, failure of $\langle \forall x\, ME(x, \eta)\rangle \forall x\, ME(x, \eta)$ fits perfectly the main theme in Dokic and Égré (2009), namely that the margin for error corresponding to Mr Magoo's perceptual *and* inferential knowledge is simply not the same as the margin for error corresponding to the purely perceptual knowledge which is Mr Magoo's initial endowment.[15] As a consequence, a correct approximation from

[15]By *perceptual and inferential* knowledge we mean here the knowledge Mr Magoo gets from what he sees and from reflecting for the first time on the limitations of his perceptive abilities. Inferential knowledge itself comes in various degrees since Mr Magoo can then make inferences grounded in his first level perceptual and inferential knowledge.

below of his *perceptual* margin can be an incorrect approximation from below of his *perceptual and inferential* margin. A nice feature of our model is that these differences in margins are not stipulated.[16] They are accounted for by the epistemic updates themselves, since the margin after reflection is nothing but the margin in the updated model.[17]

6.4.2 Discounted Margins

Epistemic updates help explain why and where things go wrong. But they can also be used to get positive results. Williamson assumes that Mr Magoo's estimate remains constant throughout the reflection process. This assumption is disputable, however. On a more realistic scenario, Mr Magoo's estimate of his current margin for error would become lower and lower as he goes through more and more rounds of reflections (intuitively, he is less and less sure about his margin, since many reflections have pushed it down). Clearly, such assumptions impact the soundness of the argument. To study exactly how, we need to take the values of Mr Magoo's successive reflections as parameters.

Let us consider *sequences of estimates* of the form $\overrightarrow{\eta} = \eta_1, \eta_2, \ldots, \eta_n, \ldots$ where η_n is Mr Magoo's estimate of his margin after the first $n-1$ rounds of reflection. An infinite sequence of iterations of the basic argument is thus determined. They are the formal rendering of the (potentially infinite) reflective process Mr Magoo engages in. Now the question is: under which conditions on $\overrightarrow{\eta}$ is it fine for us to freely iterate the argument? Or equivalently, under which conditions on $\overrightarrow{\eta}$ does Mr Magoo keep on learning things?

We are asking for a characterization of soundness conditions for an arbitrary number of iterations of the argument. A question about soundness only makes sense when a particular instance of the scenario has been chosen. So we fix d, Mr Magoo's perceptual margin for error. We shall say that a sequence of estimates $\overrightarrow{\eta}$ is *d-bounded* iff every partial sum is smaller than d, that is $\sum_{i=0}^{n} \eta_i \leq d$ for all n. We shall also say that a sequence $\overrightarrow{\eta}$ makes Mr Magoo's argument *iteratively sound* iff every iteration of the argument starting from some premise $K(s < k)$ true in the actual world and with margin estimates chosen according to $\overrightarrow{\eta}$ is sound.[18]

---

[16] On top of the use of a unified knowledge operator, this is the second advantage of our approach over the one by Dokic and Égré (2009).

[17] Note that strictly speaking the updated model is not a margin model for the Euclidean topology on the reals. This is only because $\forall x\ ME(x, \eta)$ states an asymmetric constraint on margins. To regain symmetry, and to get margin models as update models, we would need a stronger principle such as $\forall x\ (((s \geq (x - \eta)) \to \neg K(s < k)) \wedge ((s \leq (x + \eta)) \to \neg K(s > x)))$. Since nothing important hinges on this, we stick to the weaker principle. See Égré (2008) for a discussion of general margin for error principles in modal and epistemic logic.

[18] The precise definition of an iteration of the argument as parameterized by $\overrightarrow{\eta}$ and the proof of Fact 7 are given in the Appendix. Fact 7 might seem quite obvious, and indeed the proof is by no means difficult. Concluding the analysis of a paradox with a commonsensical conclusion may not be a bad thing, and one should keep in mind that the claims and proofs need to be done in CDEL instead of DEL.

Fact 7 *Let d be the margin of the margin model. A sequence of estimates $\overrightarrow{\eta}$ makes Mr Magoo's argument iteratively sound iff $\overrightarrow{\eta}$ is d-bounded.*

This fact should come as no surprise. Intuitively, the one shot version of the argument is sound if and only if Mr Magoo's estimate does not exceed his actual margin for error. Now, Mr Magoo might go for a sequence of cautious consecutive estimates instead of a one shot daring estimate. But in any case, he should not be allowed to outrange his initial perceptual margin for error. Which is just to say that consecutive estimates should not add up to more than d – the bounding condition on $\overrightarrow{\eta}$. What if Mr Magoo is overconfident? At some point in the reflective process, his estimates add up to more than his perceptual margin and he ends up with a false belief about the size of the tree. In terms of our iterated argument, the conclusion that he knows the size of the tree to be less than his initial estimate minus further improvements ends up being false. This happens when the premise about the margin left at this stage is false.

Williamson considers that the margin estimate can be assumed to remain constant, if it is taken to be small enough. Fact 7 shows that this is too strong an assumption. If $\overrightarrow{\eta}$ is of the form $\eta_1, \eta_1, \eta_1 \ldots,$ for some non-zero η_1, there is no d such that $\overrightarrow{\eta}$ is d-bounded. So no matter what the exact situation is, Mr Magoo is going to reach a false conclusion. By contrast, more realistic choices for $\overrightarrow{\eta}$ can yield iterative soundness. For example, if each estimate is no greater than it should be (considering the remaining margin at the current stage), that is if $\eta_n < d - \sum_{i=0}^{n-1} \eta_i$, then $\overrightarrow{\eta}$ is d-bounded (actually this is equivalent to $\overrightarrow{\eta}$ being d-bounded). This suggests an easy generalization of Fact 7. Let us say that a sequence of estimates $\overrightarrow{\eta}$ makes Mr Magoo's argument *iteratively coherent* if there is some situation (that is some margin model) such that the argument is iteratively sound in that situation. We get:

Fact 8 *A sequence of estimates $\overrightarrow{\eta}$ makes Mr Magoo's argument iteratively coherent iff $\overrightarrow{\eta}$ is d-bounded for some d.*

In this perspective, Williamson's account of Mr Magoo's scenario is merely wrong – there is no situation in which his assumptions about the ways of reflection yield an argument which remains sound when it is iterated. Using Quine's taxonomy (see Quine (1961)), the Margin for Error paradox may be classified as a *falsidical* paradox rather than as an antinomy. The false premise in this case is the assumption of constancy of the margin estimates. Why does this falsidical paradox tend to look like an antinomy? We have the strong intuition that some sequences of estimates yield an iteratively coherent argument. This is indeed true. Because Williamson insists on the fixed estimate for the margin being arbitrarily small, we are misled into thinking that his choice is one of the choices yielding an iteratively coherent argument. In that case, we would have had an antinomy. But we do not.

6.4.3 The Surprise Examination

To conclude, we would like to briefly compare our solution with the thorough dynamic analysis of the Surprise Examination paradox due to Gerbrandy (2007). The story is as follows. A teacher announces to her class on Monday that they will have a surprise examination this week. Clever Marilyn, a student in the class, starts thinking about it. First, the exam cannot be on Friday, because it would be known on Thursday evening that it will take place on Friday. Since it cannot be on Friday, it cannot be on Thursday either, because it would be known on Wednesday evening that it will take place on Thursday. By repeating the argument the student can conclude that there can be no surprise exam, which sounds like a plain contradiction. Let S be the statement that, for every day X in the week, if the exam is on day X, then the student does not know that it is on day X. Using epistemic updates, Gerbrandy shows that "If S correctly paraphrases the teacher's announcement, then Marylin's reasoning is cut short after having excluded the last day as the day of the exam" (p. 27). This is because there is no guarantee that S is true after the teacher has announced it (if the exam is on Thursday, S is true but $\langle S \rangle S$ is false). Gerbrandy suggests various ways to strengthen the teacher's statement. The teacher could explicitly say that the exam will still be a surprise after she has announced that it is a surprise. $S \wedge [S]S$ gives Marilyn enough information to exclude the last two days of the week, but it is still the case that $S \wedge [S]S$ need not be true after it has been announced, so that the exam could be scheduled on Wednesday. One might wish to go for an even stronger announcement δ that would state its own success, so that the equation $\delta \leftrightarrow S \wedge \langle \delta \rangle S$ holds. Gerbrandy shows that no formula of Dynamic Epistemic Logic satisfies this equation, and that δ would be "contingently paradoxical", in the sense that it would be both true and false in some situations.

Here is one way to look at the structural similarity of the Surprise Examination paradox and the Margin for Error paradox. The teacher's announcement that the exam will be a surprise allows the student to eliminate a day of the week as the day of the exam. After that, the announcement is not guaranteed to be true. If the teacher repeats her announcement, it is indeed bound to be false at some stage. Similarly, realizing that $ME(k, \eta)$ holds allows Mr Magoo to eliminate a certain range of heights as the height of the tree. After that, the principle is not guaranteed to be true. If Mr Magoo keeps on updating his knowledge according to the same estimate, the estimate will indeed be inaccurate at some stage. However, the Surprise Examination paradox involves a discrete scale, the consecutive days of the week, whereas the Margin for Error paradox involves a dense scale, the possible heights of the tree. This confers a more subtle status to iterations. The fact that δ is (contingently) paradoxical essentially says that arbitrarily many iterations are not sound. Arbitrarily many iterations with a fixed estimate are not sound either, but Fact 7 states the conditions under which arbitrarily many iterations with a variable estimate are (iteratively) sound. Thus, in the somewhat richer setting offered by Williamson, epistemic updates can be applied to yield more fine-grained results concerning the demarcation line between paradoxical and non-paradoxical scenarios.[19]

6.5 Conclusion

We are finite creatures, endowed with limited perceptual abilities. Our knowledge obeys a margin for error. But we can explore and push our limits to some extent. What is the lesson to draw from Mr Magoo's story in this perspective? Well, there is good news and there is bad news. Here is the bad news first: properly speaking, we cannot *know* our limits. Knowing (an approximation of) our margin for error would make the notion either vacuous, or inconsistent. Just as in the case of Moorean sentences, our limitation is something we can *realize*, but this is not something stable for us to know. The good news is we can always improve on our limits. As long as our sequence of estimations is adequate, there is room for further improvement, and Fact 8 characterizes the conditions under which this may happen.

The attention given to the dynamics of knowledge is an essential component of our account. In this respect, we have followed the path opened by Van Benthem (2004) and taken up by Gerbrandy (2007). One novelty here is our use of Centered Dynamic Epistemic Logic instead of plain DEL, in a context in which DEL cannot satisfactorily handle the intended models for the paradoxical scenarios. We also put forward a notion of iterative soundness, in order to tease apart paradoxical and non-paradoxical versions of Williamson's scenario, in particular to distinguish between one application of Williamson's premises, and their iteration. In this respect, Williamson's paradox is a genuine sorites, since the core question is whether – or when – it is fine to repeat the argument. Because of that, we may wonder whether the dynamic approach may be extended to deal with other sorites more generally.

Acknowledgments We are grateful to O. Roy for detailed comments. The first author also wishes to thank the participants in the Logic and Language Seminar in Stockholm and in the Logic Seminar in Gothenburg. The second author thanks participants in the Formal Epistemology Workshop 2009 in Pittsburgh. We acknowledge the ANR project "Cognitive Origins of Vagueness" (ANR-07-JCJC-0070) for financial support.

[19] In his discussion of the Surprise Examination paradox, Williamson (2000) considers a whole range of closely related paradoxes, so as to gradually turn the Margin for Error paradox into the Surprise examination paradox (to get started, think of a scenario in which Marilyn has a glimpse of a calendar on which the teacher has ringed the examination date, so that she knows it does not take place on Friday). According to Williamson, it is not exactly introspection which is to blame in the Surprise Examination paradox, but a similar assumption with respect to ascriptions of iterated knowledge. Williamson is certainly right that the similarities between the two paradoxes call for similar solutions. In this respect, the similarity between our solution to the Margin for Error paradox and Gerbrandy's solution to the Surprise Examination paradox is quite welcome.

Appendix

Completeness Proof for CDEL

We recall the main Theorem of Section 6.2 and give the proof for **K45** and the class of all frames. The proofs for **S5** and the class of reflexive frames would be similar. We extend the strategy used in Bonnay and Égré (2009) to capture epistemic updates as well as K operators.

Theorem 6.2 K45 *(resp.* **S5***) plus the recursion axioms is sound and complete with respect to Centered Semantics with updates on the class of all frames (resp. of all reflexive frames).*

Proof. Soundness We only prove soundness of the recursion axiom for knowledge. Soundness of **K45** can be proven along the lines of Bonnay and Égré (2009), and correctness for the first three recursion axioms is immediate. So for an arbitrary model $\mathcal{M}$ and a world w in $\mathcal{M}$, what we want is $\mathcal{M}, w \models_{CS} [\varphi]K\psi$ iff $\mathcal{M}, w \models_{CS} \varphi \to K(\varphi \to [\varphi]\psi)$. We get it easily by the following chain of equivalences:

$\mathcal{M}, w \models_{CS} [\varphi]K\psi$
iff $\mathcal{M}, (w, w) \models_{CS} [\varphi]K\psi$ (by definition of $\models_{CS}$)
iff if $\mathcal{M}, (w, w) \models_{CS} \varphi$, then $\mathcal{M}|\varphi_w, (w, w) \models_{CS} K\psi$ (by definition of $[-]$)
iff if $\mathcal{M}, (w, w) \models_{CS} \varphi$, then for all $w' \in \mathcal{M}|\varphi_w$ s.t. wRw',
$\mathcal{M}|\varphi_w, (w, w') \models_{CS} \psi$ (by definition of K)
iff if $\mathcal{M}, (w, w) \models_{CS} \varphi$, then for all $w' \in \mathcal{M}$ s.t. wRw' and $\mathcal{M}, (w, w') \models_{CS} \varphi$,
$\mathcal{M}|\varphi_w, (w, w') \models_{CS} \psi$ (by definition of $|\varphi_w$)
iff if $\mathcal{M}, (w, w) \models_{CS} \varphi$, then for all $w' \in \mathcal{M}$ s.t. wRw' and $\mathcal{M}, (w, w') \models_{CS} \varphi$,
$\mathcal{M}, (w, w') \models_{CS} [\varphi]\psi$ (by definition of $[-]$)
iff if $\mathcal{M}, (w, w) \models_{CS} \varphi$, then for all $w' \in \mathcal{M}$ s.t. wRw',
$\mathcal{M}, (w, w') \models_{CS} \varphi \to [\varphi]\psi$
iff if $\mathcal{M}, (w, w) \models_{CS} \varphi$, then $\mathcal{M}, (w, w) \models_{CS} K(\varphi \to [\varphi]\psi)$ (by definition of K)
iff $\mathcal{M}, (w, w) \models_{CS} \varphi \to K(\varphi \to [\varphi]\psi)$

Completeness. The proof is by contraposition. Assume that a formula does not follow from **K45** plus the recursion axioms, we want to show that φ is not valid with respect to CS. We already know that **K45** plus the recursion axioms is complete with respect to standard semantics over the class of transitive and euclidian frames. So there is a model $\mathcal{M}$, based on such a frame, and a world w in it such that $\mathcal{M}, w \not\models \varphi$. It is sufficient to show that $\mathcal{M}, w \not\models_{CS} \varphi$. This follows from a more general fact, namely that CS and Kripke semantics agree on transitive and euclidian models.

More precisely, for any transitive and euclidian model $\mathcal{M}$ and $w, w' \in \mathcal{M}$, for any formula φ in the language of Dynamic Epistemic Logic, we have that $\mathcal{M}, w' \models \varphi$ iff $\mathcal{M}, (w, w') \models_{CS} \varphi$. The proof is by induction on the complexity of the formula. We only give the case for $\varphi = [\psi]\chi$ (again, the other cases are taken care of in Bonnay and Égré (2009)).

By definition, we have that, on the side of Kripke semantics,
$\mathcal{M}, w' \models [\psi]\chi$ iff if $\mathcal{M}, w' \models \psi$, then $\mathcal{M}|\psi, w' \models \chi$
and, on the side of centered semantics,
$\mathcal{M}, (w, w') \models_{CS} [\psi]\chi$ iff if $\mathcal{M}, (w, w') \models_{CS} \psi$, then $\mathcal{M}|\psi_w, (w, w') \models_{CS} \chi$

By induction hypothesis on ψ, we immediately have that $\mathcal{M}, w' \models \psi$ iff $\mathcal{M}, (w, w') \models_{CS} \psi$. So we just need to prove that $\mathcal{M}|\psi, w \models \chi$ iff $\mathcal{M}|\psi_w, (w, w') \models_{CS} \chi$. But the induction hypothesis on ψ also tells us that $\mathcal{M}|\psi$ are $\mathcal{M}|\psi_w$ are the same. Therefore, by induction hypothesis on χ, $\mathcal{M}|\psi, w \models \chi$ iff $\mathcal{M}|\psi_w, (w, w') \models_{CS} \chi$.

Positive Introspection as Dynamic Closure

In Section 6.3, we claimed that epistemic updates stand to our recasting of Williamson's argument as positive introspection stands to the original argument, modulo closure of knowledge under logical consequence and the recursion axioms. More precisely, the claim was:

Fact 9 $Kp \to KKp \vdash_{Cl,RA} Kp \to [Kp \to q]Kq$

Proof First, we show:

$$\begin{aligned} &Kp \to [Kp \to q]Kq \\ \equiv_{RA} &Kp \to ((Kp \to q) \to K((Kp \to q) \to q)) \end{aligned}$$

By the recursion axiom for K,

$$\begin{aligned} &Kp \to [Kp \to q]Kq \\ \equiv_{RA} &Kp \to ((Kp \to q) \to K((Kp \to q) \to [(Kp \to q)]q)) \end{aligned}$$

Then by the recursion axiom for atoms:

$$\begin{aligned} &Kp \to ((Kp \to q) \to K((Kp \to q) \to [(Kp \to q)]q)) \\ \equiv_{RA} &Kp \to ((Kp \to q) \to K((Kp \to q) \to ((Kp \to q) \to q)) \end{aligned}$$

Finally, $(Kp \to q) \to ((Kp \to q) \to q)$ is of the form $A \to (A \to B)$ which is equivalent to $A \to B$.

Now it is sufficient to show

$$Kp \to KKp \vdash_{Cl} Kp \to ((Kp \to q) \to K((Kp \to q) \to q))$$

Note that $Kp \to ((Kp \to q) \to q)$ is a tautology. But we have $Kp \to KKp$, so by closure under logical consequence $Kp \to K((Kp \to q) \to q)$. *A fortiori*, $Kp \to ((Kp \to q) \to K((Kp \to q) \to q))$.

Note that as may happen in DEL, the above entailment cannot be generalized to arbitrary formulae instead of the atoms.

Sequences of Estimates

We recall the main Fact of Section 6.4, provide a precise characterization of what we mean by 'repeating the argument' and give a detailed proof. We work with margin models. In what follows d is Mr Magoo's perceptual margin for error and w_r is the actual world, so that epistemically possible worlds are worlds $w_{r'}$ with $|r - r'| \leq d$.

Fact 10 *A sequence of estimates $\overrightarrow{\eta}$ makes Mr Magoo's argument iteratively sound iff $\overrightarrow{\eta}$ is d-bounded.*

We start with some definitions. Let ζ_n be the sum of the first n margin estimates, that is $\zeta_n = \sum_{i=0}^{n} \eta_i$ and let $\langle ME(k, \overrightarrow{\eta})\rangle^n$ be short for the first n reflections, that is $\langle ME(k, \overrightarrow{\eta})\rangle^n = \langle ME(k, \eta_1)\rangle \ldots \langle ME(k - \zeta_{n-1}, \eta_n)\rangle$. In other words, $\langle ME(k, \overrightarrow{\eta})\rangle^n$ is the sequence of the first n updates according to Mr Magoo's reflective powers *as given by* $\overrightarrow{\eta}$. The premises at round $n + 1$ are the same as at round 1, but for the fact that n consecutive reflections have taken place. So we get the premises for round $n + 1$ essentially by prefixing $\langle ME(k, \overrightarrow{\eta})\rangle^n$ and adapting the parameters accordingly:

(3n+1') $\langle ME(k, \overrightarrow{\eta})\rangle^n K(s < k - \zeta_n)$
(3n+2') $\langle ME(k, \overrightarrow{\eta})\rangle^n ME(k - \zeta_n, \eta_{n+1})$
(3n+3') $\langle ME(k, \overrightarrow{\eta})\rangle^n$
$(K(s < (k - \zeta_n)) \to [ME(k - \zeta_n, \eta_{n+1})]K(s < (k - \zeta_{n+1})))$

Proof Assume that $K(s < k - \zeta_{n-1})$ is true at w_r, the net effect of $\langle ME(k - \zeta_{n-1}, \eta_n)\rangle$ at w_r is to delete all worlds $w_{r'}$ such that $r' \in [k - \zeta_n, k - \zeta_{n-1}]$, as can be proven by an easy induction on n. Let us now prove both directions.

($\Leftarrow$) If $\overrightarrow{\eta}$ is d-bounded, the argument is iteratively sound. We prove by induction on n that (3n+1'), (3n+2') and (3n+3') are true. First note that according to $\models_{CS}$, introspection always holds, so that (3n+3') is always true and we only need to check the first two premises. For $n = 1$, (1') is true by definition of iterative-soundness. Since $\overrightarrow{\eta}$ is d-bounded, we have in particular $\eta_1 \leq d$, so (2') is true.[20] For $n + 1$, (3(n+1)+1') is true by induction hypothesis, since it follows from (3n+1'), (3n+2')

[20]See statement (i) on p. 114.

and (3n+3') which are true. It only remains to be shown that $\langle ME(k, \vec{\eta})\rangle^n ME(k-\zeta_n, \eta_{n+1})$. It amounts to showing that when the space of worlds is restricted to $[0, k-\zeta_n[$, $ME(k-\zeta_n, \eta_{n+1})$ is still true in the actual world, that is $\mathcal{M}_d|[0, k-\zeta_n[, w_r \vDash_{CS} ME(k-\zeta_n, \eta_{n+1})$. We have just seen that $\mathcal{M}_d|[0, k-\zeta_n[, w_r \vDash_{CS} K(s < (k-\zeta_n))$, so we want $\mathcal{M}_d|[0, k-\zeta_n[, w_r \vDash_{CS} s < (k-\zeta_{n+1})$. Since η was d-bounded, we know that $\zeta_{n+1} \leq d$ and since we had $\mathcal{M}_d, w_r \vDash_{CS} K(s < k), r < k-d$. Therefore $r < k - \zeta_{n+1}$ as needed.

($\Rightarrow$) If $\vec{\eta}$ is not d-bounded, the argument is not iteratively sound. By hypothesis, there is n such that $\zeta_n \leq d$ and $\zeta_{n+1} > d$. Let k be such that $r + d < k < r+\zeta_{n+1}$. We have $\mathcal{M}_d, w_r \vDash_{CS} K(s < k)$. It is then sufficient to show that $\mathcal{M}_d$-$[k-\zeta_n, +\infty[, w_r \vDash_{CS} K(s < (k-\zeta_n)) \wedge (s \geq (k-\zeta_{n+1}))$. Since $\zeta_n \leq d$, $\mathcal{M}_d$-$[k-\zeta_n, +\infty[, w_r \vDash_{CS} K(s < (k-\zeta_n))$ follows from the proof for the other direction. We want $r \geq k - \zeta_{n+1}$. We know $k < r + \zeta_{n+1}$ hence $k - \zeta_{n+1} < r$.

References

Van Benthem J (2004) What one may come to know. Analysis 64(2):95–105

van Benthem J, Gerbrandy J, Hoshi T, Pacuit E (2009) Merging frameworks of interaction. Journal of Philosophical Logic 38:491–526

Bonnay D, Égré P (2009) Inexact knowledge with introspection. Journal of Philosophical Logic 38(2):179–228

van Ditmarsch H, van der Hoek W, Kooi B (2007) Dynamic epistemic logic. Synthese library, Springer, New York

Dokic J, Égré P (2009) Margin for error and the transparency of knowledge. Synthese 166:1–20

Égré P (2008) Reliability, margin for error and self-knowledge. In: Pritchard D, Hendricks V (eds) New waves in epistemology, Palgrave McMillan, New York, pp 215–250

Gerbrandy JD (2007) The surprise examination in dynamic epistemic logic. Synthese 155(1):21–33

Gillies A (2001) A new solution to Moore's paradox. Philosophical Studies 105:237–250

Leitgeb H (2002) Review of Timothy Williamson, *Knowledge and its Limits*. Grazer Philosophische Studien 65:195–205

Quine WVO (1961) The ways of paradox. In: The ways of paradox and other essays, third edition, Harvard University Press, Cambridge, MA pp 1–18

Williamson T (1992) Inexact knowledge. Mind 101:217–242

Williamson T (2000) Knowledge and its Limits. Oxford University Press, Oxford

Chapter 7
Simple Evidence Elimination in Justification Logic

Bryan Renne

7.1 Introduction

Suppose that the prosecutor presents the jury with exhibit x_1, an audio recording of a boss ordering his subordinate to falsify the accounting ledgers so as to deceive the investors into thinking that his insolvent company is not actually insolvent. Suppose further that the judge provides the jury with oral instructions x_2 stating that the jury may use the following principle in reaching its verdict: "if the boss ordered his subordinate to falsify the ledgers, then the boss is guilty of fraud." Using the principle described by the judge's instructions x_2, the recording x_1 provides the jury with sufficient evidence to find the boss guilty of fraud.

But now suppose that the boss' attorney challenges the authenticity of x_1 (the recording) by presenting further evidence that succeeds in convincing the jury that x_1 (the recording) is not authentic and so should be set aside. This challenge has the effect of *eliminating* the evidence x_1; that is, the challenge makes it so that the jury no longer considers x_1 as evidence that the boss ordered his subordinate to falsify the ledgers. So while the jury still has the judge's instructions x_2 for use in reaching its verdict, it will no longer use x_1 (the recording) as evidence. Assuming that there is no further evidence that the boss ordered his subordinate to falsify the ledgers, the jury will then find the boss not guilty of fraud.

This simplistic example of courtroom evidence presents two important features of evidence. First, evidence is something that can be combined according to logical principles in order to draw conclusions. Second, in drawing conclusions on the basis of evidence, one is sometimes required to set aside (or *eliminate*) certain pieces of evidence and then determine which conclusions can still be drawn using only the evidence that remains.

In this chapter, we study a logic for reasoning about these and other issues of evidence. Our logic is an extension of a basic system of *Justification Logic*, a

B. Renne (✉)
Faculty of Philosophy, University of Groningen, Oude Boteringestraat 52, 9712 GL Groningen, The Netherlands
e-mail: bryan@renne.org

P. Girard et al. (eds.), *Dynamic Formal Epistemology*, Synthese Library 351,
DOI 10.1007/978-94-007-0074-1_7,

family of logics for reasoning about evidence and justification for rational agents (Artemov 2008, Fitting 2009, Kuznets 2008, Renne 2008). Justification Logic originated in the proof-theoretic studies of Gödel, who sought an exact provability semantics for the modal logic S4 (Gödel 1995). Artemov later discovered the *Logic of Proofs* as this long-sought connection between S4 and Gödel's intended S4 provability semantics (Artemov 2001), and a number of authors (including S. Artemov, M. Fitting, R. Iemhoff, N. Krupski, V. Krupski, R. Kuznets, R. Milnikel, B. Renne, and others) have since grown the study of the Logic of Proofs into a broader research project – *Justification Logic* – whose purpose is to investigate a wide-ranging family of logics of evidence and justification for rational agents.

In this chapter, we present a system of Justification Logic for reasoning about evidence and evidence elimination. Our theory is called SEE (for Simple Evidence Elimination). We will describe the syntax and semantics of SEE, prove the theory sound and complete with respect to its semantics, and then use our simplistic courtroom evidence example to show how SEE can be used to reason about evidence and evidence elimination.

7.2 Syntax

The language of SEE allows us to describe propositional truth, the evidence a rational individual holds for a given assertion, and the elimination of such evidence.

Definition 7.1 $\mathfrak{L}$(SEE) (pronounced "el-ess-e-e"), the *language of Simple Evidence Elimination*, consists of the *terms* t and the *formulas* φ formed by the following grammar.

$$
\begin{aligned}
t &::= c_k \mid x_k \mid t_1 \cdot_\varphi t_2 \mid t_1 + t_2 \mid\, !t \mid t^{k,\varphi} \\
\varphi &::= p_k \mid \bot \mid \top \mid \varphi_1 \star \varphi_2 \mid \neg\varphi \mid t{:}\, \varphi \mid [k, \varphi_1]\varphi_2 \mid t{:}\, {}^{k,\varphi_1}\varphi_2 \\
& \quad\;\; k \in \mathbb{N},\ \star \in \{\supset, \wedge, \vee, \equiv\}
\end{aligned}
$$

A term of the form c_k is called a *constant* and a term of the form x_k is called a *variable*; the constants and variables make up the *atomic terms*. To say that a term t is *variable-free* means that each non-superscript, non-subscript occurrence of an atomic term in t is a constant. (Examples: $c_0 \cdot_{x_1 : p_5} c_2$ is variable-free, whereas $c_0 \cdot_{x_1 : p_5} x_2$ is not; $c_7^{3, x_8 : p_1}$ is variable-free, whereas $x_7^{3, x_8 : p_1}$ is not.)[1] The p_k's make up the set of *propositional letters*. $\bot$ is the propositional constant for falsity and $\top$ is the propositional constant for truth. Both above and throughout the chapter, we use the symbol $\star$ as a metavariable ranging over the binary logical connectives $\supset$ (implication), $\wedge$ (conjunction), $\vee$ (disjunction), and $\equiv$ (equivalence). A formula of

[1] For Justification Logic aficionados: we will describe the reason we use a subscript formula φ in forming the term $t \cdot_\varphi s$ from terms t and s later in the chapter.

the form $t:\varphi$ is called an *evidence assertion* and is assigned the informal reading "t is evidence that φ." A formula of the form $[k,\varphi]\psi$ is called an *update assertion* and is assigned the informal reading "after [elimination] (k,φ), φ [is true]." Modals of the form $[k,\varphi]$ are called *update modals.* A formula of the form $t{:}^{x_k,\varphi}\ \psi$ is called an *elimination assertion* and is assigned the informal reading "[elimination] (k,φ) eliminates evidence t for φ."

Notation 7.1 We let $\mathcal{T}$ denote the set of all terms in $\mathfrak{L}(\mathsf{SEE})$. Whenever it is convenient, we will identify $\mathfrak{L}(\mathsf{SEE})$ with the set of formulas in the language $\mathfrak{L}(\mathsf{SEE})$.

In $\mathfrak{L}(\mathsf{SEE})$, terms play the role of abstract pieces of evidence that may be combined using the term-forming operations given in the grammar of Definition 7.1. The idea is that these term-forming operations represent logical operations of evidence formation. As an example, we will see shortly that $t+s$ is evidence for everything that one or both of t or s evidences. In this way, $t\mapsto t+s$ and $s\mapsto t+s$ each indicate the operation on evidence that takes one piece of evidence and combines it monotonically with another, thereby evidencing all things that were evidenced by one or more of the two constituent pieces.

Formulas of the form $t:\varphi$ express the statement that t is evidence for φ. So we see that the monotonic combination of evidence $t\mapsto t+s$ can be described by the principle $(t{:}\,\varphi)\supset(t+s):\varphi$, which says, "if t is evidence that φ, then $t+s$ is [also] evidence that φ." By writing down a number of principles describing the behavior of the term-forming operations, one can describe a system of evidence satisfying desirable properties.

Since constants will play a special role described later, we will restrict our notion of evidence elimination so as to eliminate evidence assertions of the form $x_k:\varphi$. To make things simple, we will only eliminate one variable at a time, and we will use the formula

$$[k,\varphi]\psi$$

to mean that ψ is true after we eliminate the evidence x_k that φ. This elimination, which we will write as (k,φ), has the effect of making the formula $x_k:\varphi$ false.

The elimination (k,φ) can also have consequences for other pieces of evidence built using x_k. As an example, we saw how the jury's combined evidence (consisting of the judge's instructions combined with the recording) had to be eliminated as a result of eliminating a part of the combination (the recording). So we see that our theory will also need to reason about how the elimination (k,φ) can lead to the elimination of assertions $t:\psi$ in which t is built using x_k. To specify the consequences of the elimination (k,φ) on more complicated pieces of evidence, we will use elimination assertions.

The elimination assertion $t{:}\ ^{k,\varphi}\psi$ says that the elimination (k,φ) will have the consequence of eliminating $t{:}\,\psi$, thereby making it so that $t{:}\,\psi$ is false. This allows us to provide schematic descriptions of how an elimination (k,φ) can affect the truth of evidence assertions $t{:}\,\psi$ for more complicated terms t.

In the next section, we present the intended semantics for our language $\mathfrak{L}(\mathsf{SEE})$. This semantics describes the conventions we will adopt with regard to evidence behavior in defining our system for reasoning about evidence and evidence elimination.

7.3 Semantics

Our semantics is based the semantics developed by Fitting (2005) and Mkrtychev (1997) for the Logic of Proofs. This semantics introduces what we call an *evidence labeling* in order to directly regulate the truth of evidence assertions $t:\varphi$. Placing certain properties of evidence closure on an evidence labeling yields what we call an *evidence function*.

Definition 7.2 An *evidence labeling* is a subset of $\mathcal{T} \times \mathfrak{L}(\mathsf{SEE})$. For a set S of $\mathfrak{L}(\mathsf{SEE})$-formulas, an *S-evidence function* is an evidence labeling $\mathcal{A}$ that satisfies each of the following schematic properties.

- *Constant Specification S*: if $k \in \mathbb{N}$ and $\varphi \in S$, then $(c_k, \varphi) \in \mathcal{A}$.
- *Application*: if $(t, \varphi \supset \psi) \in \mathcal{A}$ and $(s, \varphi) \in \mathcal{A}$, then $(t \cdot_\varphi s, \psi) \in \mathcal{A}$.
- *Sum*: if $(t, \varphi) \in \mathcal{A}$ or $(s, \varphi) \in \mathcal{A}$, then $(t + s, \varphi) \in \mathcal{A}$.
- *Checker*: if $(t, \varphi) \in \mathcal{A}$, then $(!t, t:\varphi) \in \mathcal{A}$.
- *Update*: if $(t, \varphi) \in \mathcal{A}$, then $(t^{k,\psi}, [k, \psi]\varphi) \in \mathcal{A}$.

If it is convenient and unlikely to cause confusion, we may drop the prefix "S-" in referring to an S-evidence function.

Remark 7.1 We will use an evidence labeling $\mathcal{A}$ to determine the truth of evidence assertions $t:\varphi$ in the following way: $(t, \varphi) \in \mathcal{A}$ will mean that $t:\varphi$ is true. Under this reading, the defining properties of an S-evidence function give us the following connection between term-forming operations and evidence closure principles.

- Constant Specification S tells us that c_k is evidence for φ whenever φ is in S. Thinking of the set S as a collection of "basic statements" that are to be accepted without detailed justification, this property has us use the constants as evidence for the statements that have been identified as "basic." We will later take S as the set of axioms in our to-be-defined axiomatic theory for simple evidence elimination, thereby identifying the axioms of the theory as the "basic statements" that will be evidenced by a constant.
- Application tell us that $t \cdot_\varphi s$ is evidence for ψ whenever t is evidence for $\varphi \supset \psi$ and s is evidence for φ. Thus $t \cdot_\varphi s$ represents the combination of the evidence t for $\varphi \supset \psi$ with the evidence s for φ so as to evidence ψ according to the rule of *Modus Ponens*:

$$\frac{\varphi \supset \psi \quad \varphi}{\psi},$$

which is read, "from assumptions $\varphi \supset \psi$ and φ, conclude ψ." The subscript φ in $t \cdot_\varphi s$ indicates the important rule φ plays as the antecedent of $\varphi \supset \psi$ in the above application of Modus Ponens.

- Sum tells us that $t + s$ is evidence for everything evidenced by one or more of t and s. So $t + s$ is the monotonic combination of evidence t with evidence s.
- Checker tells us that in case t is evidence for φ, then $!t$ checks that t is evidence for φ. So $!t$ provides a means of verifying an evidence assertion.
- Update tells us that in case t is evidence for φ, then $t^{k,\psi}$ is evidence that φ is true after elimination (k, ψ). To make sense of this, if we think of t as very strong evidence that φ is always true, then we ought to be able to use t in an argument showing that φ is true after elimination (k, ψ) by virtue of the fact that φ is always true. We use the term $t^{k,\psi}$ to represent this argument.

While an evidence labeling (and thus an evidence function) will allow us to determine the truth of evidence assertions $t : \varphi$, we still need a way to determine the truth of propositional letters. This is the purpose of a *valuation*.

Definition 7.3 A *valuation* is a set of propositional letters.

We will use a valuation V to determine propositional truth in the following way: $p_k \in V$ means that p_k is true. This is all we need to determine propositional truth.

Taken together, an evidence labeling $\mathcal{A}$ and a valuation V make up a pair $(\mathcal{A}, V)$ that we call an *evidenced valuation*. (An evidenced valuation whose evidence labeling is in fact an evidence function is what we will call a *model*.) Evidenced valuations provide all the ingredients we need to define a notion of truth for $\mathfrak{L}(\mathsf{SEE})$-formulas.

Definition 7.4 Let S be a set of $\mathfrak{L}(\mathsf{SEE})$-formulas. An *evidenced valuation* is a pair $(\mathcal{A}, V)$ consisting of an evidence labeling $\mathcal{A}$ and a valuation V. An S-*model* is an evidenced valuation $(\mathcal{A}, V)$ satisfying the property that $\mathcal{A}$ is an S-evidence function. If it is convenient and unlikely to cause confusion, we may drop the prefix "S-" in referring to an S-model.

Definition 7.5 (Truth) Let $(\mathcal{A}, V)$ be an evidenced valuation. For an $\mathfrak{L}(\mathsf{SEE})$-formula φ, we write $\mathcal{A}, V \models \varphi$ to mean that φ is *true in* $(\mathcal{A}, V)$, and we write $\mathcal{A}, V \not\models \varphi$ to mean that φ is not true in $(\mathcal{A}, V)$. We define the notion of truth for an $\mathfrak{L}(\mathsf{SEE})$-formula in the evidenced valuation $(\mathcal{A}, V)$ by the following induction on $\mathfrak{L}(\mathsf{SEE})$-formula construction.

- $\mathcal{A}, V \models p_k$ means that $p_k \in V$.
- $\mathcal{A}, V \not\models \bot$ and $\mathcal{A}, V \models \top$.
- $\mathcal{A}, V \models \varphi_1 \star \varphi_2$ means that $\mathcal{A}, V \models \varphi_1$ star $\mathcal{A}, V \models \varphi_2$ for $\star \in \{\supset, \wedge, \vee, \equiv\}$.[2]

[2]The word "star" is to be replaced by the English reading for the binary logical connective $\star$; in particular, $\supset$ is read "implies", $\wedge$ is read "and", $\vee$ is read "or", and $\equiv$ is read "if and only if." Note that the connectives $\supset$ and $\equiv$ are to be understood as being defined in the appropriate way in terms of the *material conditional*, which is given by saying that "φ implies ψ" is true exactly when φ is false or ψ is true.

AXIOM SCHEME

EV. $x_k : \varphi$

RULES

$$\frac{\vdash t:(\psi \supset \chi)}{\vdash (t \cdot_\psi s):\chi}\text{(EAL)} \qquad \frac{\vdash s:\psi}{\vdash (t \cdot_\psi s):\chi}\text{(EAR)}$$

$$\frac{\vdash t:\psi \qquad \vdash s:\psi}{\vdash (t+s):\psi}\text{(ES)}$$

$$\frac{\vdash t:\psi}{\vdash\, !t:(t:\psi)}\text{(EC)}$$

$$\frac{\vdash t:\psi}{\vdash t^{j,\chi}:[j,\chi]\psi}\text{(EU)}$$

Fig. 7.1 The theory $\mathsf{E}(k, \varphi)$

- $\mathcal{A}, V \models \neg\varphi$ means that $\mathcal{A}, V \not\models \varphi$.
- $\mathcal{A}, V \models t : \varphi$ means that $(t, \varphi) \in \mathcal{A}$.
- $\mathcal{A}, V \models t\!: {}^{k,\varphi}\psi$ means that $\mathsf{E}(k, \varphi) \vdash t : \psi$.
  The theory $\mathsf{E}(k, \varphi)$ is defined in Fig. 7.1. We will write $\mathsf{E}(k, \varphi) \vdash t : \psi$ to mean that the $\mathfrak{L}(\mathsf{SEE})$-formula $t : \psi$ is derivable in $\mathsf{E}(k, \varphi)$, and we will write $\mathsf{E}(k, \varphi) \nvdash t : \psi$ to mean that $t : \psi$ is not derivable in $\mathsf{E}(k, \varphi)$. Our reason for using the theory $\mathsf{E}(k, \varphi)$ will be explained in a moment.
- $\mathcal{A}, V \models [k, \varphi]\psi$ means that $\mathcal{A}[k, \varphi], V \models \psi$, where

$$\mathcal{A}[k, \varphi] := \{(t, \chi) \in \mathcal{A} \mid \mathsf{E}(k, \varphi) \nvdash t : \chi\} \ .$$

The definition of truth (Definition 7.5) identifies the truth of evidence assertions $t : \varphi$ in an evidenced valuation $(\mathcal{A}, V)$ with the contents of the evidence labeling $\mathcal{A}$, in the sense that $\mathcal{A}, V \models t : \varphi$ if and only if $(t, \varphi) \in \mathcal{A}$. Thus we see that if an evidence valuation $(\mathcal{A}, V)$ happens to be a model (Definition 7.4), which means that $\mathcal{A}$ is an evidence function (Definition 7.2), then the truth of evidence assertions $t : \varphi$ is regulated in a way that respects the intended meanings of the term-forming operations (described in Remark 7.1).

Before we describe the other key clauses within our definition of truth, let us first take a moment to recall and then flesh out the motivating ideas behind our notion of evidence elimination. First, we represent eliminations using a pair (k, φ) consisting of a natural number $k \in \mathbb{N}$ and a formula $\varphi \in \mathfrak{L}(\mathsf{SEE})$. An elimination (k, φ) is to eliminate certain evidence assertions $t : \psi$, in the sense that the occurrence of the elimination (k, φ) will make it the case that $t : \psi$ is false for certain evidence assertions $t : \psi$. As for determining the evidence assertions $t : \psi$ that ought to be eliminated under the elimination (k, φ), we use the following principles.

- *Elimination Base*: (k, φ) eliminates $x_k : \varphi$.
- *Elimination Triggers*. For each of the evidence function properties (Definition 7.2) other than Constant Specification S, use the inverse of the property to trigger eliminations of evidence assertions $t : \psi$ based on the evidence assertions that have already been eliminated. (Note: in reading the inverse of an evidence function property from Definition 7.2 for the purpose of this principle, we interpret the negation of an assertion $(t, \psi) \in \mathcal{A}$ as saying, "elimination (k, φ) eliminates $t : \psi$.") Written in detail, this principle specifies the following elimination triggers.

  - *Inverse Application Trigger*: if (k, φ) eliminates $t : (\psi \supset \chi)$ or (k, φ) eliminates $s : \psi$, then (k, φ) also eliminates $(t \cdot_{\psi} s) : \chi$.
  - *Inverse Sum Trigger*: if (k, φ) eliminates $t : \psi$ and (k, φ) eliminates $s : \psi$, then (k, φ) also eliminates $(t + s) : \psi$.
  - *Inverse Checker Trigger*: if (k, φ) eliminates $t : \psi$, then (k, φ) also eliminates $!t : (t : \psi)$.
  - *Inverse Update Trigger*: if (k, φ) eliminates $t : \psi$, then (k, φ) also eliminates $t^{j,\chi} : [j, \chi]\psi$.

The idea behind the elimination principles is that the evidence function properties describe logical principles of evidence closure that intuitively connect the veracity of one or more evidence assertions $s_1 : \chi_1$ and $s_2 : \chi_2$ with the veracity of an evidence assertion $t(s_1, s_2) : \chi$ whose evidence $t(s_1, s_2)$ is built out of s_1 and s_2 using one of the term-forming operations (Definition 7.1). In essence, the term-forming operation that allows us to construct $t(s_1, s_2)$ out of the terms s_1 and s_2 is to be identified with a certain logical principle for constructing the more complicated piece of evidence $t(s_1, s_2)$ for χ out of the simpler pieces of evidence s_1 (for χ_1) and s_2 (for χ_2) according to our description in Remark 7.1. So when we eliminate one or more of the evidence assertions $s_1 : \chi_1$ and $s_2 : \chi_2$, thereby undermining the veracity of each assertion that we eliminate, we may end up undermining the veracity of the assertion $t(s_1, s_2) : \chi$ because the veracity of this assertion intuitively depends on the veracity of one or more of $s_1 : \chi_1$ and $s_2 : \chi_2$. Whether this happens depends on whether the elimination (k, φ) has falsified the antecedent of the evidence function property governing the term-forming operation that lets us form $t(s_1, s_2)$ from s_1 and s_2. Illustrative example: if the elimination (k, φ) eliminates $t : \psi$ and $s : \psi$, then this has the effect of falsifying the antecedent of the Sum property ("if $t : \psi$ or $s : \psi$, then $(t + s) : \psi$"). But falsifying the antecedent of the Sum property has the intuitive effect of undermining the veracity of the evidence assertion $(t + s) : \psi$ because $t + s$ is supposed to evidence all those things that are evidenced by one or more of t and s (see Remark 7.1). Therefore, if (k, φ) eliminates $t : \psi$ and $s : \psi$, then (k, φ) should also eliminate $(t + s) : \psi$. (Note that the statement in the previous sentence is just the inverse of the Sum property, where we use the reading of inverses specified above in the description of the Elimination Triggers principle). As this illustrative example has shown, the inverse of an evidence function property tells us when the elimination of certain evidence assertions intuitively ought to trigger the elimination of another evidence assertion. So this is why we specified the Elimination Triggers property as we did above.

As an example of how an elimination affects the truth of evidence assertions, let us name a few of the evidence assertions $s:\chi$ that are to be eliminated by an occurrence of the elimination $(1,\varphi)$. First, the elimination $(1,\varphi)$ will obviously eliminate $x_1:\varphi$ due to the principle of Elimination Base. But since $(1,\varphi)$ eliminates $x_1:\varphi$, the Inverse Application Trigger says that $(1,\varphi)$ must also eliminate $(t\cdot_\varphi x_1):\psi$. So the elimination $(1,\varphi)$ eliminates both $x_1:\varphi$ and $(t\cdot_\varphi x_1):\psi$. But these eliminations trigger further eliminations, including the elimination of $(x_1+x_1):\varphi$ (by the Inverse Sum Trigger), the elimination of $!(t\cdot_\varphi x_1):((t\cdot_\varphi x_1):\psi)$ (by the Inverse Checker Trigger), and the elimination of $x_1^{2,\psi}:[2,\psi]\varphi$ (by the Inverse Update Trigger), along with many other eliminations. This is how the elimination $(1,\varphi)$ brings about the elimination of a wide variety of evidence assertions $s:\chi$.

We now examine the way in which the definition of truth handles elimination assertions $t:\ ^{k,\varphi}\psi$. Our intention is that $t:\ ^{k,\varphi}\psi$ is true in an evidenced valuation $(\mathcal{A},V)$ if and only if the elimination (k,φ) eliminates $t:\psi$. (It is in this way that elimination assertions allow us to describe the effects of the elimination (k,φ) within our formal language.) So we see that the truth of an elimination assertion $t:\ ^{k,\varphi}\psi$ is identified with the action of the elimination (k,φ) on the evidence assertion $t:\psi$. Since we have said that we want this action to follow the logical closure principles described by the principles of Elimination Base and Elimination Triggers, whether the action of the elimination (k,φ) affects the evidence assertion $t:\psi$ is a question of logical consequence and it is not hard to see that this notion of logical consequence is encapsulated by the simple axiomatic theory $\mathsf{E}(k,\varphi)$ in Fig. 7.1. This is the reason why the truth of an elimination assertion $t:\ ^{k,\varphi}\psi$ in an evidenced valuation has been identified with the derivability of $t:\psi$ in the theory $\mathsf{E}(k,\varphi)$.

Some readers may find this reliance on the axiomatic theory $\mathsf{E}(k,\varphi)$ within our definition of truth a bit strange because we are connecting the notion of derivability in the axiomatic theory $\mathsf{E}(k,\varphi)$, a syntactic notion, with our definition of truth, a semantic notion. Unfortunately, some such dependence is unavoidable in our framework because we insist that the action of the elimination (k,φ) on evidence assertions ensures that whenever evidence assertions $s_1:\psi_1$ and $s_2:\psi_2$ are eliminated, then so are the evidence assertions $t(s_1,s_2):\psi$ whose evidence $t(s_1,s_2)$ has its veracity intuitively dependent on the veracity of the evidence of one or more of s_1 (for ψ_1) and s_2 (for ψ_2) in the way we described above. Since this notion of dependence is of an essentially logical nature, the notion of truth must somehow utilize the notion of logical consequence encapsulated by the theory $\mathsf{E}(k,\varphi)$. But note that such a notion of logical dependence is just what we want: think of our example of simplistic courtroom evidence, where the elimination of evidence x_1 (the recording of the boss ordering his subordinate to falsify the ledgers) is to bring about the elimination of the logical combination of evidence obtained by joining evidence x_1 (the recording) with evidence x_2 (the judge's instructions that the boss is guilty of fraud if he orders his subordinate to falsify the ledgers).

The reader who is still suspicious of the above connection between our notion of truth and $\mathsf{E}(k, \varphi)$-derivability will hopefully find some comfort in the fact that the theory $\mathsf{E}(k, \varphi)$ is extremely simple and well-behaved. In particular, notice that the conclusion of each rule produces an evidence assertion $t' : \varphi'$ with a term t' that is more complex (contains more symbols) than any term t occurring in a hypothesis $t : \varphi$ of the rule. Further, there is a one-to-one correspondence between the syntactic term-forming operations (from Definition 7.1) and the rules of $\mathsf{E}(k, \varphi)$. Also, the lone Axiom EV of $\mathsf{E}(k, \varphi)$ pertains to atomic terms (in fact, to the single variable x_k for a fixed $k \in \mathbb{N}$). It is thus not difficult to see that this theory is decidable.[3]

Finally, let us look at the definition of truth for update assertions $[k, \varphi]\psi$. Our intention is to have $[k, \varphi]\psi$ true in an evidence valuation $(\mathcal{A}, V)$ if and only if ψ is true after we alter the evidence labeling $\mathcal{A}$ according to the action of the elimination (k, φ) on evidence assertions. But since we used the theory $\mathsf{E}(k, \varphi)$ to characterize those evidenced assertions $t : \chi$ that are to be eliminated, in the sense that (k, φ) eliminates $t : \chi$ if and only if $\mathsf{E}(k, \varphi) \vdash t : \chi$, then we alter $\mathcal{A}$ by removing all term-formula pairs $(t, \chi) \in \mathcal{A}$ such that $\mathsf{E}(k, \varphi) \vdash t : \chi$. Thus we see that in defining $\mathcal{A}[k, \varphi]$ by setting

$$\mathcal{A}[k, \varphi] := \{(t, \chi) \in \mathcal{A} \mid \mathsf{E}(k, \varphi) \nvdash t : \chi\},$$

the evidence labeling $\mathcal{A}[k, \varphi]$ ensures that the evidence assertions $t : \chi$ true in $(\mathcal{A}[k, \varphi], V)$, the evidence labeling that obtains after the occurrence of the elimination (k, φ), are just those evidence assertions $t : \chi$ that were true in $(\mathcal{A}, V)$ before the occurrence of (k, φ) and were also left intact by the action of (k, φ) on evidence assertions.

The notion of formula validity in which we will be interested is given relative to a set S of "basic assertions" that are to be justified using a constant.

Definition 7.6 (Validity) Let S be a set of $\mathfrak{L}(\mathsf{SEE})$-formulas and φ be an $\mathfrak{L}(\mathsf{SEE})$-formula. To say that φ is S-*valid*, written $S \models \varphi$, means that $\mathcal{A}, V \models \varphi$ for each S-model $(\mathcal{A}, V)$. We write $S \not\models \varphi$ to mean that φ is not S-valid.

Since our notion of formula validity is given relative a set S of "basic assertions," it will be important to see that the semantic elimination operation

$$(\mathcal{A}, V) \mapsto (\mathcal{A}[k, \varphi], V)$$

from Definition 7.5 maps S-models to S-models. Said informally, we wish to see that this operation preserves the property of being an S-model.

Lemma 7.1 (S-Model Preservation) *Let S be a set of $\mathfrak{L}(\mathsf{SEE})$-formulas. If $(\mathcal{A}, V)$ is an S-model, then $(\mathcal{A}[k, \varphi], V)$ is also an S-model.*

---

[3] In particular, determining whether $\mathsf{E}(k, \varphi) \vdash t : \psi$ is $O(2^{|t|})$, where $|t|$ is equal to the number of occurrences of term-forming operations that were used in constructing t out of variables and constants according to the grammar in Definition 7.1.

Proof It suffices for us to show that $\mathcal{A}[k,\varphi]$ is an S-evidence function under the assumption that $\mathcal{A}$ is an S-evidence function. According to the definition of S-evidence functions (Definition 7.2), $\mathcal{A}[k,\varphi]$ is an S-evidence function if and only if it satisfies each of Constant Specification S, Application, Sum, Checker, and Update.

Let us check that $\mathcal{A}[k,\varphi]$ satisfies Constant Specification S. We observe that the axiomatics of $\mathsf{E}(k,\varphi)$ in Fig. 7.1 ensures that $\mathsf{E}(k,\varphi) \nvdash c_j : \varphi$ for every $j \in \mathbb{N}$. Applying the definition of $\mathcal{A}[k,\varphi]$ (Definition 7.5), it follows that $(c_j, \psi) \in \mathcal{A}$ if and only if $(c_j, \psi) \in \mathcal{A}[k,\varphi]$ for each $j \in \mathbb{N}$. But then the fact that $\mathcal{A}$ satisfies Constant Specification S implies that $\mathcal{A}[k,\varphi]$ satisfies Constant Specification S.

Let us check that $\mathcal{A}[k,\varphi]$ satisfies Application. We observe that the axiomatics of $\mathsf{E}(k,\varphi)$ in Fig. 7.1 ensures that $\mathsf{E}(k,\varphi) \nvdash (t \cdot_\psi s) : \chi$ if and only if $\mathsf{E}(k,\varphi) \nvdash t : (\psi \supset \chi)$ and $\mathsf{E}(k,\varphi) \nvdash s : \psi$. Applying the definition of $\mathcal{A}[k,\varphi]$ (Definition 7.5), we have that $(t, \psi \supset \chi) \in \mathcal{A}[k,\varphi]$ and $(s,\psi) \in \mathcal{A}[k,\varphi]$ together imply that $(t, \psi \supset \chi) \in \mathcal{A}$, $\mathsf{E}(k,\varphi) \nvdash t : (\psi \supset \chi)$, $(s,\psi) \in \mathcal{A}$, and $\mathsf{E}(k,\varphi) \nvdash s : \psi$. Since $\mathcal{A}$ satisfies Application, $(t, \psi \supset \chi) \in \mathcal{A}$ and $(s,\psi) \in \mathcal{A}$ together imply that $(t \cdot_\psi s, \chi) \in \mathcal{A}$. And the result from the second sentence of this paragraph shows that $\mathsf{E}(k,\varphi) \nvdash t : (\psi \supset \chi)$ and $\mathsf{E}(k,\varphi) \nvdash s : \psi$ together imply that $\mathsf{E}(k,\varphi) \nvdash (t \cdot_\psi s) : \chi$. Applying again the definition of $\mathcal{A}[k,\varphi]$, we have shown that $(t, \psi \supset \chi) \in \mathcal{A}[k,\varphi]$ and $(s,\psi) \in \mathcal{A}[k,\varphi]$ together imply that $(t \cdot_\psi s, \chi) \in \mathcal{A}[k,\varphi]$. It follows that $\mathcal{A}[k,\varphi]$ satisfies Application.

The argument that $\mathcal{A}[k,\varphi]$ satisfies each of Sum, Checker, and Update is shown similarly. Conclusion: $\mathcal{A}[k,\varphi]$ is an S-evidence function.

7.4 Axiomatics

We are now in a position to describe the axiomatics of our theory of Simple Evidence Elimination, SEE.

Definition 7.7 SEE (pronounced "ess-e-e"), the *Theory of Simple Evidence Elimination*, is defined in Fig. 7.2. For each $\mathfrak{L}$(SEE)-formula φ, we write SEE $\vdash \varphi$ to mean that φ is derivable in SEE and we write SEE $\nvdash \varphi$ to mean that φ is not derivable in SEE.

SEE an extension of a basic theory of Justification Logic.[4] Like other Justification Logics, SEE satisfies Artemov's *Internalization Theorem*, which provides a sense in which the structure of terms can mirror reasoning within the theory.

Theorem 7.1 (Artemov's Internalization Theorem; Artemov (2001)) SEE $\vdash \varphi$ *implies* SEE $\vdash t : \varphi$ *for a variable-free term* $t \in \mathcal{T}$.

---

[4]This theory was called J4 in Renne (2008), though we note that the languages of J4 and SEE vary slightly. In particular, while the language of SEE, $\mathfrak{L}$(SEE), uses a subscript formula φ in forming the term $t \cdot_\varphi s$ from terms t and s, the language of J4 forms the term $t \cdot s$ from terms t and s without a subscript formula. We require such a subscript formula in $\mathfrak{L}$(SEE) in order to be able to express that the formula $(t \cdot_\psi s)\colon^{k,\varphi} \chi$ is equivalent to some Boolean combination of formulas of the form $t\colon^{k,\varphi} \chi_1$ and $s\colon^{k,\varphi} \chi_2$ for appropriate χ_1 and χ_2. The equivalence we want is Axiom X3 of SEE (Fig. 7.2).

CLASSICAL LOGIC AND EVIDENCE

CL. Axiom schemes for classical propositional logic

E1. $(t:(\varphi \supset \psi)) \supset ((s:\varphi) \supset (t \cdot_{\varphi} s):\psi)$

E2. $(t:\varphi) \supset (t+s):\varphi$

$(s:\varphi) \supset (t+s):\varphi$

E3. $(t:\varphi) \supset !t:(t:\varphi)$

E4. $(t:\varphi) \supset t^{k,\psi}:[k,\psi]\varphi$

UPDATE AND ELIMINATION

U1. $[k,\varphi]q \quad \equiv \quad q$

U2. $[k,\varphi](\psi \star \chi) \quad \equiv \quad [k,\varphi]\psi \star [k,\varphi]\chi$

U3. $[k,\varphi]\neg\psi \quad \equiv \quad \neg[k,\varphi]\psi$

U4. $[k,\varphi](t:\psi) \quad \equiv \quad (t:\psi) \wedge \neg(t:^{k,\varphi}\psi)$

U5. $[k,\varphi](t:^{j,\chi}\psi) \quad \equiv \quad (t:^{j,\chi}\psi)$

X1. $(c_j:^{k,\varphi}\psi) \quad \equiv \quad \bot$

X2. $(x_j:^{k,\varphi}\psi) \quad \equiv \quad \begin{cases} \top & \text{if } (j,\psi) = (k,\varphi) \\ \bot & \text{otherwise} \end{cases}$

X3. $((t \cdot_{\psi} s):^{k,\varphi}\chi) \quad \equiv \quad (t:^{k,\varphi}(\psi \supset \chi)) \vee (s:^{k,\varphi}\psi)$

X4. $((t+s):^{k,\varphi}\psi) \quad \equiv \quad (t:^{k,\varphi}\psi) \wedge (s:^{k,\varphi}\psi)$

X5. $(!t:^{k,\varphi}\theta) \quad \equiv \quad \begin{cases} t:^{k,\varphi}\chi & \text{if } \theta = (t:\chi) \\ \bot & \text{otherwise} \end{cases}$

X6. $(t^{j,\chi}:^{k,\varphi}\theta) \quad \equiv \quad \begin{cases} t:^{k,\varphi}\psi & \text{if } \theta = [j,\chi]\psi \\ \bot & \text{otherwise} \end{cases}$

Note: $q \in \{p_k, \bot, \top\}$ and $\star \in \{\supset, \wedge, \vee, \equiv\}$.

RULES

$$\frac{k \in \mathbb{N} \quad \varphi \text{ an axiom}}{\vdash c_k:\varphi}\,(\text{CN})$$

$$\frac{\vdash \varphi \supset \psi \quad \vdash \varphi}{\vdash \psi}\,(\text{MP}) \qquad \frac{k \in \mathbb{N} \quad \vdash \varphi}{\vdash [k,\psi]\varphi}\,(\text{UN})$$

Fig. 7.2 The theory SEE

Proof By induction on the length of the SEE-derivation of φ. In the base case, φ is an axiom, and it follows from Rule CN that SEE $\vdash c_0:\varphi$. Taking $t := c_0$, a variable-free term, the result follows. In the induction step, φ follows by a rule of inference. We consider each rule of inference in turn.

- Induction Case: $\varphi = (c_k : \psi)$ follows from ψ by Rule CN.
  Take $t := !c_k$, a variable-free term. We observe that SEE $\vdash t : \varphi$ by Axiom E3 and Rule MP.
- Induction Case: φ follows from $\psi \supset \varphi$ and ψ by Rule MP.
  By the induction hypothesis, there are variable-free terms s_1 and s_2 such that SEE $\vdash s_1 : (\psi \supset \varphi)$ and SEE $\vdash s_2 : \psi$. Applying Axiom E1 and Rule MP, it follows that SEE $\vdash (s_1 \cdot_\psi s_2) : \varphi$. Taking $t := s_1 \cdot_\psi s_2$, we observe that t is variable-free.
- Induction Case: $\varphi = [k, \psi]\chi$ follows from χ by Rule UN.
  By the induction hypothesis, there is a variable-free term s such that SEE $\vdash s : \chi$. It follows by Axiom E4 and Rule MP that SEE $\vdash s^{k,\psi} : [k, \psi]\chi$. Taking $t := s^{k,\psi}$, we observe that t is variable-free.

The Internalization Theorem provides a sense in which rational agents can formulate specific arguments describing the process of logical deduction; that is, in deducing φ using a specific SEE-deduction, the term t yielded by the proof of the Internalization Theorem provides an explicit description of the step-by-step reasoning that took place in the deduction. This bolsters the sense in which terms serve as pieces of evidence in theories of Justification Logic .

Our Soundness Theorem says that if we take S to be the set of SEE-axioms, thereby equating these axioms with the "basic statements" that are to be justified by a constant, then all SEE-theorems are S-valid.

Theorem 7.2 (Soundness) *Let S be the set of* SEE*-axioms.* SEE $\vdash \varphi$ *implies* $S \models \varphi$.

Proof We show by induction on the length of derivation in SEE that each SEE-theorem is S-valid. In the base case of this induction, we must show that each SEE-axiom is S-valid.

- Base Case: Axiom CL is S-valid.
  This follows from the usual truth-table arguments for classical propositional logic.
- Base Case: Axioms E1–E4 are S-valid.
  Let $(\mathcal{A}, V)$ be an S-model. That E1 is true in $(\mathcal{A}, V)$ follows from the definition of truth (Definition 7.5) and the fact that $\mathcal{A}$ satisfies Application. Similarly, E2 is true in $(\mathcal{A}, V)$ because $\mathcal{A}$ satisfies Sum, E3 is true in $(\mathcal{A}, V)$ because $\mathcal{A}$ satisfies Checker, and E4 is true in $(\mathcal{A}, V)$ because $\mathcal{A}$ satisfies Update. Since $(\mathcal{A}, V)$ was an arbitrarily chosen S-model, we have shown that E1–E4 are each S-valid.
- Base Case: Axioms U1–U5 are S-valid.
  That each of Axioms U1–U5 is S-valid follows directly from the definition of truth (Definition 7.5). The most interesting case is Axiom U4, so let us write out the argument for this axiom as a paradigmatic example for the others.
  Let $(\mathcal{A}, V)$ be an S-model. To have $\mathcal{A}, V \models [k, \varphi](t : \psi)$ means that $\mathcal{A}[k, \varphi], V \models t : \psi$, which itself means that $(t, \psi) \in \mathcal{A}[k, \varphi]$. By the definition of $\mathcal{A}[k, \varphi]$ (Definition 7.5), $(t, \psi) \in \mathcal{A}[k, \varphi]$ is equivalent to $(t, \psi) \in \mathcal{A}$ and $\mathsf{E}(k, \varphi) \nvdash t : \psi$. By the definition of truth, the latter conjunction is itself

equivalent to the statement that $\mathcal{A}, V \models t:\psi$ and $\mathcal{A}, V \not\models t:{}^{k,\varphi}\psi$. Again applying the definition of truth, the latter conjunction is equivalent to $\mathcal{A}, V \models (t:\psi) \wedge \neg(t:{}^{k,\varphi}\psi)$. We therefore have shown that $\mathcal{A}, V \models [k,\varphi](t:\psi)$ is equivalent to $\mathcal{A}, V \models (t:\psi) \wedge \neg(t:{}^{k,\varphi}\psi)$, and so it follows by the definition of truth that U4 is true in $(\mathcal{A}, V)$. Since $(\mathcal{A}, V)$ was an arbitrarily chosen S-model, we have shown that U4 is S-valid.

- Base Case: Axiom X1 is S-valid.
  Let $(\mathcal{A}, V)$ be an S-model. By the definition of truth, $\mathcal{A}, V \models (c_j:{}^{k,\varphi}\psi) \equiv \bot$ is equivalent to $\mathsf{E}(k,\varphi) \nvdash c_j:\psi$. By an examination of the axiomatics of $\mathsf{E}(k,\varphi)$ from Fig. 7.1, the latter is simply true. Since $(\mathcal{A}, V)$ was an arbitrarily chosen S-model, we have shown that X1 is S-valid.
- Base Case: Axiom X2 is S-valid.
  Let $(\mathcal{A}, V)$ be an S-model. By the definition of truth, $\mathcal{A}, V \models (x_k:{}^{k,\varphi}\varphi) \equiv \top$ is equivalent to $\mathsf{E}(k,\varphi) \vdash x_k:\varphi$. By an examination of the axiomatics of $\mathsf{E}(k,\varphi)$ from Fig. 7.1, the latter is simply true.

  Also by the definition of truth, $\mathcal{A}, V \models (x_j:{}^{k,\varphi}\psi) \equiv \bot$ for $(j,\psi) \neq (k,\varphi)$ is equivalent to $\mathsf{E}(k,\varphi) \nvdash x_j:\psi$. By an examination of the axiomatics of $\mathsf{E}(k,\varphi)$ from Fig. 7.1, it follows from our assumption $(j,\psi) \neq (k,\varphi)$ that $\mathsf{E}(k,\varphi) \nvdash x_j:\psi$ is simply true.

  Since $(\mathcal{A}, V)$ was an arbitrarily chosen S-model, we have shown that X2 is S-valid.
- Base Case: Axiom X3 is S-valid.
  Let $(\mathcal{A}, V)$ be an S-model. By the definition of truth, $\mathcal{A}, V \models (t \cdot_\psi s):{}^{k,\varphi}\chi$ is equivalent to $\mathsf{E}(k,\varphi) \vdash (t \cdot_\psi s):\chi$. By an examination of the axiomatics of $\mathsf{E}(k,\varphi)$ from Fig. 7.1, the latter is equivalent to the statement that $\mathsf{E}(k,\varphi) \vdash t:(\psi \supset \chi)$ or $\mathsf{E}(k,\varphi) \vdash s:\psi$. But the latter disjunction is what it means to say that $\mathcal{A}, V \models \big(t:{}^{k,\varphi}(\psi \supset \chi)\big) \vee (s:{}^{k,\varphi}\psi)$. We therefore have shown that $\mathcal{A}, V \models (t \cdot_\psi s):{}^{k,\varphi}\chi$ is equivalent to $\mathcal{A}, V \models \big(t:{}^{k,\varphi}(\psi \supset \chi)\big) \vee (s:{}^{k,\varphi}\psi)$, and so it follows from the definition of truth that X3 is true in $(\mathcal{A}, V)$. Since $(\mathcal{A}, V)$ was an arbitrarily chosen S-model, we have shown that X3 is S-valid.
- Base Case: Axioms X4–X6 are S-valid.
  These are shown by arguments similar to the above argument for Axiom X3.

This completes the base cases of the induction. For the induction cases, we are to show that the SEE-rules preserve S-validity. We consider each rule in turn.

- Induction Case: $c_k:\varphi$ was derived from φ using Rule CN.
  Let $(\mathcal{A}, V)$ be an S-model. φ is an SEE-axiom and therefore $\varphi \in S$. It follows that $(c_k,\varphi) \in \mathcal{A}$ by the fact that $\mathcal{A}$ satisfies Constant Specification S. But then $\mathcal{A}, V \models c_k:\varphi$. Since $(\mathcal{A}, V)$ was an arbitrarily chosen S-model, we have shown that $c_k:\varphi$ is S-valid.
- Induction Case: ψ was derived from $\varphi \supset \psi$ and φ using Rule MP.
  By the induction hypothesis, each of $\varphi \supset \psi$ and φ is S-valid. Let $(\mathcal{A}, V)$ be an S-model. It follows from the S-validity of $\varphi \supset \psi$ and φ that $\mathcal{A}, V \models (\varphi \supset \psi) \wedge \varphi$

and thus that $\mathcal{A}, V \models \psi$. Since $(\mathcal{A}, V)$ was an arbitrarily chosen S-model, we have shown that ψ is S-valid.

- Induction Case: $[k, \psi]\varphi$ was derived from φ using Rule UN.
  By the induction hypothesis, φ is S-valid. Let $(\mathcal{A}, V)$ be an S-model. It follows from the S-Model Preservation Lemma (Lemma 7.1) that $(\mathcal{A}[k, \psi], V)$ is an S-model. Applying the S-validity of φ, we then have that $\mathcal{A}[k, \psi], V \models \varphi$. But the latter is what it means to have $\mathcal{A}, V \models [k, \psi]\varphi$. Since $(\mathcal{A}, V)$ was an arbitrarily chosen S-model, we have shown that $[k, \psi]\varphi$ is S-valid.

The converse of the Soundness Theorem (Theorem 7.2) is the Completeness Theorem. To prove the Completeness Theorem, we introduce a notion of *depth* for $\mathfrak{L}$(SEE)-formulas that will come up later.

Definition 7.8 ($\mathfrak{L}$(SEE)-Depth) The $\mathfrak{L}$(SEE)-*depth function* is a function $d : \mathfrak{L}(\mathsf{SEE}) \to \mathbb{N}$ that maps each formula $\varphi \in \mathfrak{L}(\mathsf{SEE})$ to a natural number $d(\varphi)$ according to the definition in Fig. 7.3. We call $d(\varphi)$ the *depth* of φ.

As it turns out, each $\mathfrak{L}$(SEE)-formula φ is provably equivalent in SEE to an $\mathfrak{L}$(SEE)-formula φ° with $d(\varphi^\circ) = 0$, which says that φ° does not contain occurrences of update modals within the scope of a term.[5] The formula φ°, called the *reduction* of φ, is defined as follows.

Definition 7.9 The $\mathfrak{L}$(SEE)-*reduction function* is a function $\circ : \mathfrak{L}(\mathsf{SEE}) \to \mathfrak{L}(\mathsf{SEE})$ that maps each formula $\varphi \in \mathfrak{L}(\mathsf{SEE})$ to the formula $\varphi^\circ \in \mathfrak{L}(\mathsf{SEE})$ according to the definition in Fig. 7.4. We call φ° the *reduction* of φ.

Lemma 7.2 (Reduction Lemma) *$d(\varphi^\circ) = 0$ and* SEE $\vdash \varphi \equiv \varphi^\circ$.

Proof By an induction on the construction of φ. Abbreviations: $\vdash \gamma$ abbreviates SEE $\vdash \gamma$, "SEE" abbreviates "reasoning in SEE", "IH" abbreviates "induction hypothesis."

$$
\begin{array}{lcl}
d(q) & := & 0 \\
d(\varphi \star \psi) & := & \max\{d(\varphi), d(\psi)\} \\
d(\neg\varphi) & := & d(\varphi) \\
d(t:\varphi) & := & 0 \\
d(t:^{k,\varphi}\psi) & := & 0 \\
d\big([k,\varphi]\psi\big) & := & 1 + d(\psi)
\end{array}
$$

Note: $q \in \{p_k, \bot, \top\}$ and $\star \in \{\supset, \wedge, \vee, \equiv\}$.

Fig. 7.3 Definition of a function $d : \mathfrak{L}(\mathsf{SEE}) \to \mathbb{N}$

[5] To say that an $\mathfrak{L}$(SEE)-formula θ contains a piece of syntax *within the scope of a term* means that there is a subformula $t : \psi$ of θ such that there is an occurrence of the piece of syntax in ψ.

$$
\begin{array}{lll}
q^\circ & := & q \\
(\varphi \star \psi)^\circ & := & \varphi^\circ \star \psi^\circ \\
(\neg\varphi)^\circ & := & \neg\varphi^\circ \\
(t : \varphi)^\circ & := & t : \varphi \\
(t :^{k,\varphi} \psi)^\circ & := & t :^{k,\varphi} \psi \\
([k,\varphi]q)^\circ & := & q \\
([k,\varphi](\psi \star \chi))^\circ & := & ([k,\varphi]\psi)^\circ \star ([k,\varphi]\chi)^\circ \\
([k,\varphi]\neg\psi)^\circ & := & \neg([k,\varphi]\psi)^\circ \\
([k,\varphi](t : \psi))^\circ & := & (t : \psi) \wedge \neg(t :^{k,\varphi} \psi) \\
([k,\varphi](t :^{j,\chi} \psi))^\circ & := & t :^{j,\chi} \psi \\
([k,\varphi][j,\psi]\chi)^\circ & := & ([k,\varphi]([j,\psi]\chi)^\circ)^\circ
\end{array}
$$

Note: $q \in \{p_k, \bot, \top\}$ and $\star \in \{\supset, \wedge, \vee, \equiv\}$.

Fig. 7.4 Definition of a function $\circ : \mathfrak{L}(\mathsf{SEE}) \to \mathfrak{L}(\mathsf{SEE})$

- Base Case: $\varphi = q$ for $q \in \{p_k, \bot, \top\}$.

$$
\begin{aligned}
d(q^\circ) &= d(q) && \text{Fig. 7.4} \\
&= 0 && \text{Fig. 7.3}
\end{aligned}
$$

$$
\begin{array}{ll}
1. \vdash q \equiv q & \text{by SEE} \\
2. \vdash q \equiv q^\circ & \text{by 1, Fig. 7.4}
\end{array}
$$

- Induction Case: $\varphi = (\psi \star \chi)$.

$$
\begin{aligned}
& d((\psi \star \chi)^\circ) \\
&= d(\psi^\circ \star \chi^\circ) && \text{by Fig. 7.4} \\
&= \max\{d(\psi^\circ), d(\chi^\circ)\} && \text{by Fig. 7.3} \\
&= 0 && \text{by IH}
\end{aligned}
$$

$$
\begin{array}{ll}
1. \vdash \psi \equiv \psi^\circ & \text{by IH} \\
2. \vdash \chi \equiv \chi^\circ & \text{by IH} \\
3. \vdash (\psi \star \chi) \equiv (\psi^\circ \star \chi^\circ) & \text{by 1, 2, SEE} \\
4. \vdash (\psi \star \chi) \equiv (\psi \star \chi)^\circ & \text{by 3, Fig. 7.4}
\end{array}
$$

- Induction Case: $\varphi = \neg\psi$.

$$\begin{array}{ll} & d((\neg\psi)^\circ) \\ = d(\neg\psi^\circ) & \text{by Fig. 7.4} \\ = d(\psi^\circ) & \text{by Fig. 7.3} \\ = 0 & \text{by IH} \end{array}$$

$$\begin{array}{ll} 1. \vdash \psi \equiv \psi^\circ & \text{by IH} \\ 2. \vdash \neg\psi \equiv \neg\psi^\circ & \text{by 1, SEE} \\ 3. \vdash \neg\psi \equiv (\neg\psi)^\circ & \text{by 2, Fig. 7.4} \end{array}$$

- Induction Case: $\varphi = (t : \psi)$.

$$\begin{array}{ll} & d((t : \psi)^\circ) \\ = d(t : \psi) & \text{by Fig. 7.4} \\ = 0 & \text{by Fig. 7.3} \end{array}$$

$$\begin{array}{ll} 1. \vdash (t : \psi) \equiv (t : \psi) & \text{by SEE} \\ 2. \vdash (t : \psi) \equiv (t : \psi)^\circ & \text{by 1, Fig. 7.4} \end{array}$$

- Induction Case: $\varphi = (t :^{k,\psi} \chi)$.
  Similar to the previous induction case ($\varphi = t : \psi$).
- Induction Case: $\varphi = [k, \psi]\theta$.
  By a sub-induction on the depth $d(\theta)$ of θ with a sub-sub-induction on the construction of θ. Abbreviations: "SIH" abbreviates "sub-induction hypothesis" and "SSIH" abbreviates "sub-sub-induction hypothesis." A reference to an SEE axiom indicates that the result is by reasoning in SEE that makes crucial use of the axiom in question.
  - Sub-Base Case: $d(\theta) = 0$; Sub-Sub-Base Case: $\theta = q$, where $q \in \{p_k, \bot, \top\}$.

$$\begin{array}{rll} d(([k, \psi]q)^\circ) & = d(q) & \text{by Fig. 7.4} \\ & = 0 & \text{by Fig. 7.3} \end{array}$$

$$\begin{array}{ll} 1. \vdash [k, \psi]q \equiv q & \text{by Axiom U1} \\ 2. \vdash [k, \psi]q \equiv ([k, \psi]q)^\circ & \text{by 1, Fig. 7.4} \end{array}$$

  - Sub-Base Case: $d(\theta) = 0$; Sub-Sub-Induction Case: $\theta = (\chi \star \omega)$.

$$\begin{array}{ll} & d(([k, \psi](\chi \star \omega))^\circ) \\ = d(([k, \psi]\chi)^\circ \star ([k, \psi]\omega)^\circ) & \text{by Fig. 7.4} \\ = \max\{d(([k, \psi]\chi)^\circ), d(([k, \psi]\omega)^\circ)\} & \text{by Fig. 7.3} \\ = 0 & \text{by SSIH} \end{array}$$

$$
\begin{array}{ll}
1. \vdash [k, \psi]\chi \equiv ([k, \psi]\chi)^\circ & \text{by SSIH} \\
2. \vdash [k, \psi]\omega \equiv ([k, \psi]\omega)^\circ & \text{by SSIH} \\
3. \vdash ([k, \psi]\chi \star [k, \psi]\omega) \equiv ([k, \psi]\chi)^\circ \star ([k, \psi]\omega)^\circ & \text{by 1, 2, SEE} \\
4. \vdash ([k, \psi]\chi \star [k, \psi]\omega) \equiv ([k, \psi](\chi \star \omega))^\circ & \text{by 3, Fig. 7.4} \\
5. \vdash [k, \psi](\chi \star \omega) \equiv ([k, \psi](\chi \star \omega))^\circ & \text{by 4, Axiom U2}
\end{array}
$$

– Sub-Base Case: $d(\theta) = 0$; Sub-Sub-Induction Case: $\theta = \neg\chi$.

$$
\begin{aligned}
d(([k, \psi]\neg\chi)^\circ) &= d(\neg([k, \psi]\chi)^\circ) && \text{by Fig. 7.4} \\
&= d(([k, \psi]\chi)^\circ) && \text{by Fig. 7.3} \\
&= 0 && \text{by SSIH}
\end{aligned}
$$

$$
\begin{array}{ll}
1. \vdash [k, \psi]\chi \equiv ([k, \psi]\chi)^\circ & \text{by SSIH} \\
2. \vdash \neg[k, \psi]\chi \equiv \neg([k, \psi]\chi)^\circ & \text{by 1, SEE} \\
3. \vdash \neg[k, \psi]\chi \equiv ([k, \psi]\neg\chi)^\circ & \text{by 2, Fig. 7.4} \\
4. \vdash [k, \psi]\neg\chi \equiv ([k, \psi]\neg\chi)^\circ & \text{by 3, Axiom U3}
\end{array}
$$

– Sub-Base Case: $d(\theta) = 0$; Sub-Sub-Induction Case: $\theta = (t : \chi)$.

$$
\begin{aligned}
&d(([k, \psi](t : \chi))^\circ) \\
&= d((t : \chi) \wedge \neg(t\colon {}^{k,\psi} \chi)) && \text{by Fig. 7.4} \\
&= 0 && \text{by Fig. 7.3}
\end{aligned}
$$

$$
\begin{array}{ll}
1. \vdash [k, \psi](t : \chi) \equiv (t : \chi) \wedge \neg(t\colon {}^{k,\psi} \chi) & \text{by Axiom U4} \\
2. \vdash [k, \psi](t : \chi) \equiv ([k, \psi](t : \chi))^\circ & \text{by Fig. 7.4}
\end{array}
$$

– Sub-Base Case: $d(\theta) = 0$; Sub-Sub-Induction Case: $\theta = (t\colon {}^{j,\chi} \omega)$.

$$
\begin{aligned}
&d(([k, \psi](t\colon {}^{j,\chi} \omega))^\circ) \\
&= d(t\colon {}^{j,\chi} \omega) && \text{by Fig. 7.4} \\
&= 0 && \text{by Fig. 7.3}
\end{aligned}
$$

$$
\begin{array}{ll}
1. \vdash [k, \psi](t\colon {}^{j,\chi} \omega) \equiv (t\colon {}^{j,\chi} \omega) & \text{by Axiom U5} \\
2. \vdash [k, \psi](t\colon {}^{j,\chi} \omega) \equiv ([k, \psi](t\colon {}^{j,\chi} \omega))^\circ & \text{by Fig. 7.4}
\end{array}
$$

– Sub-Induction Case: $d(\theta) > 0$; Sub-Sub-Base and Sub-Sub-Induction Case: $\theta = [j, \chi]\omega$.
By the SIH, we have that $d(([j, \chi]\omega)^\circ) = 0$. It therefore follows that

$$d([k, \psi]([j, \chi]\omega)^\circ) = 1$$

by Fig. 7.3. Applying the fact that $d([k, \psi][j, \chi]\omega) \geq 2$, we have shown that the SIH also applies to the formula $[k, \psi]([j, \chi]\omega)^\circ$, which gives us the following.

$$\begin{aligned} & d(([k, \psi][j, \chi]\omega)^\circ) \\ &= d(([k, \psi]([j, \chi]\omega)^\circ)^\circ) && \text{by Fig. 7.4} \\ &= 0 && \text{by SIH} \end{aligned}$$

$$\begin{array}{ll} 1. \vdash [j, \chi]\omega \equiv ([j, \chi]\omega)^\circ & \text{by SIH} \\ 2. \vdash [k, \psi][j, \chi]\omega \equiv [k, \psi]([j, \chi]\omega)^\circ & \text{by 1, SEE} \\ 3. \vdash [k, \psi]([j, \chi]\omega)^\circ \equiv ([k, \psi]([j, \chi]\omega)^\circ)^\circ & \text{by SIH} \\ 4. \vdash [k, \psi][j, \chi]\omega \equiv ([k, \psi]([j, \chi]\omega)^\circ)^\circ & \text{by 2, 3, SEE} \\ 5. \vdash [k, \psi][j, \chi]\omega \equiv ([k, \psi][j, \chi]\omega)^\circ & \text{by 4, Fig. 7.4} \end{array}$$

– Sub-Induction Case: $d(\theta) > 0$, Sub-Sub-Induction Cases $\theta = (\chi \star \omega)$, $\theta = \neg\chi$, $\theta = (t : \chi)$, and $\theta = (t :\ ^{j,\chi}\omega)$ are handled as in the corresponding sub-sub-induction cases of Sub-Induction Case $d(\theta) = 0$. (See above.)

In addition to the Reduction Lemma (Lemma 7.2), we will need one more lemma to facilitate our proof of the forthcoming Completeness Theorem. This lemma is as follows.

Lemma 7.3 SEE $\vdash t\colon {}^{k,\varphi}\psi$ *if and only if* $\mathsf{E}(k, \varphi) \vdash t : \psi$.

Proof Let S be the set of SEE-axioms. It is not difficult to see that $(\mathcal{T} \times \mathfrak{L}(\mathsf{SEE}), \emptyset)$ is an S-model.[6] It therefore follows by soundness (Theorem 7.2) and the definition of validity (Definition 7.6) that SEE $\vdash t\colon {}^{k,\varphi}\psi$ implies $\mathcal{T} \times \mathfrak{L}(\mathsf{SEE}), \emptyset \models t\colon {}^{k,\varphi}\psi$. Applying the definition of truth, the latter implies $\mathsf{E}(k, \varphi) \vdash t : \psi$. So we see that SEE $\vdash t\colon {}^{k,\varphi}\psi$ implies $\mathsf{E}(k, \varphi) \vdash t : \psi$. To prove the converse of this implication, we argue by induction on the length of derivation in $\mathsf{E}(k, \varphi)$ that $\mathsf{E}(k, \varphi) \vdash t : \psi$ implies SEE $\vdash t\colon {}^{k,\varphi}\psi$. (The axiomatics of $\mathsf{E}(k, \varphi)$ are defined in Fig. 7.1, and the axiomatics of SEE are defined in Fig. 7.2.)

- Base Case: $\mathsf{E}(k, \varphi) \vdash x_k : \varphi$ by Axiom EV of $\mathsf{E}(k, \varphi)$.
  We have SEE $\vdash x_k\colon {}^{k,\varphi}\varphi$ by Axiom X2 of SEE.
- Induction Case: $\mathsf{E}(k, \varphi) \vdash (t \cdot_\psi s) : \chi$ using Rule EAL of $\mathsf{E}(k, \varphi)$.
  By the induction hypothesis, SEE $\vdash t\colon {}^{k,\varphi}(\psi \supset \chi)$. Reasoning in SEE, it follows that SEE $\vdash \big(t\colon {}^{k,\varphi}(\psi \supset \chi)\big) \vee (s\colon {}^{k,\varphi}\psi)$. Applying Axiom X3 of SEE, the result follows.
- Induction Case: $\mathsf{E}(k, \varphi) \vdash (t \cdot_\psi s) : \chi$ using Rule EAR of $\mathsf{E}(k, \varphi)$.
  Similar to the previous induction case.

[6] It is easy to see that $\mathcal{A}^* := \mathcal{T} \times \mathfrak{L}(\mathsf{SEE})$ satisfies each of the defining properties of S-evidence functions (Definition 7.2) by the fact that $(t, \psi) \in \mathcal{A}^*$ holds for every $(t, \psi) \in \mathcal{T} \times \mathfrak{L}(\mathsf{SEE})$.

- Induction Case: $\mathsf{E}(k, \varphi) \vdash (t + s) : \psi$ using Rule ES of $\mathsf{E}(k, \varphi)$.
  By the induction hypothesis, $\mathsf{SEE} \vdash t\colon {}^{k,\varphi}\psi$ and $\mathsf{SEE} \vdash s\colon {}^{k,\varphi}\psi$. Reasoning in SEE, it follows that $\mathsf{SEE} \vdash (t\colon {}^{k,\varphi}\psi) \wedge (s\colon {}^{k,\varphi}\psi)$. Applying Axiom X4 of SEE, the result follows.
- Induction Case: $\mathsf{E}(k, \varphi) \vdash\, !t : (t : \psi)$ using Rule EC of $\mathsf{E}(k, \varphi)$.
  By the induction hypothesis, $\mathsf{SEE} \vdash t\colon {}^{k,\varphi}\psi$. Applying Axiom X5 of SEE, the result follows.
- Induction Case: $\mathsf{E}(k, \varphi) \vdash t^{j,\chi} : [j, \chi]\psi$ using Rule EU of $\mathsf{E}(k, \varphi)$.
  Similar to the previous induction case, except that we use Axiom X6 of SEE.

Our Completeness Theorem says that if we take S to be the set of SEE-axioms, thereby equating these axioms with the "basic statements" that are to be justified by a constant , then the S-valid $\mathfrak{L}(\mathsf{SEE})$-formulas are SEE-theorems.

Theorem 7.3 (Completeness) *Let S be the set of* SEE*-axioms. $S \models \varphi$ implies* $\mathsf{SEE} \vdash \varphi$.

Proof Let us make a few definitions leading up to a canonical model argument. A *conjunction* of a finite set of $\mathfrak{L}(\mathsf{SEE})$-formulas is the conjunction whose conjuncts are the members of the finite set. To say that a formula φ *implies* $\bot$ means that $\mathsf{SEE} \vdash \varphi \supset \bot$. To say that a set of $\mathfrak{L}(\mathsf{SEE})$-formulas is *consistent* means that no conjunction of a finite subset implies $\bot$. To say that a set of $\mathfrak{L}(\mathsf{SEE})$-formulas is *inconsistent* means that the set is not consistent. To say that a set of $\mathfrak{L}(\mathsf{SEE})$-formulas is *maximal consistent* means that the set is consistent and adding any $\mathfrak{L}(\mathsf{SEE})$-formula not already in the set would produce an inconsistent set. Any consistent set of $\mathfrak{L}(\mathsf{SEE})$-formulas may be extended to a maximal consistent set of $\mathfrak{L}(\mathsf{SEE})$-formulas using a Lindenbaum Argument.

Assume that $\mathsf{SEE} \nvdash \varphi$. It follows from the Soundness Theorem (Theorem 7.2)[7] that $\mathsf{SEE} \nvdash \bot$ and hence that $\{\neg\varphi\}$ is consistent and so may be extended to a maximal consistent set T^φ. Define the evidence labeling $\mathcal{A}^\varphi$ by setting

$$\mathcal{A}^\varphi := \{(t, \psi) \in \mathcal{T} \times \mathfrak{L}(\mathsf{SEE}) \mid t : \psi \in T^\varphi\}$$

and define the valuation V^φ by setting

$$V^\varphi := \{p_k \in \mathfrak{L}(\mathsf{SEE}) \mid k \in \mathbb{N} \text{ and } p_k \in T^\varphi\}.$$

We wish to show that the evidenced valuation $(V^\varphi, \mathcal{A}^\varphi)$ is an S-model. Definition 7.4 implies that it suffices for us to show that $\mathcal{A}^\varphi$ is an S-evidence function. By Definition 7.2, to say that $\mathcal{A}^\varphi$ is an S-evidence function means that $\mathcal{A}^\varphi$ satisfies each of Constant Specification S, Application, Sum, Checker, and Update. So let us check that $\mathcal{A}^\varphi$ indeed satisfies each of these properties.

[7] In particular, in the proof of Lemma 7.3, we pointed out that $(\mathcal{T} \times \mathfrak{L}(\mathsf{SEE}), \emptyset)$ is an S-model. Therefore, were it the case that $\mathsf{SEE} \vdash \bot$, it would follow by soundness (Theorem 7.2) that $\mathcal{T} \times \mathfrak{L}(\mathsf{SEE}), \emptyset \models \bot$. Since the definition of truth (Definition 7.5) says that the latter is impossible, we conclude that $\mathsf{SEE} \nvdash \bot$.

- $\mathcal{A}^\varphi$ satisfies Constant Specification S.
  Choose $\psi \in S$. It follows by Rule CN and the maximal consistency of T^φ that $(c_k : \psi) \in T^\varphi$. Applying the definition of $\mathcal{A}^\varphi$, we have that $(c_k, \psi) \in \mathcal{A}^\varphi$. It follows that $\mathcal{A}^\varphi$ satisfies Constant Specification S.
- $\mathcal{A}^\varphi$ satisfies Application.
  Assume $(t, \psi \supset \chi) \in \mathcal{A}^\varphi$ and $(s, \psi) \in \mathcal{A}^\varphi$. Applying the definition of $\mathcal{A}^\varphi$, we have that $\big(t : (\psi \supset \chi)\big) \in T^\varphi$ and $(s : \psi) \in T^\varphi$. But then it follows by Axiom E1 and the maximal consistency of T^φ that $\big((t \cdot_\psi s) : \chi\big) \in T^\varphi$. Applying the definition of $\mathcal{A}^\varphi$, we have that $(t \cdot_\psi s, \chi) \in \mathcal{A}^\varphi$. It follows that $\mathcal{A}^\varphi$ satisfies Application.
- $\mathcal{A}^\varphi$ satisfies Sum, Checker, and Update by arguments similar to that for Application (though we use Axioms E2, E3, and E4, respectively).

  Conclusion: $\mathcal{A}^\varphi$ is an S-evidence function, and $(\mathcal{A}^\varphi, V)$ is an S-model.

We now wish to prove a property of $(\mathcal{A}^\varphi, V^\varphi)$ called the *Truth Lemma*: $\theta \in T^\varphi$ if and only if $\mathcal{A}^\varphi, V^\varphi \models \theta$. We first prove the Truth Lemma for $\mathfrak{L}(\mathsf{SEE})$-formulas θ with $d(\theta) = 0$ (Definition 7.8) by an induction on the construction of θ.

- Base Case: $\theta = q$, where $q \in \{p_k, \bot, \top\}$.
  If $\theta \in \{\bot, \top\}$, then the result follows by the maximal consistency of T^φ and the definition of truth (Definition 7.5). If $\theta = p_k$, then the result follows by the definition of V^φ and the definition of truth.
- Induction Case: $\theta = (\psi \star \chi)$, where $d(\psi) = d(\chi) = 0$.
  This case follows easily from the induction hypothesis.
- Induction Case: $\theta = \neg\psi$, where $d(\psi) = 0$.
  This case also follows easily from the induction hypothesis.
- Induction Case: $\theta = (t : \psi)$.
  We have $(t : \psi) \in T^\varphi$ if and only if $(t, \psi) \in \mathcal{A}^\varphi$. But the latter is what it means to have that $\mathcal{A}^\varphi, V^\varphi \models t : \psi$.
- Induction Case: $\theta = (t{:}\ ^{j,\chi} \psi)$.
  By an easy inductive argument (with the induction on the construction of t), we see that $\mathsf{E}(j, \chi) \vdash t : \psi$ or $\mathsf{E}(j, \chi) \nvdash t : \psi$. (This argument really is easy: just look at the axiomatics of $\mathsf{E}(j, \chi)$ in Fig. 7.1.) Applying Lemma 7.3, we have that $\mathsf{SEE} \vdash t{:}\ ^{j,\chi} \psi$ or $\mathsf{SEE} \nvdash t{:}\ ^{j,\chi} \psi$. It therefore follows by the maximal consistency of T^φ that $(t{:}\ ^{j,\chi} \psi) \in T^\varphi$ if and only if $\mathsf{SEE} \vdash t{:}\ ^{j,\chi} \psi$. But the latter is equivalent to $\mathsf{E}(j, \chi) \vdash t : \psi$ by Lemma 7.3. Since $\mathsf{E}(j, \chi) \vdash t : \psi$ is equivalent to $\mathcal{A}^\varphi, V^\varphi \models t{:}\ ^{j,\chi} \psi$ by the definition of truth, the result follows.

This completes our argument that the Truth Lemma holds of $\mathfrak{L}(\mathsf{SEE})$-formulas θ with $d(\theta) = 0$. Let us now argue that the Truth Lemma holds for all $\mathfrak{L}(\mathsf{SEE})$-formulas θ; that is, we show that $\theta \in T^\varphi$ if and only if $\mathcal{A}^\varphi, V^\varphi \models \theta$.

It follows from the Reduction Lemma (Lemma 7.2) and the maximal consistency of T^φ that $\theta \in T^\varphi$ if and only if $\theta^\circ \in T^\varphi$. But $d(\theta^\circ) = 0$ by the Reduction Lemma (Lemma 7.2), and so it follows by what we showed above that $\theta^\circ \in T^\varphi$ if and only if $\mathcal{A}^\varphi, V^\varphi \models \theta^\circ$. But since we showed that $(\mathcal{A}^\varphi, V^\varphi)$ is an S-model, it follows from the Reduction Lemma (Lemma 7.2) and soundness (Theorem 7.2) that $\mathcal{A}^\varphi, V^\varphi \models \theta^\circ$ is equivalent to $\mathcal{A}^\varphi, V^\varphi \models \theta$. All together, we have shown that $\theta \in T^\varphi$ if and only if $\mathcal{A}^\varphi, V^\varphi \models \theta$, which is the statement of the Truth Lemma.

Having proved the Truth Lemma, we may finish the overall proof. First, since $\neg\varphi \in T^\varphi$, it follows from the Truth Lemma that $\mathcal{A}^\varphi, V^\varphi \models \neg\varphi$. Applying the definition of truth, $\mathcal{A}^\varphi, V^\varphi \not\models \varphi$. Since $(\mathcal{A}^\varphi, V^\varphi)$ is an S-model, we have $S \not\models \varphi$ by the definition of validity (Definition 7.6). Thus we have shown that from the assumption $\mathsf{SEE} \nvdash \varphi$, which we made near the beginning of this proof, we may conclude that $S \not\models \varphi$. The statement of the theorem follows.

Soundness (Theorem 7.2) and completeness (Theorem 7.3) show that when we take S to be the set of SEE-axioms, thereby equating these axioms with the "basic statements" that are to be justified by a constant, then the set of SEE-theorems is equal to the set of S-valid $\mathfrak{L}$(SEE)-formulas. Accordingly, SEE exactly characterizes the S-valid formulas for the set S of SEE-axioms.

7.5 The Courtroom Evidence Example Formalized

Our simplistic example of courtroom evidence was described in the introduction of this chapter. In this example, the jury begins with two pieces of evidence.

1. $x_1 : O$
   In words: x_1 (the recording) is evidence that the boss ordered his subordinate to falsify the ledgers. (We used O as a mnemonic for "ordered." O is to be understood as an abbreviation for a propositional letter.)
2. $x_2 : (O \supset G)$
   In words: x_2 (the judge's instructions) is evidence that "if the boss ordered his subordinate to falsify the ledgers, then the boss is guilty of fraud." (We used G as a mnemonic for "guilty." G is to be understood as an abbreviation for a propositional letter.)

Using the symbol X to denote the conjunction $(x_1 : O) \wedge \big(x_2 : (O \supset G)\big)$, it follows by Axiom E1 of SEE (Fig. 7.2) that

$$\mathsf{SEE} \vdash X \supset (x_2 \cdot_O x_1) : G.$$

In words: "[given assumptions X,] $x_2 \cdot_O x_1$ is evidence that the boss is guilty of fraud." The combined evidence $x_2 \cdot_O x_1$ represents the jury using its evidence x_2 that $O \supset G$ and its evidence x_1 that O to conclude that G using the principle of Modus Ponens:

$$\frac{O \supset G \quad O}{G}\,.$$

This application of Modus Ponens may be read, "from assumptions $O \supset G$ and O, conclude G." Here the subscript O in the evidence $x_2 \cdot_O x_1$ indicates the important role O plays as the antecedent of the implication $O \supset G$ in this application of Modus Ponens.

But now let us examine the effect of the boss' attorney's successful challenge as to the authenticity of evidence x_1 (the recording), in which the boss' attorney presents further evidence that succeeds in convincing the jury that x_1 (the recording) is not authentic and so should be set aside. We may equate this successful challenge with the elimination $(1, O)$ because this is the elimination that will eliminate the evidence assertion $x_1 : O$ in accord with the boss' attorney's successful challenge of the evidence x_1 (that the boss ordered his subordinate to falsify the ledgers). Proceeding, we have the following derivation in $\mathsf{E}(1, O)$ (Fig. 7.1).

1. $x_1 : O$ Axiom EV
2. $(x_2 \cdot_O x_1) : G$ by 1, Rule EAR

That is, the elimination $(1, 0)$ has the effect of eliminating the evidence assertions $x_1 : O$ and $(x_2 \cdot_O x_1) : G$. Applying Lemma 7.3, it follows that

$$\mathsf{SEE} \vdash x_1 :^{1,O} O \quad \text{and} \quad \mathsf{SEE} \vdash (x_2 \cdot_O x_1) :^{1,O} G.$$

Using Axioms U4 and U3 of SEE (Fig. 7.2), it follows that

$$\mathsf{SEE} \vdash [1, O]\neg(x_1 : O) \quad \text{and} \quad \mathsf{SEE} \vdash [1, O]\neg\big((x_2 \cdot_O x_1) : G\big);$$

that is, "after elimination $(1, O)$, x_1 is not evidence that the boss ordered his subordinate to falsify the ledgers" and "after elimination $(1, O)$, $x_2 \cdot_O x_1$ is not evidence that the boss is guilty of fraud."

All together, we have shown that

$$\mathsf{SEE} \vdash X \supset (x_2 \cdot_O x_1) : G \quad \text{and} \quad \mathsf{SEE} \vdash X \supset [1, O]\neg\big((x_2 \cdot_O x_1) : G\big).$$

So while the jury could at first combine its evidence x_1 (the recording) with evidence x_2 (the judge's instructions) to produce evidence $x_2 \cdot_O x_1$ that the boss is guilty of fraud, the boss' attorney's successful challenge of the evidence x_1 that O ("the boss ordered his subordinate to falsify the ledgers") eliminates evidence x_1 for O, leaving the jury without the evidence $x_2 \cdot_O x_1$ that the boss is guilty of fraud.

7.6 Conclusion

We have presented SEE, the Theory of Simple Evidence Elimination, and we showed that it is sound and complete with respect to its intended semantics. Using a simplistic example of courtroom evidence, we showed how SEE can be used to reason about evidence and evidence elimination. In future work, we plan to extend our theory to one that not only allows for evidence elimination but also evidence *introduction*, whereby a piece t of evidence may be introduced for an assertion φ, which has the effect of making the evidence assertion $t : \varphi$ true. Such a joint theory of evidence, evidence elimination, and evidence introduction would provide a much fuller account in Justification Logic of the dynamics of evidence held by a rational individual.

Acknowledgments The author expresses his deep gratitude to Melissa Nicolls, Paul Warnett, and Toireasa Jespersen for helpful criticisms and suggestions concerning the courtroom evidence example. But do note that the author is solely responsible for the content of this chapter, including but not limited to any deficiencies, inaccuracies, or absurdities that the courtroom evidence example may contain or suggest.

References

Artemov S (2008) The logic of justification. The Review of Symbolic Logic 1(4):477–513
Artemov SN (2001) Explicit provability and constructive semantics. The Bulletin of Symbolic Logic 7(1):1–36
Fitting M (2005) The Logic of Proofs, semantically. Annals of Pure and Applied Logic 132(1):1–25
Fitting M (2009) Reasoning with justifications. In: Makinson D, Malinowski J, Wansing H (eds) Towards mathematical philosophy, trends in logic, vol 28, Springer, Netherlands, pp 107–123
Gödel K (1995) Vortrag bei Zilsel/Lecture at Zilsel's (*1938a). In: Feferman S, Dawson JW Jr, Goldfarb W, Parsons C, Solovay RM (eds) Unpublished essays and lectures, Kurt Gödel Collected Works, vol III, Oxford University Press, Oxford, pp 86–113
Kuznets R (2008) Complexity issues in justification logic. PhD thesis, The City University of New York
Mkrtychev A (1997) Models for the Logic of Proofs. In: Adian S, Nerode A (eds) Logical Foundations of Computer Science, Springer, New York, LNCS, vol 1234, pp 266–275
Renne B (2008) Dynamic epistemic logic with justification. Ph.D. thesis, The City University of New York

Chapter 8
Belief Update as Social Choice

Johan van Benthem

8.1 Introduction

The purpose of this technical Note is to make a link between two areas, dynamic logics for belief revision, and social choice theory. Our primary motivation is that, in this way, a further underpinning may be found for current belief update rules on the logic side, viewing agents, so to speak, as communities of signals. Conversely, with the proposed link, ideas from dynamic logics of information might percolate to social choice theory as well. We will state our results for two specific frameworks, dynamic epistemic-doxastic logic as in Baltag and Smets (2008), and relation merge in the style of Andréka et al. (2002). We will not explain the two frameworks in any great depth: this is a text for experts.

8.2 Dynamic-Doxastic Belief Change

We start with the doxastic-dynamic framework of Baltag and Smets (2008), van Benthem (2007), that treats belief revision in the spirit of dynamic epistemic logic (Baltag et al. 1998, van Ditmarsch et al. 2007, van Benthem 2010). Here are some basic notions – for convenience, stated for the case where epistemic accessibility is an equivalence relation among worlds, and plausibility is a pre-order inside the equivalence classes.

Definition 8.1 (Epistemic-doxastic models) Epistemic-doxastic models are structures $\boldsymbol{M} = (W, \{\sim_i\}_{i \in I}, \{\leq_i\}_{i \in I}, V, s)$ where the relations $\sim_i$ stand for epistemic accessibility, for each agent i in the group I, and the relations $\leq_i$ compare worlds as follows, $\leq_i xy$ if agent i considers world x at least as plausible as y. The V is a valuation map for proposition letters, and finally, s stands for the actual world.

J. van Benthem (✉)
Institute for Logic, Language, & Computation (ILLC), University of Amsterdam, PO Box 94242, 1090 GE, Amsterdam, The Netherlands; Department of Philosophy, Stanford University, Stanford, CA 94305, USA
e-mail: johan@science.uva.nl; johan.vanbenthem@uva.nl

P. Girard et al. (eds.), *Dynamic Formal Epistemology*, Synthese Library 351,
DOI 10.1007/978-94-007-0074-1_8,

Such models interpret knowledge as truth in all epistemically accessible worlds, and belief as truth in all most plausible worlds among the latter. Conditional beliefs can also be interpreted, but no such formal language is needed here.

Next, the dynamics is brought about by the following representation of events, with $<$ standing for the strict version of the plausibility order $\leq$ (that is, we set $x < y$ iff $x \leq y \wedge \neg\, y \leq x$).

Definition 8.2 (Event models) An *event model* $\boldsymbol{E} \equiv (E, \{\sim_i\}_{i \in G}, \{\leq_i\}_{i \in I}, \{Pre_e\}_{e \in E}, s)$ having a set of *events* E with (a) epistemic relations $\sim_i$ and plausibility orders $\leq_i$, for each agent i, (b) a map *Pre* assigning *preconditions* Pre_e to events e, stating just when these are executable, and finally, (c) an *actual event e*.

Definition 8.3 (Product update) For any epistemic-doxastic model $(\boldsymbol{M}, s)$ and event model $(\boldsymbol{E}, e)$, the *product model* $(\boldsymbol{M} \times \boldsymbol{E}, (s, e))$ is an epistemic-doxastic model with the following main components (where we drop agent subscripts for convenience):

(a) the domain is the set $\{(s, e) \mid s$ a world in $\boldsymbol{M}$, e an event in $\boldsymbol{E}$, $(M, s) \models PRE_e\}$,
(b) $(s, e) \sim (t, f)$ iff $s \sim t \wedge e \sim f$,
(c) $(s, e) \leq (t, f)$ iff $(eIf \wedge s \leq t) \vee e < f$, where eIf abbreviates $e \leq f \wedge f \leq e$,
(d) the valuation for atoms p at (s, e) is the same as that at s in $\boldsymbol{M}$.

The construction adds one layer of events to the worlds in the current information model, subject to the preconditions: information flows when these rule out some combinations. The epistemic update rule says that agents can only learn if they could distinguish worlds before (and have perfect memory for this) or if the new observation told them (that is, they only learn through observation). Next, the doxastic update rule is reminiscent of complex Jeffrey-style update with new probabilistic information: the last events seen decides the ordering if agents have a strong plausibility opinion, otherwise, the old plausibility order gets copied.

Here, we focus on the plausibility update part, also called the Priority Update Rule. This looks like just one stipulation out of many, but the mechanism is quite general because different event models generate different belief revision policies.[1]

The idea in what follows is really simple. We view the new plausibility relation in $\boldsymbol{M} \times \boldsymbol{E}$ as a result of social choice between "actors" $\boldsymbol{M}$ (the agents and all their signals received so far) and a new complex signal event $\boldsymbol{E}$. Getting ahead of ourselves, the conditions that we will propose on this process capture Priority Update, if we take the actors in a hierarchy of authority. But they leave a little more room if we treat the two actors as equally important, resulting in something more like epistemic update. Such latitude is an asset to the analysis, and we could find still more if we relaxed some of our social choice conditions even further.

[1]An alternative mechanism are the dynamic logic programs of van Benthem and Liu (2007).

8.3 "Social Choice" as Preference Merge for Groups

In this Note, I will not work with standard social choice theory (starting from the classic (Arrow 1951)),[2] but with a mathematical framework for relation merge among agents from Andréka et al. (2002). Motivations include social choice, but also aggregating criteria in linguistic analysis, or closer to dynamic epistemic logic: *belief merge* in groups of agents that meet, and need to arrive at a shared model. There are two central ideas in the set-up. First, most bare accounts of creating collective relations from individual ones work with an input that is too poor, viz. just a set of relations. In general, however, we need a richer input, in the form of a "priority graph" that also indicates dominance order in the group of argument relations. Moreover, for this to work, relations are to be reflexive transitive *pre-orders*, not necessarily connected. The existence of cases of incomparability is crucial for obtaining an elegant mathematical theory.

Definition 8.4 (Prioritized relation merge) Given an acyclic strictly ordered *priority graph* $\mathbf{G} = (G, <)$ of indices, that may be thought of as standing for agents i in some set I,[3] the merged *group priority relation* is defined as follows:

$$x \leq_G y \text{ iff for all indices } i \in \mathbf{G}, \text{ either } x \leq_i y, \text{ or there is a } j > i \text{ in } \mathbf{G} \text{ with } x <_j y$$

The intuitive explanation of this particular aggregation mechanism is the following. Either x weakly dominates y for all relations, or if not, then x *compensates* for this failure by doing strictly better than y on some comparison relation that has a higher priority in the graph. In special cases like linear priority graphs, this fits the well-known hierarchical lexicographic ordering.

In Andréka et al. (2002) ("ARS") prove "universality" of their aggregation procedure, and completeness with algebraic equations.[4] Girard (2008) and Liu (2008) show how this elegant set-up fits well with dynamic epistemic logic. It generalizes belief merge, priority-based preference, and ceteris paribus "agendas" (van Benthem et al. (2009)), while Girard (2008) also provides a new complete axiomatization in a matching hybrid modal language.

8.4 Belief Change as Social Choice: The Motivating Analogy

Next we turn the tables, and analyze belief revision itself as a process of social choice in a group – not of agents, but of "signals" in the loose sense explained. Here is how we cast the mathematics of the earlier product update mechanism.

---

[2] Franz Dietrich has pointed out with concrete suggestions that interesting connections might be made with more standard social choice literature but this will have to wait till another occasion.

[3] Indices may have multiple occurrences in the graph, but we will ignore this subtlety here.

[4] Priority graphs have natural operations of *sequential composition* (put one graph above another) and *parallel composition* (take a disjoint union of graphs). These yield an elegant calculus: (a) disjoint union leads to *intersection* of relations, (b) sequential composition to lexicographic order.

Abstract setting: ordering pair objects given component orders Two ordered sets (A, R) and (B, S) are given, with possibly different domains A, B: for instance, think of a doxastic model $\boldsymbol{M}$ and an event model $\boldsymbol{E}$ with their separate domains and plausibility orders. Now we seek *to order the product* $A \times B$ by a suitable relation $O(R, S)$ over pairs (a, b).[5,6] I will think in terms of plausibility *pre-orders* henceforth, in line with the *ARS* preference merge.

The main analogy The Priority Update Rule takes the event model $\boldsymbol{E}$ to rank "above" the doxastic model $\boldsymbol{M}$, and defines the following weak order in $\boldsymbol{M} \times \boldsymbol{E}$:

$$(s, e) \leq (t, f) \text{ iff } (s \leq t \ \wedge \ eIf) \ \vee \ e < f.^{7}$$

But this is reminiscent of the earlier priority graph merge. By some simple manipulation in propositional logic, this formulation is easily seen to be equivalent to

$$(s, e) \leq (t, f) \text{ iff } (s \leq t \ \wedge \ e \leq f) \ \vee \ e < f.$$

And that is just the idea of *ARS*: a pair (x, y) must satisfy both component relations R, but y can "compensate failures" where $\neg Rxy$ by doing strictly better than x with respect to some relation R' with higher priority than R. Hence, it makes sense to analyze belief update in terms of choice principles for relation merge.

Note: epistemic entanglement Actual Priority Update works entangled with epistemic accessibility, as we only compare plausibility links inside epistemic equivalence classes. We omit this feature here, thereby also freeing up the earlier accessibility symbol $\sim$ for re-use in a moment.

In what follows, we will not bother about precise analogies with *ARS*. Our aim is to state an analysis that speaks for itself. Of course, while working with pre-orders, we have to read the following intuitions and keep in mind four distinct cases:

$$x < y, x > y, x \sim y \text{ (indifferent, } x \leq y \ \wedge \ y \leq x),$$
$$\text{and also } x \# y \text{ (incomparable, } \neg x \leq y \ \wedge \ \neg y \leq x).$$

For greater vividness, we mark these cases graphically as follows, in the given order:

$$\rightarrow \qquad \leftarrow \qquad \sim \qquad \#$$

8.5 Finding Intuitive Conditions on Plausibility Update

What sort of process are we trying to capture? I will first choose a very restrictive set to zoom in exclusively on priority update. Later on I relax this, to get greater

[5] In general, we may only need to order a *subset* of this full product space, since some relevant constraint may rule out pairs: as we have seen already with the DEL event preconditions.

[6] We can also rephrase things over one set of "pair objects". First lift component relations to pairs: $(a, c)R_1(b, d)$ iff aR_1c, $(a, c)R_2(b, d)$ iff cR_2d, then merge in single domain style. As the referee points out, in social choice, this would be a rather ego-centric ordering of bundles of objects.

[7] By a simple computation, we also get the equivalence $(s, e) < (t, f)$ iff $(s < t \wedge e \leq f) \vee e < f$.

variety in update rules. The first condition is familiar from model theory, generalized quantifier theory for natural language, and many other areas. It says that the choice should not depend on individual features of objects, only their ordering pattern:

Condition (a): Permutation invariance

Consider any two permutations of A and B. Thinking of A, B as disjoint sets, without loss of generality, we can see this as one permutation π. We require:

$$O(\pi[R], \pi[S]) = \pi[O(R, S)].$$

This standard invariance condition imposes a strong uniformity on possible formats of definition (cf. the structural invariance accounts of "logicality" discussed in Benthem (2002)). But without pursuing this in its generality here, we at once impose one more constraint:

Condition (b): Locality

$$O(R, S)((a, b), (a', b')) \text{ iff } O(R|\{a, a'\}, S|\{b, b'\})((a, b), (a', b')).$$

This says that we will only compare in terms of just the objects explicitly mentioned in the comparison. This is a very strong version of context-independence, akin to "Independence from Irrelevant Alternatives" in social choice.

Digression Locality in the above intuitive sense holds for belief revision policies like radical update $\Uparrow A$ (van Benthem 2007), where we place all A-objects above all $\neg A$-objects in the new ordering, while keeping the old ordering inside these two zones. And indeed, this update can be modeled through Priority Update using a two-point event model with an A-signal more plausible than a $\neg A$-signal. But Locality fails intuitively for *conservative* update $\uparrow A$ where we place only *the best A-worlds* on top in the new ordering. This requires a check if worlds to be "promoted" are *maximal* in A in the given order, and this involves running through other worlds. From the technical viewpoint of Priority Update, this is still no objection, since we can *re-encode* a conservative revision rule through the radical one, by shifting some relevant information to another location. Say, we can change the input event model to a new one with two artificial "signals" *best* A and $\neg$ *best* A, making the former more plausible than the latter. But it has been objected that this construal is ad-hoc, and if we reject such artificial signals, we may want to allow for other non-local update mechanisms, too.

Table format Taken together, Permutation Invariance and Locality force any operation O to be definable by its behaviour in the following 4×4-Table:

		S on b, b'			
		$\rightarrow$	$\leftarrow$	$\sim$	#
R on a, a'	$\rightarrow$	-	-	-	-
	$\leftarrow$	-	-	-	-
	$\sim$	-	-	-	-
	#	-	-	-	-

Here entries stand for the 4 isomorphism classes on two objects: all that matters given the invariance condition. Under some conditions, table entries are forced.[8] We now fill in the same four types of entry in the Table, subject to further conditions.

Caveat Strictly speaking, one might just want to put *YES*/*NO* in the slots marking whether the relation $\leq$ holds. In using the four types one should check that all intuitions to be stated hold for Priority Product Update as defined above.

Choice conditions on the aggregation procedure Now we state some conditions on how the component relations are going to be used in the final result. Even though we will only be using these conditions for a choice involving two actors, they make sense more generally. The names we are using have been chosen for their vividness, but nothing is claimed for them in naturalistic terms:

Condition (c): Abstentions

If a subgroup votes indifferent ($\sim$), then the others determine the outcome.

Condition (d): Closed agenda

The social outcome always occurs among the opinions of the voters.

This implies *Unanimity*: "if all group members agree, take the shared outcome". But it is much stronger. Finally, consider agents who care, and are not indifferent about outcomes. An "over-rule" is a case where one opinion wins over the other.

Condition (e): Overruling

If an agent's opinion ever overrules that of another, then her opinion always does.

This goes against the spirit of democracy and letting everyone win once in a while. But we should not hide the fact that this is indeed what the Priority Rule does, given its bias toward the last event.

8.6 Characterizing Priority Update

Our main result now captures Priority Update, though with one twist. We will indeed derive that the ordering of the inputs must be hierarchical. But we will not force the "authority" to be the second argument – or in the above setting: the event model $\boldsymbol{E}$. This seems a somewhat extraneous decision beyond global choice analysis.[9] With this understanding, the result to follow speaks of "a", not "the", Priority Update:

---

[8] For instance, if singletons are reflexive, agents will be indifferent between x and x itself, and by the Abstentions principle below, we only need to look at the other pair relations.

[9] In fact, the other option of giving priority to the first argument: say, the initial model $\boldsymbol{M}$, is an interesting conservative anti-Jeffreyan variant where little learning takes place.

Theorem 8.1 *A preference aggregation function is a Priority Update iff it satisfies Permutation Invariance, Locality, Abstentions, Closed Agenda, and Overruling.*

Proof First, Priority Update satisfies all previously stated conditions. Here one needs to check that the original formulation boils down to the case format in our Table. For instance, if event arguments are incomparable, this will block any comparison between the pairs, whence the last column. Also, if $e < f$, and $s < t$, it is easy to check that then $(s, e) < (t, f)$. Etcetera.

Conversely, we analyze Table entries subject to our conditions. The diagonal is clear by Unanimity, and the row and column for indifference by Abstentions:

R on a, a' \ S on b, b'	$\rightarrow$	$\leftarrow$	$\sim$	#
$\rightarrow$	$\rightarrow$	1	$\rightarrow$	2
$\leftarrow$	3	$\leftarrow$	$\leftarrow$	4
$\sim$	$\rightarrow$	$\leftarrow$	$\sim$	#
#	5	6	#	#

This leaves six slots to be filled. But there are really only three choices, by symmetry considerations. E.g., an entry for $\rightarrow\leftarrow$ automatically induces one for $\leftarrow\rightarrow$.

Now consider slot 1. By Closed Agenda, this must be either $\leftarrow$ or $\rightarrow$. Without loss of generality, consider the latter: S overrules R. Using Overruling to fill the other cases with S's opinion, and applying Permutation Invariance, our Table is:

R on a, a' \ S on b, b'	$\rightarrow$	$\leftarrow$	$\sim$	#
$\rightarrow$	$\rightarrow$	$\leftarrow$	$\rightarrow$	#
$\leftarrow$	$\rightarrow$	$\leftarrow$	$\leftarrow$	#
$\sim$	$\rightarrow$	$\leftarrow$	$\sim$	#
#	$\rightarrow$	$\leftarrow$	#	#

It is easy to see that this final diagram is precisely that for Priority Update in its original sense. The other possibility would give preference to the ordering on $\boldsymbol{M}$.[10]

8.7 Weaker Conditions: Additional Update Rules

Now we relax our conditions to allow a democratic variant where both arguments count equally – in our scenario, a "flat" epistemic-style product update $\boldsymbol{M} \times \boldsymbol{E}$ where

$$(s, e) \leq (t, f) \text{ iff } s \leq t \ \wedge \ e \leq f.$$[11]

[10] Both are instances of the so-called "But" operator of ARS, i.e., a "Leader/Follower" pattern.

[11] This intersection of relations is the second basic operation of ARS: "And", instead of "But".

Now Closed Agenda fails. For instance, clearly, with this rule, the above "clash" of cases $\rightarrow \leftarrow$ ends up in #. Instead, we formulate two new principles:

Condition (f): Unanimity

If voters all agree, then their vote is the social outcome.

Condition (g): Alignment

If anyone changes their vote to get closer to the current group outcome, then that group outcome does not change.

Here is an instance of a more liberal characterization result for update rules:

Theorem 8.2 *A preference merge function satisfies Permutation Invariance, Locality, Abstentions, Overruling, Unanimity, and Alignment if and only if it is either (a) a priority update, or (b) flat product update.*[12]

Proof The crucial step is now that, without Closed Agenda, Slot 6 in our diagram

		S on b, b'			
		$\rightarrow$	$\leftarrow$	$\sim$	#
R on a, a'	$\rightarrow$	$\rightarrow$	6	$\rightarrow$	2
	$\leftarrow$	3	$\leftarrow$	$\leftarrow$	4
	$\sim$	$\rightarrow$	$\leftarrow$	$\sim$	#
	#	5	6	#	#

may also have entries $\sim$ or #. However, the former outcome can be ruled out by Alignment. If S were to change its opinion to $\sim$, the outcome would still have to be $\sim$, but it is $\rightarrow$. So, the outcome must be #. But then, using Alignment for both voters (plus some Permutation Invariance), we see that all remaining slots must be #:

		S on b, b'			
		$\rightarrow$	$\leftarrow$	$\sim$	#
R on a, a'	$\rightarrow$	$\rightarrow$	#	$\rightarrow$	#
	$\leftarrow$	#	$\leftarrow$	$\leftarrow$	#
	$\sim$	$\rightarrow$	$\leftarrow$	$\sim$	#
	#	#	#	#	#

This is clearly the table corresponding to the flat update.

8.8 Further Questions, and Conclusion

The preceding results are extremely simple, and just the start of a general line of thinking. For instance, we may get new update rules when we relax the choice

[12] Christian List has suggested that the results in this chapter are close to a characterization of "lexicographic dictatorships" by Luce and Raiffa (1957): cf. D'Aspremont (1985). These links are yet to be explored, again looking at social choice postulates for their belief revision content. Another result that List has suggested as an alternative belief revision mechanism is May's Theorem capturing the essence of democratic majority voting: cf. Goodin and List (2006).

conditions.[13] For instance, as said before, Overruling goes against the grain of democracy, and dropping it would allow for *mixtures* of influence for the two arguments, in particular, $\boldsymbol{M}$ and $\boldsymbol{E}$, in line with rules for *inductive learning*. Also, we noted that giving up Locality would give independent status to conservative update rules, on a par with our radical priority upgrade. Finally, it would be of interest to extend our analysis to postulates for more signals and belief merge, though *ARS* do a good job there already.[14]

Also, further themes from social choice theory might make sense in our logical setting. For instance, what would be the belief revision counterpart of having systematic restrictions on the set of individual preference profiles that can occur?[15] My answer would be the following: assumptions on the "continuity" of the information streams that we encounter in the world. It has often been observed hat we only learn well if the universe is reasonably kind to us.

But perhaps the main benefit of the perspective offered here is the following. I find the idea natural that update and revision over time amounts to integrating signals. And I am positively intrigued by the idea that, as an agent, "I" am "we": the social aggregation of my original self plus all signals received, depending on the manner in which I took them.

Acknowledgment I thank Christian List for his useful feedback suggesting deeper forays into social choice theory – and especially, Franz Dietrich for many informative critical comments, only part of which I have been able to process in this short note.

References

Andréka H, Ryan M, Schobbens P (2002) Operators and laws for combining preference relations. Journal of Logic and Computation 12(1):13–53

Arrow K (1951) Social choice and individual value. Cowles Foundations and Wiley, New York

Baltag A, Smets S (2008) A qualitative theory of dynamic interactive belief revision. Proceedings LOFT 2007 (Logic and foundations of game theory and decision theory). Texts in Logic and Games 3:9–58

Baltag A, Moss LS, Solecki S (1998) The logic of public announcements, common knowledge, and private suspicions. In: Gilboa I (ed) Proceedings of the 7th conference on theoretical aspects of rationality and knowledge (TARK 98), pp 43–56

Benthem J (2002) Invariance and definability: two faces of logical constants. Reflections on the Foundations of Mathematics Essays in Honor of Sol Feferman. pp 426–446

---

[13] Franz Dietrich notes that our condition of Permutation Invariance is very strong, and that the literature on Arrow-style theorems suggests better results using only weaker versions of it.

[14] As for more technical issues, more might be said about relative power of different update rules in achieving new relational patterns on models. Compare Priority Update versus Flat Product Update. Which rule is more general in its dynamic effects, if we allow *re-encoding of arguments*? Priority Update can never make an established strict preference for x over y "indifferent" again, while I think Democracy can mimic any effect of Priority by suitably re-encoded sequences of events.

[15] Again this follows up on a question from Franz Dietrich.

van Benthem J (2007) Dynamic logic for belief revision. Journal of Applied Non-Classical Logics 17(2):129–155
van Benthem J (2010) Logical dynamics of information and interaction. Monograph to appear with Cambridge University Press, Cambridge, 366 pp.
van Benthem J, Liu F (2007) Dynamic logic of preference upgrade. Journal of Applied Non-Classical Logics 17(2):157–182
van Ditmarsch H, van der Hoek W, Kooi B (2007) Dynamic epistemic logic. Synthese library, Springer, New York
van Benthem J, Girard P, Roy O (2009) Everything else being equal: A modal logic approach to ceteris paribus preferences. Journal of Philosophical Logic 38(1):83–125
D'Aspremont C (1985) Axioms for social welfare orderings. Social goals and social organization: Essays in memory of Elisha Pazner, pp 19–76
Girard P (2008) Modal logic for belief and preference change. PhD thesis, Stanford University and ILLC, University of Amsterdam
Goodin R, List C (2006) A conditional defense of plurality rule: generalizing May's theorem in a restricted informational environment. American Journal of Political Science 50(4):940–949
Liu F (2008) Changing for the better: Preference dynamics and agent diversity. Dissertation, ILLC, University of Amsterdam

Chapter 9
Revision with Conditional Probability Functions: Two Impossibility Results

François Lepage and Charles Morgan

9.1 Context and Background

It is nowadays a commonplace idea to take the probability of a counterfactual $A{>}B$ to be the probability of the consequent B after some revision of the probability function that shifts the probability of A to 1. The underlying intuition is that the belief which a rational agent gives to a counterfactual seems to accord with the belief this agent would give to the consequent if the antecedent were true.

The first serious and disastrous attempt to provide a probabilistic interpretation to an extension of the classical propositional logic containing a counterfactual along these lines was that of Stalnaker (1968, 1970) who suggested the use of Popper's two-place probability functions and conditionalization. Lewis (1976) showed that only trivial probability functions satisfy Stalnaker's constraints. In the same paper, Lewis presented a completely different way (see Gärdenfors 1988, for a comparison) to shift from Pr to Pr^A; he called his technique *Imaging*. His technique is quite simple: Lewis assumes to be given some possible worlds structure, consisting of a set of possible worlds and a linear ordering relation of a certain sort on the set of possible worlds; the ordering relation is interpreted as a relation of proximity between worlds. In addition, Lewis assumes to be given an a priori probability distribution over the set of worlds. A proposition is taken to be a set of possible words and the probability of a proposition is the sum of the probability of the worlds where the proposition is true. In this framework, the probability of a counterfactual $A{>}B$ is the probability of the consequent B after a shift of the probability of any non-A-world onto the nearest A-world according to the proximity relation. This process is however very specific. Not only it is defined in the framework of possible world semantics, but it also assumes the existence of a linear ordering over the set of worlds. Furthermore, it uses the framework of absolute probability functions which is not the most general approach for probabilistic interpretations.

F. Lepage (✉)
Département de Philosophie, Université du Québec à Montréal, C.P. 6128, succursale Centre-ville, Montréal, Québec H3C 3J7, Canada
e-mail: francois.lepage@umontreal.ca

P. Girard et al. (eds.), *Dynamic Formal Epistemology*, Synthese Library 351,
DOI 10.1007/978-94-007-0074-1_9,

The results that will be presented below begin by considering a generalization of the revision process in two fundamental ways: (1) we propose to abandon the dependence on possible worlds structures and rely on purely probabilistic interpretations; and (2) we use the general framework of conditional probability functions.

9.1.1 Conditional Probability Functions

It is Popper (1934) who first presented an axiomatization of a probability calculus which takes as a primitive notion that of two-place probability functions. The two important features of Popper's approach are (1) conditionalization on sentences that have a 0 probability is always defined and (2) the classical absolute probability functions can be considered as a limiting case of conditionalization on a tautology.

However, the great breakthrough in the probabilistic interpretation of the logical calculus is Hartry Field's (1977) axiomatization of the notion of probabilistic validity *without making use* of the notion of truth value and more generally without using any classical semantic notion that makes use of truth functions. Field's axiomatizations of conditional probability both for propositional calculus and for first-order predicate calculus can be used for proofs of soundness and completeness of the usual axiomatizations of these calculi.

There are many ways to provide axiomatizations of conditional probability functions. The most general is probably the following from Morgan (2003) which uses two-place probability functions where the second place is a set of sentences instead of a single sentence.

In what follows, we will use theses "full blood" conditional probability functions, functions that take two arguments, a sentence and a set of sentences (not necessarily finite) and give a real number between 0 and 1. The problem is then to define a revision function, i.e., a function that assigned to each conditional probability function a *revised* conditional probability function.

This sets the limits of our triviality results. For example, a theory like that of Boutilier (1995) which uses absolute probability functions escape our results *but* if we replace absolute probability functions by conditional probability functions, our triviality results hold. The same is true for Joyce's causal decision theory (Joyce 1999). However, both Dubois and Prade Possibility logic (Dubois and Prade 1988) and Spohn Ordinal conditional functions (Spohn 1988) seem immunized against our results.

Let L be the language of the classical propositional logic (L is the set of wff's defined as usual).

A conditional probability function is any function Pr from $L \times \wp(L)$ into the unit interval [0,1], subject to the following constraints:

NP.1 $0 \leq \Pr(A, \Gamma) \leq 1$
NP.2 If $A \in \Gamma$ then $\Pr(A, \Gamma) = 1$
NP.3 $\Pr(A \vee B, \Gamma) = \Pr(A, \Gamma) + \Pr(B, \Gamma) - \Pr(A \wedge B, \Gamma)$
NP.4 $\Pr(A \wedge B, \Gamma) = \Pr(A, \Gamma) \times \Pr(B, \Gamma \cup \{A\})$
NP.5 $\Pr(\neg A, \Gamma) = 1 - \Pr(A, \Gamma)$ unless $\Pr(B, \Gamma) = 1$ for all B.

NP.6 $\Pr(A \wedge B, \Gamma) = \Pr(B \wedge A, \Gamma)$
NP.7 $\Pr(C, \Gamma \cup \{A \wedge B\}) = \Pr(C, \Gamma \cup \{A, B\})$
NP.8 $\Pr(A \vee \neg A, \Gamma) = 1$

(Pr-abnormal, Pr-normal): Γ is Pr-abnormal iff for all sentences A, $\Pr(A, \Gamma) = 1$. Γ is Pr-normal iff for at least one sentence A, $\Pr(A, \Gamma) < 1$.

The natural way to define the notion of *Semantic consequence* is:

A is a *semantic consequence* of a set of sentences Γ (in symbols $\Gamma \Vdash A$) iff for all Pr and all Δ, $\Pr(A, \Gamma \cup \Delta) = 1$.

Let $\vdash$ be the symbol of derivability of the classical calculus. Then it can be proved that

$$\Gamma \vdash A \text{ iff } \Gamma \Vdash A$$

This result can be generalized. We know since (Morgan 1982, 2003) (see also Field 1977, Leblanc 1979, Van Fraassen 1981) that any extension of classical propositional calculus is sound and complete for probabilistic interpretations.

We will need the following classical lemmas. The proofs are in the Appendix.

Lemma 9.1 *If* Γ *is Pr-normal, then* $\Pr(A \wedge \neg A, \Gamma) = 0$.

Lemma 9.2 *If* Γ *is Pr-normal but* $\Gamma \cup \{A\}$ *is Pr-abnormal, then* $\Pr(A, \Gamma) = 0$.

Lemma 9.3 $\Pr(A \supset B, \Gamma) = 1 - \Pr(A, \Gamma) + \Pr(A \wedge B, \Gamma)$.

Lemma 9.4 *If* $\Pr(A \supset B, \Gamma) = 1$, *then* $\Pr(A, \Gamma) \leq \Pr(B, \Gamma)$.

Lemma 9.5 *If* $\Pr(A, \Delta) \leq \Pr(B, \Delta)$ *for all* Δ, *then* $\Pr(A \supset B, \Gamma) = 1$ *for all* Γ.

Lemma 9.6 *Disjunction is the least upper bound, in the following sense:*

(A) $\Pr(A, \Gamma) \leq \Pr(A \vee B, \Gamma)$ *for all* Γ.
(B) $\Pr(B, \Gamma) \leq \Pr(A \vee B, \Gamma)$ *for all* Γ.
(C) *Let* C *be any sentence such that:*

  (i) $\Pr(A, \Delta) \leq \Pr(C, \Delta)$ *for all* Δ, *and*
  (ii) $\Pr(B, \Delta) \leq \Pr(C, \Delta)$ *for all* Δ.

Then for all Δ, $\Pr(A \vee B, \Delta) \leq \Pr(C, \Delta)$.

Lemma 9.7 *If* $\Pr(A, \Gamma) = 1$, *then* $\Pr(A \wedge B, \Gamma) = \Pr(B, \Gamma)$.

Lemma 9.8 *If* $\Pr(A, \Gamma) = 1$, *then* $\Pr(B, \Gamma) = \Pr(B, \Gamma \cup \{A\})$

9.2 First Result

Let $L_{>}$ be the extension of L obtained by adding a new binary connective $>$. Formally,

(i) $L \subseteq L_>$;
(ii) for any $A, B \in L_>$, $\neg A$, $A \wedge B$, $A > B \in L_>$;
(iii) nothing else is in $L_>$.

(The other classical connectives are introduced by the usual definitions.)

Proposition 9.1 *Let A be a wff and let* $(\ )^A$ *be a function mapping the set of conditional probability functions into the set of conditional probability functions. We write* Pr^A *as shorthand for* $(\mathrm{Pr})^A$. *Suppose that for all formulas A the function* $(\ )^A$ *satisfies the following conditions:*

(1) For any Γ and any Pr, $\mathrm{Pr}^A(A, \Gamma) = 1$
(2) For any Γ, any Pr and any wff B, $\mathrm{Pr}(A > B, \Gamma) = \mathrm{Pr}^A(B, \Gamma)$

then $\mathrm{Pr}(A > (B > A), \Gamma) = 1$

Proof

1. $\mathrm{Pr}(A > (B > A), \Gamma) = \mathrm{Pr}^A(B > A, \Gamma)$	By (2)
2. $= \mathrm{Pr}^A(B > A, \Gamma \cup \{A\})$	By Lemma 9.8
3. $= \mathrm{Pr}^{AB}(A, \Gamma \cup \{A\})$	By (2)
4. $= 1$	By NP.2

Such revision functions do exist: just take $(\ )^A$ to be conditionalization on A and "$>$" to be "$\supset$".

This simple result trivializes any semantics of counterfactuals for which (1) and (2) hold: no logic of counterfactuals should validate $A > (B > A)$ (just think about $A > (\neg A > A)$).

Moreover, if we assume that $(A > B) \supset (A \supset B)$ is a theorem of a sound axiomatization of $L_>$ (notice that $(A > B) \supset (A \supset B)$ is a theorem of all Lewis' V-logics (Lewis 1973)), then

(3) $\mathrm{Pr}((A > B) \supset (A \supset B), \Gamma) = 1$ for all Pr, A, B, Γ

But if (3) holds, it is trivial to prove that "$>$" and "$\supset$" are probabilistically indiscernible.

Proposition 9.2 *If* (1), (2) *and* (3) *as above hold, then*

$$\mathrm{Pr}((A > B) \equiv (A \supset B), \Gamma) = 1 \text{ for all Pr}, A, B, \Gamma$$

The proof of Proposition 9.2 needs the following lemmas. The proofs are in the Appendix.

Lemma 9.9 $\mathrm{Pr}(A > B, \Gamma) \leq \mathrm{Pr}(A \supset B, \Gamma)$, *for all* Γ.

Lemma 9.10 $\mathrm{Pr}(B, \Gamma) \leq \mathrm{Pr}(A > B, \Gamma)$, *for all* Γ.

Lemma 9.11 $\Gamma \cup \{\neg A\}$ is Pr^A-*abnormal.*

Lemma 9.12 $\mathrm{Pr}(\neg A > (A > B), \Gamma) = 1$, *for all* Γ.

Lemma 9.13 $\Pr(\neg A, \Gamma) \leq \Pr(A > B, \Gamma)$, *for all* Γ.

Lemma 9.14 $\Pr(\neg A \vee B, \Gamma) \leq \Pr(A > B, \Gamma)$, *for all* Γ.

Proof of Proposition 9.2.

This is a direct consequence of Lemma 9.9, the definition of material implication in terms of $\vee$ and $\neg$ and Lemma 9.14.

9.3 Second Result

Let us consider a slightly more refined definition of the probability of a counterfactual. In addition to a shift from Pr to $\Pr^A$ we consider a shift from Γ to another background Γ^*_A. This approach is absolutely general. In what follows, we put no constraints on $\Pr^A$ (it could even be Pr) nor on Γ^*_A. The only general constraints are $\Pr(A > B, \Gamma) = \Pr^A(B, \Gamma^*_A)$ and $\Pr(A > A, \Gamma) = \Pr^A(A, \Gamma^*_A) = 1$. We need two things. The first is the postulate that there is a *background revision function* such that for any set of sentences Γ

$$( )^*_A : \wp(L_>) \to \wp(L_>)$$
$$\Gamma \mapsto \Gamma^*_A$$

Secondly, for any Pr and Γ, let $f_{\Pr,\Gamma} : L_> \to [0, 1]$ be such that $f_{\Pr,\Gamma}(X) = \Pr(X, \Gamma)$. $f_{\Pr,\Gamma}$ is just the one place function obtained from Pr by setting Γ to a constant value.

Using the background revision function, we can describe the shift of conditional probability as a function $S( )^A$ that takes $f_{\Pr,\Gamma}$ as argument to give $S(f_{\Pr,\Gamma})^A = f_{\Pr,\Gamma^*_A}$.

We know that *CPF* is closed under conditionalization, i.e., if Pr is a *CPF* then $\Pr'$ which is such that for a given Δ, any X and any Γ, $\Pr'(X, \Gamma) = \Pr(X, \Gamma \cup \Delta)$ is also a *CPF* (it is very easy to check that $\Pr'$ satisfies NP.1-NP.8). Indeed, if the set of conditional probability functions were not closed under conditionalization, then the whole notion of Bayesian updating of beliefs would have to be abandoned. We do not necessarily advocate Bayesian updating as the only method of updating belief sets, but it would be bizarre to rule it out as a possible method of updating beliefs at least in some cases.

Proposition 9.3 *Let* Pr *and* Γ *be any conditional probability function and any background. Let* Δ *be a set of wff's and* $\Pr'$ *the conditional probability function obtained from* Pr *by conditionalization on* Δ. *Then, for any* wff X,

$$\Pr(X, \Gamma^*_A \cup \Delta) = \Pr(X, (\Gamma \cup \Delta)^*_A)$$

i.e., $( )^*_A$ and conditionalization commute.

Proof $f_{\Pr',\Gamma} = f_{\Pr,\Gamma\cup\Delta}$ because by hypothesis $\Pr'$ is obtained from Pr by conditionalization on Δ. Thus $S(f_{\Pr',\Gamma})^A = S(f_{\Pr,\Gamma\cup\Delta})^A$ and $f_{\Pr',\Gamma^*_A} = f_{\Pr,(\Gamma\cup\Delta)^*_A}$ which implies that for any wff X, $f_{\text{Pr}',\Gamma^*_A}(X) = f_{\text{Pr},(\Gamma\cup\Delta)^*_A}(X)$ and thus $\text{Pr}'(X, \Gamma^*_A) = \text{Pr}(X, (\Gamma \cup \Delta)^*_A)$ and finally $\text{Pr}(X, \Gamma^*_A \cup \Delta) = \text{Pr}(X, (\Gamma \cup \Delta)^*_A)$.

This trivializes any probabilistic interpretation of counterfactuals.

Proposition 9.4 *There is no* $(\,)^*_A$ *such that*

(i) *for any* Pr, Γ, B *and* A, $\text{Pr}(A > B, \Gamma) = \text{Pr}^A(B, \Gamma^*_A)$
(ii) *for any* Pr, Γ, *and* A, $\text{Pr}(A > A, \Gamma) = \text{Pr}^A(A, \Gamma^*_A) = 1$
(iii) *for some* Pr, Γ *and* A *such that* A *is not a contradiction,* $\neg A \in \Gamma$ *and* Γ^*_A is Pr^A*-normal.*

Proof Let A be a non contradiction, and suppose that $\neg A \in \Gamma$ and Γ^*_A is Pr^A-normal.
By (ii) $\text{Pr}(A > A, \Gamma) = \text{Pr}^A(A, \Gamma^*_A) = 1$.
Because $\neg A \in \Gamma$, $\text{Pr}(A > A, \Gamma) = \text{Pr}(A > A, \Gamma \cup \{\neg A\}) = 1$. $\text{Pr}^A(A, (\Gamma \cup \{\neg A\})^*_A) = 1$ by (i).
By (iii), NP.2 and NP.5, $\text{Pr}^A(\neg A, (\Gamma \cup \{\neg A\})^*_A) = \text{Pr}^A(\neg A, \Gamma^*_A) = 0$.
By Proposition 9.3 $\text{Pr}^A(\neg A, (\Gamma \cup \{\neg A\})^*_A) = \text{Pr}^A(\neg A, \Gamma^*_A \cup \{\neg A\})$. By NP.2 $\text{Pr}^A(\neg A, \Gamma^*_A \cup \{\neg A\}) = 1$. So $0 = 1$.

So for any background revision function that respects (i) and (ii) and commutes with conditionalization, if $\neg A \in \Gamma$ then for any A which is not a contradiction and any Pr, Γ^*_A is Pr^A-abnormal. Notice that conditionalization commutes with itself, i.e., if $\Gamma^*_A = \Gamma \cup \{A\}$ then $\Gamma^*_A \cup \{A\} = (\Gamma \cup \{A\})^*_A$. However, if $\neg A \in \Gamma$, $\Gamma^*_A = \Gamma \cup \{A\}$ is inconsistent and thus Pr-abnormal for any Pr.

9.4 Closing Remarks

Although at our starting point we focused on Lewis' imaging for motivation, our results do not rely on any process like imaging, nor on possible worlds structures of any kind. Once again, our triviality results apply to any revision process associated with the evaluation of a counterfactual such that (1) the antecedent of the counterfactual has a probability of one for the revised conditional probability function and (2) the probability of the counterfactual for the former conditional probability function is that of the consequent for the revised conditional probability function. For sure, any kind of imaging in the spirit of Lewis will satisfy (1) and (2) and thus cannot escape the triviality results. In short, there is no way to generalize the Lewis approach beyond possible worlds constructs in such a way as to apply to general conditional probability distributions.

Acknowledgment The authors would like to thank the anonymous referee for very helpful comments and suggestions.

Appendix

Lemma 9.1 *If* Γ *is Pr-normal, then* $\Pr(A \wedge \neg A, \Gamma) = 0$.

Proof Suppose Γ is Pr-normal. From NP.3 and NP.8 we have:

$$1 = \Pr(A \vee \neg A, \Gamma) = \Pr(A, \Gamma) + \Pr(\neg A, \Gamma) - \Pr(A \wedge \neg A, \Gamma).$$

Applying NP.5 *to the right we have:*

$$1 = \Pr(A, \Gamma) + 1 - \Pr(A, \Gamma) - \Pr(A \wedge \neg A, \Gamma).$$

Simple arithmetic gives the desired result.

Lemma 9.2 *If* Γ *is* Pr-*normal but* $\Gamma \cup \{A\}$ *is* Pr-*abnormal, then* $\Pr(A, \Gamma) = 0$.

Proof Suppose Γ is Pr-normal but $\Gamma \cup \{A\}$ is Pr-abnormal. From Lemma 9.1 and NP.4, we have:

$$0 = \Pr(A \wedge \neg A, \Gamma) = \Pr(A, \Gamma) \times \Pr(\neg A, \Gamma \cup \{A\}).$$

Since $\Gamma \cup \{A\}$ is Pr-abnormal, $\Pr(\neg A, \Gamma \cup \{A\}) = 1$ and thus $\Pr(A, \Gamma) = 0$.

Lemma 9.3 $\Pr(A \supset B, \Gamma) = 1 - \Pr(A, \Gamma) + \Pr(A \wedge B, \Gamma)$.

Proof If Γ is Pr-abnormal, the result is immediate, so suppose that Γ is Pr-normal. By the definition of $\supset$ and applying NP.3, we have:

$$\Pr(A \supset B, \Gamma) = \Pr(\neg A \vee B, \Gamma) = \Pr(\neg A, \Gamma) + \Pr(B, \Gamma) - \Pr(\neg A \wedge B, \Gamma).$$

Applying NP.6 and NP.4 yields:

$$\Pr(A \supset B, \Gamma) = \Pr(\neg A, \Gamma) + \Pr(B, \Gamma) - \Pr(B, \Gamma) \times \Pr(\neg A, \Gamma \cup \{B\}).$$

Since Γ is Pr-normal, we can apply NP.5 to obtain:

$$\Pr(A \supset B, \Gamma) = 1 - \Pr(A, \Gamma) + \Pr(B, \Gamma) - \Pr(B, \Gamma) \times \Pr(\neg A, \Gamma \cup \{B\}). \quad (9.1)$$

On one hand, suppose that $\Gamma \cup \{B\}$ is Pr-abnormal. Then applying Lemma 9.2 to (9.1) yields:

$$\Pr(A \supset B, \Gamma) = 1 - \Pr(A, \Gamma).$$

But Lemma 9.2 and NP.6 and NP.4 ensure that $\Pr(A \wedge B, \Gamma) = 0$. So we have:

$$\Pr(A \supset B, \Gamma) = 1 - \Pr(A, \Gamma) + \Pr(A \wedge B, \Gamma).$$

On the other hand, suppose that $\Gamma \cup \{B\}$ is Pr-normal. Then applying NP.5 to (9.1) gives:

$$\begin{aligned}\Pr(A \supset B, \Gamma) &= 1 - \Pr(A, \Gamma) + \Pr(B, \Gamma) - \Pr(B, \Gamma) \times (1 - \Pr(A, \Gamma \cup \{B\})) \\ &= 1 - \Pr(A, \Gamma) + \Pr(A \wedge B, \Gamma)\end{aligned}$$

So, in either case, the result follows.

Lemma 9.4 If $\Pr(A \supset B, \Gamma) = 1$, *then* $\Pr(A, \Gamma) \leq \Pr(B, \Gamma)$.

Proof Suppose $\Pr(A \supset B, \Gamma) = 1$. Then from Lemma 9.3 we have:

$$1 = 1 - \Pr(A, \Gamma) + \Pr(A \wedge B, \Gamma)$$

So by elementary arithmetic we have:

$$\Pr(A, \Gamma) = \Pr(A \wedge B, \Gamma)$$

But by NP.1 and NP.4 and NP.6, we know:

$$\Pr(A \wedge B, \Gamma) \leq \Pr(B, \Gamma)$$

Hence the desired result follows.

Lemma 9.5 *If* $\Pr(A, \Delta) \leq \Pr(B, \Delta)$ *for all* Δ, *then* $\Pr(A \supset B, \Gamma) = 1$ *for all* Γ.

Proof Suppose $\Pr(A, \Delta) \leq \Pr(B, \Delta)$ for all Δ. Let Γ be arbitrary. Then by NP.1 and NP.2 we have:

$$\Pr(B, \Gamma \cup \{A\}) = 1$$

But by Lemma 9.3 and NP.4 we have:

$$\Pr(A \supset B, \Gamma) = 1 - \Pr(A, \Gamma) + \Pr(A, \Gamma) \times \Pr(B, \Gamma \cup \{A\})$$

The desired result follows immediately from these two equations.

Lemma 9.6 *Disjunction is the least upper bound, in the following sense:*

(A) $\Pr(A, \Gamma) \leq \Pr(A \vee B, \Gamma)$ *for all* Γ.
(B) $\Pr(B, \Gamma) \leq \Pr(A \vee B, \Gamma)$ *for all* Γ.
(C) *Let* C *be any sentence such that:*

 (i) $\Pr(A, \Delta) \leq \Pr(C, \Delta)$ *for all* Δ, *and*
 (ii) $\Pr(B, \Delta) \leq \Pr(C, \Delta)$ *for all* Δ.

Then for all Γ, $\Pr(A \vee B, \Delta) \leq \Pr(C, \Delta)$.

Proof For the (B) part, by NP.1 we know the maximum value for $\Pr(B, \Gamma \cup \{A\})$ is 1, so by NP.3 and NP.4 we have:

$$\begin{aligned}\Pr(A \vee B, \Gamma) &= \Pr(A, \Gamma) + \Pr(B, \Gamma) - \Pr(A, \Gamma) \times \Pr(B, \Gamma \cup \{A\}) \\ &\geq \Pr(A, \Gamma) + \Pr(B, \Gamma) - \Pr(A, \Gamma)\end{aligned} \tag{9.2}$$

and the right hand side just reduces to $\Pr(B, \Gamma)$, as desired. Part (A) follows in a similar fashion by first applying NP.6 to the conjunction. For the (C) part of the Lemma, suppose both of the following hold:

$$\Pr(A, \Delta) \leq \Pr(C, \Delta) \text{for all} \Delta. \tag{9.3}$$

$$\Pr(B, \Delta) \leq \Pr(C, \Delta) \text{for all} \Delta. \tag{9.4}$$

Taking Δ to be of the form $\Gamma \cup \{A\}$ for arbitrary Γ, and using NP.1 and NP.2, we then obtain the following:

$$\Pr(C, \Gamma \cup \{A\}) = 1 \text{ for all} \Gamma. \tag{9.5}$$

$$\Pr(C, \Gamma \cup \{B\}) = 1 \text{ for all} \Gamma. \tag{9.6}$$

Next we multiply $\Pr(A \vee B, \Gamma \cup \{C\})$ by $\Pr(C, \Gamma)$ and apply NP.3 and NP.4:

$$\begin{aligned}&\Pr(C, \Gamma) \times \Pr(A \vee B, \Gamma \cup \{C\}) = \\ &\quad \Pr(C, \Gamma) \times \Pr(A, \Gamma \cup \{C\}) + \Pr(C, \Gamma) \times \Pr(B, \Gamma \cup \{C\}) \\ &\quad - \Pr(C, \Gamma) \times \Pr(A, \Gamma \cup \{C\}) \times \Pr(B, \Gamma \cup \{A, C\})\end{aligned} \tag{9.7}$$

Then applying NP.4 and NP.6 to the right hand side of (9.7), we obtain:

$$\begin{aligned}&\Pr(C, \Gamma) \times \Pr(A \vee B, \Gamma \cup \{C\}) = \\ &\quad \Pr(A, \Gamma) \times \Pr(C, \Gamma \cup \{A\}) + \Pr(B, \Gamma) \times \Pr(C, \Gamma \cup \{B\}) \\ &\quad - \Pr(A, \Gamma) \times \Pr(B, \Gamma \cup \{A\}) \times \Pr(C, \Gamma \cup \{A, B\})\end{aligned} \tag{9.8}$$

But then using (9.5) and (9.6) on the right of (9.8) we have:

$$\begin{aligned}&\Pr(C, \Gamma) \times \Pr(A \vee B, \Gamma \cup \{C\}) = \\ &\quad \Pr(A, \Gamma) + \Pr(B, \Gamma) - \Pr(A, \Gamma) \times \Pr(B, \Gamma \cup \{A\})\end{aligned} \tag{9.9}$$

Hence from (9.9), NP.1 and NP.3 we have:

$$\Pr(A, \Gamma) + \Pr(B, \Gamma) - \Pr(A, \Gamma) \times \Pr(B, \Gamma \cup \{A\}) \leq \Pr(C, \Gamma) \tag{9.10}$$

Finally the desired result follows by applying NP.3 and NP.4 to the left hand side of (9.10)

Lemma 9.7 *If* $\Pr(A, \Gamma) = 1$, *then* $\Pr(A \wedge B, \Gamma) = \Pr(B, \Gamma)$.

Proof Suppose $\Pr(A, \Gamma) = 1$. Then by Thm 1.8, $\Pr(A \vee B, \Gamma) = 1$. So, by NP.3, we have:

$$1 = \Pr(A \vee B, \Gamma) = \Pr(A, \Gamma) + \Pr(B, \Gamma) - \Pr(A \wedge B, \Gamma)$$

Hence it follows that:

$$1 = 1 + \Pr(B, \Gamma) - \Pr(A \wedge B, \Gamma)$$

The desired result then follows by elementary arithmetic.

Lemma 9.8 *If* $\Pr(A, \Gamma) = 1$, *then* $\Pr(B, \Gamma) = \Pr(B, \Gamma \cup \{A\})$

Proof Suppose $\Pr(A, \Gamma) = 1$. From NP.4, we have:

$$\begin{aligned} \Pr(A \wedge B, \Gamma) &= \Pr(A, \Gamma) \times \Pr(B, \Gamma \cup \{A\}) \\ &= \Pr(B, \Gamma \cup \{A\}) \end{aligned}$$

The desired result then follows immediately by Lemma 9.7.

Lemma 9.9 $\Pr(A > B, \Gamma) \leq \Pr(A \supset B, \Gamma)$, *for all* Γ.

Proof By (3). and soundness, we have:

$$\Pr((A > B) \supset (A \supset B), \Gamma) = 1$$

The desired result then follows by Lemma 9.4.

Lemma 9.10 $\Pr(B, \Gamma) \leq \Pr(A > B, \Gamma)$, *for all* Γ.

Proof This result follows immediately from Lemma 9.4, Lemma 9.9, and Proposition 9.1.

Lemma 9.11 $\Gamma \cup \{\neg A\}$ is $\Pr^A$*-abnormal.*

Proof By elementary arithmetic, we have:

$$\Pr^A(A, \Gamma \cup \{\neg A\}) = 1 - (1 - \Pr^A(A, \Gamma \cup \{\neg A\}))$$

For proof by contradiction, suppose $\Gamma \cup \{\neg A\}$ is $\Pr_A$-normal. Then applying NP.5 to the right hand side, we obtain:

$$\Pr^A(A, \Gamma \cup \{\neg A\}) = 1 - \Pr^A(\neg A, \Gamma \cup \{\neg A\})$$

Applying NP.2 to the right hand side yields:

$$\Pr^A(A, \Gamma \cup \{\neg A\}) = 1 - 1 = 0$$

But we know by (1)

$$\Pr^A(A, \Gamma \cup \{\neg A\}) = 1$$

Thus the supposition is in error, and it must be the case that $\Gamma \cup \{\neg A\}$ is $\Pr^A$-abnormal.

Lemma 9.12 $\Pr(\neg A > (A > B), \Gamma) = 1$, *for all* Γ.

Proof Applying (2), we have:

$$\Pr(\neg A > (A > B), \Gamma) = \Pr^{\neg A}(A > B, \Gamma), \text{ for all } \Gamma.$$

But then applying Lemma 9.8 and (1) on the right hand side gives:

$$\Pr(\neg A > (A > B), \Gamma) = \Pr^{\neg A}(A > B, \Gamma \cup \{\neg A\}), \text{ for all } \Gamma.$$

Another application of (2) to the right hand side gives:

$$\Pr(\neg A > (A > B), \Gamma) = \Pr^{\neg AA}(B, \Gamma \cup \{\neg A\}), \text{ for all } \Gamma.$$

But then by Lemma 9.11, the right hand side must be 1.

Lemma 9.13 $\Pr(\neg A, \Gamma) \leq \Pr(A > B, \Gamma)$, *for all* Γ.

Proof This result follows immediately from Lemma 9.4, Lemma 9.9, and Lemma 9.12

Lemma 9.14 $\Pr(\neg A \vee B, \Gamma) \leq \Pr(A > B, \Gamma)$, *for all* Γ.

Proof The desired result is an immediate result of the part (C) of Lemma 9.6, along with Lemma 9.10 and Lemma 9.13.

References

Boutilier C (1995) On the revision of probabilistic belief states. Notre Dame Journal of Formal Logic 36(1):158–183

Dubois D, Prade H (1988) Possibility theory: An approach to computerized processing of uncertainty. Plenum Publishing Company, New York

Field H (1977) Logic, meaning, and conceptual role. The Journal of Philosophy, LXXXIV(7):379–409

Van Fraassen B (1981) Probabilistic semantics objectified: Postulates and logics. Journal of Philosophical Logic 10:371–394

Gärdenfors P (1988) Knowledge in flux. MIT Press, Cambridge

Joyce JM (1999) The foundations of causal decision theory. Cambridge University Press, Cambridge

Leblanc H (1979) Probabilistic semantics for first-order logic. Zeitschr. f. Logik and Grundlagen d. Math. Bd. 25:497–509

Leblanc H, Roeper P (2000) Probability theory and probability semantics. University of Toronto Press, Toronto
Lewis D (1973) Counterfactuals. Basil Blackwell, Oxford
Lewis D (1976) Probabilities of conditionals and conditional probabilities. Philosophical Review 85:297–315
Morgan C (1982) There is a probabilistic semantics for every extension of classical sentence logic. Journal of Philosophical Logic 11:431–442
Morgan C (2003) two values, three values, many values, no values. In: Fitting M, Orlowska E (eds) Beyond two: Theory and applications of multiple valued logic. Springer Verlag, Berlin, pp 348–374
Popper K (1934) Logik der Forschung. Spinger Verlag, Vienna, English translation (1959) The Logic of Scientific Discovery. Routledge, London.
Spohn W (1988) Ordinal conditional functions: A dynamic theory of epistemics states. In: Harper WL, Skyrms B (eds) Causation in Decision, Belief Change, and Statistics, vol II, Reidel, Dordrecht
Stalnaker R (1968) A theory of conditionals. In: Rescher N (ed) Studies in Logical Theory, Blackwell, Oxford, APQ Monography No. 2, reprinted in Causations and Conditionals, Sosa E (ed), Oxford University Press (1968), pp 165–179
Stalnaker R (1970) Probability and conditionals. Philosophy of Science 37(1):64–80

Chapter 10
Indeterminacy and Belief Change

Horacio Arló-Costa

10.1 Introduction

Consider the following example:

> Suppose that you are scientist facing the following problem: you have to choose between three scientific theories a, b and c. Let's suppose as well that you have conflicting criteria to evaluate the theories. For example, simplicity is one criteria and according to this criterion you have formed a relation R and according to R a is preferred to b and b is preferred to c. Suppose that you are also able to order the theories according to explanatory power. In this case you form a rival order R' and according to R' c is preferred to b and b is preferred to a. Which theory should you choose? Let's assume that each dimension of epistemic value (simplicity, explanatory power, etc) has equal weight to you.
>
> Of course there are more mundane situations that have exactly the same form. For example, instead of the theories you might compare candidates for an academic job and the rival orderings R and R' can be rankings of the candidates according to teaching ability (R) and capacity to conduct original research (R'). The same question arises here: which candidate should you choose?

Most of the existing theories of belief revision are unable to deal with examples of this sort. They assume that the agent is capable of integrating all dimensions of epistemic value in a unique weak order and that all happens as if one chooses optimal options according to this weak order. By the same token most theories of choice have equal difficulty choosing in situations of unresolved conflict. It seems, nevertheless, that in many cases of the sort illustrated above it might be complicated or directly impossible to form a unique ordering that faithfully integrates all epistemic dimensions. This chapter deals with situations of this kind. Is it possible to modify the existing theory of belief revision to deal with cases where there are multiple dimensions of epistemic value?

H. Arló-Costa (✉)
Department of Philosophy, Carnegie Mellon University, Baker Hall 135, Pittsburgh, PA 15213-3890, USA
e-mail: hcosta@andrew.cmu.edu

P. Girard et al. (eds.), *Dynamic Formal Epistemology*, Synthese Library 351,
DOI 10.1007/978-94-007-0074-1_10,

10.1.1 The Received View and Why It Does Not Work

The classical framework of *optimization* used in standard choice theory recommends choosing, among the feasible options, a *best* alternative. So, if S is the feasible set and R is a weak preference relation over S, optimization recommends focusing on the following set of best elements of S:

$$G(S, R) = \{y \in S\text{: for all } z \in S, yRz\}$$

But many economists have recently pointed out that this stringent form of maximization might not be the kind of maximization that one can apply in practical problems where information is usually incomplete and sometimes scarce. For example, the Nobel-winner Amartya Sen remarks:

> The general discipline of maximization differs from the special case of optimization in taking an alternative as choosable when it is not known to be worse that any other. [...] The basic contrast between maximization and optimization arises from the possibility that the preference ranking may be incomplete (Sen 1997, p. 767).

Consider the following simple example. Say that it is known that element a is preferred to b and that element c is preferred to d, but any other information relating these elements is unavailable. In this situation no element dominates all others. The set $\{a, b, c, d\}$ does not have optimal elements. But it is clear that that the set has two maximal elements, namely a and c. These are the un-dominated elements in the set.

To define a maximal set we can use the asymmetric part $P(R)$ of a binary acyclic relation R as follows:

$$M(S, R) = \{y \in S\text{: for no } z \in S, zP(R)y\}$$

It is easy to see that in general (for any binary relation R and any non-empty feasible set S) we have that $G(S, R) \subseteq M(S, R)$. When R is complete $G(S, R) = M(S, R)$. Moreover a maximal set $M(S, R)$ can always be replicated by optimizing a complete relation R^+ obtained from R by transforming incomparabilities into indifferences. Obviously this new relation R^+ has to be complete but it might fail to be transitive. In addition, although this new relation can mimic the maximizing behavior of $P(R)$ it is clear that it should not be used for representing knowledge. Sen warns against using this kind of relations in representing economic knowledge in particular, but it is clear that the problem is more general.

If we consider the previous example where we have a feasible set $S = \{a, b, c, d\}$ and a relation R according to which a is preferred to b and that element c is preferred to d; the relation R^+ will transform previous incomparabilities into indifferences. For example, a will now be indifferent to c and to d, and therefore a is optimal with respect to R^+. In general, the previously un-dominated elements (under R) will now be optimal (under R^+). But it is clear that the added indifferences are not part of the knowledge of the modeled agent. Maximality can be mimicked by optimality under a relation that does not represent the knowledge of the agent in question. In fact, there is a big difference between two elements being incomparable and two elements being indifferent. Especially the incomparabilites might not be removable by adding new knowledge. The nature of the problem might require to use incomparabilities rather than indifference. So, R^+ should not be confused with the basic relation R which represents the knowledge of the agent (which can be essentially incomplete).

The discipline of belief change has been dominated by the use of optimization techniques. In order to study the contraction of a theory K with a sentence A the AGM trio Alchourrón et al. (1985) has proposed to focus on the set $K \bot A$ of maximal subsets of K that fail to entail A as the feasible set from which one make choices. Then the idea is to utilize a *selection function* γ that when applied to $K \bot A$ selects a non-empty set of $K \bot A$. In particular *partial meet contraction* focuses on selection functions that are *relational*, i.e. selection functions for which there is a binary relation $\leq$ such that:

$$\gamma(K \bot A) = \{Y \in K \bot A : \text{ for all } Z \in K \bot A, Z \leq Y\}$$

Then $K \div A$, the *contraction of* K *by* A, is defined as the intersection of the elements of $\gamma(K \bot A)$. Obviously this definition relies on a process of optimization of the sort discussed above. Two main criticisms can be raised against this way of articulating contraction. One concerns the feasible set, which many see as too restrictive. The second criticism concerns the use of optimization techniques. In many applications one might not have access to the binary relation needed to optimize, a relation that imposes strong demands, like completeness.[1] In particular one might face cases of indeterminacy, which can be caused, for example, by lack of information or, alternatively, by the existence of conflicting standards of valuation (as in the introductory example).

When an agent faces indeterminacy between a set of permissible orderings one standard solution is to consider their shared agreements – that is, the *categorical* relation obtained by considering all ordered pairs shared by all the permissible orderings. This categorical relation need not be complete. So, optimizing this relation might not be possible. But, of course, one can maximize the resulting incomplete relation. This is the central idea considered in Arló-Costa (2006). In particular I considered there the problem of maximizing a quasi-transitive relation (i.e. a reflexive relation whose asymmetric component is transitive). The resulting notion of contraction is weaker than AGM.

This approach, nevertheless, has strong limitations. Consider the example presented above where the agent has two orderings R and R'. The categorical relation representing the shared agreements between these two relations is empty. So, we do not have anything to say about this simple case. But it seems that intuitively there are two options that should be selected, namely a and c.

[1]The usual procedure in the belief revision literature is to define a selection function γ as a function that returns a non-empty output when applied to a non-empty remainder set. This makes implicit requirements on the marking-off relation $\leq$. For example, the relation cannot be irreflexive.

The formal idea that articulates this intuition is to take as admissible any option that is deemed as maximal by *some* permissible ordering. This idea was introduced by Isaac Levi (in a different form) in Levi (1986). One can consider the question of what is the notion of contraction (revision) that arises when the problem of indeterminacy of epistemic value is solved via Levi's criterion. We will appeal to some results in the theory of choice to answer this question.

In our example, the application of Levi's criterion is not decisive. We end up with two options, a and c that are tied. Both of them are equally admissible. If we want a method that is decisive we can deploy a secondary notion of security that unties ties. Formally this can be represented by another weak order that discriminates between the two options selected by the first step of the method.

It is an open problem what constraints on choice fully represent this two tier method. In the last section we will report on recent results towards a solution to this problem, and we will consider applications to the theory of belief revision in conditions of indeterminacy. We will conclude by considering some applications in philosophy of science presented in an unpublished paper by Morgenbesser and Koslow (2008).

10.2 Technical Preliminaries

In the following we presuppose a propositional language $\mathcal{L}$ with the connectives $\neg, \wedge, \vee, \rightarrow, \leftrightarrow$. We let $For(\mathcal{L})$ denote the set of formulae of $\mathcal{L}$,$a, b, c, \ldots p, q, r, \ldots$ denote propositional variables of $\mathcal{L}$, and $\alpha, \beta, \delta, \ldots$, $\varphi, \psi, \chi, \ldots$ denote arbitrary formulae of $\mathcal{L}$. Sometimes we assume that the underlying language $\mathcal{L}$ is *finite*. By this we mean that $\mathcal{L}$ has only finitely many propositional variables.

As is customary, we assume that $\mathcal{L}$ is governed by a consequence operation Cn: $\mathcal{P}(For(\mathcal{L})) \rightarrow \mathcal{P}(For(\mathcal{L}))$ such that for all $A, B \subseteq For(\mathcal{L})$,

(i) $A \subseteq \mathrm{Cn}(A)$.
(ii) If $A \subseteq B$, then $\mathrm{Cn}(A) \subseteq \mathrm{Cn}(B)$.
(iii) $\mathrm{Cn}(\mathrm{Cn}(A)) \subseteq \mathrm{Cn}(A)$.
(iv) $\mathrm{Cn}_0(A) \subseteq \mathrm{Cn}(A)$, where Cn_0 is classical tautological implication.
(v) If $\varphi \in \mathrm{Cn}(A)$, then there is some finite $A_0 \subseteq A$ such that $\varphi \in \mathrm{Cn}(A_0)$.
(vi) If $\varphi \in \mathrm{Cn}(A \cup \{\psi\})$, then $\psi \rightarrow \varphi \in \mathrm{Cn}(A)$.

Conditions (i)–(vi) are respectively called *Inclusion, Monotony, Idempotence, Supraclassicality, Compactness*, and *Deduction* (Hansson 1999, p. 26). As usual, A is called *logically closed* if $\mathrm{Cn}(A) = A$, and $A \vdash \varphi$ is an abbreviation for $\varphi \in \mathrm{Cn}(A)$. We let $\mathbb{K}$ denote the collection of logically closed sets in $\mathcal{L}$.

We let $\mathcal{W}_{\mathcal{L}}$ denote the collection of all maximal consistent sets of $\mathcal{L}$ with respect to Cn. Members of $\mathcal{W}_{\mathcal{L}}$ are often called *possible worlds* or just *worlds*. For a non-empty collection of worlds W of $\mathcal{W}_{\mathcal{L}}$, let $\widehat{W}$ denote the set of sentences of $\mathcal{L}$ which are members of all worlds in W (briefly, $\widehat{W} := \bigcap_{w \in W} w$). If A is a set of sentences of $\mathcal{L}$, we let $[\![A]\!] := \{w \in \mathcal{W}_{\mathcal{L}} : A \subseteq w\}$. If φ is a sentence of $\mathcal{L}$, we write $[\![\varphi]\!]$ instead of $[\![\{\varphi\}]\!]$. Observe that for every set of sentences A of $\mathcal{L}$, $\mathrm{Cn}(A) = \widehat{[\![A]\!]}$. A member of $\mathcal{P}(\mathcal{W}_{\mathcal{L}})$ is often called a proposition, and $[\![\varphi]\!]$ is often called the *proposition expressed by* φ. Intuitively, $[\![A]\!]$ consists of those worlds in which all sentences in A hold. Finally, let $\mathcal{E}_{\mathcal{L}}$ be the set of all elementary subsets of $\mathcal{W}_{\mathcal{L}}$, i.e., $\mathcal{E}_{\mathcal{L}} := \{W \in \mathcal{P}(\mathcal{W}_{\mathcal{L}}) : W = [\![\varphi]\!]$ for some $\varphi \in For(\mathcal{L})\}$.

10.3 Some Results from the Theory of Choice

Choice can be analyzed in a rather abstract framework by appealing to standard techniques used in the pure theory of consumer choice. We can start with a universal set X which remains fixed throughout the analysis. Let then $\mathcal{S}$ be a distinguished non-empty collection of non-empty subsets of X. The pair $(X, \mathcal{S})$ will be called the *choice space* of the agent.

Now we can introduce a new notion that will be useful below. A *choice function* on a choice space $(X, \mathcal{S})$ is a function C defined on $\mathcal{S}$ that assigns a non-empty subset (*choice set*) $C(S)$ of S to each and every S in $\mathcal{S}$. Intuitively, a selection function $C\colon \mathcal{S} \to \mathcal{P}(X)$ chooses the "best" elements of each S in $\mathcal{S}$.

The following condition demarcates a special class of choice functions. Often in the literature it is assumed that choice functions satisfy this condition.

Regularity. For each $S \in \mathcal{S}$, $C(S) \neq \emptyset$.

We call a choice function satisfying Regularity *regular*.

A binary relation $\geq$ on X *rationalizes* (or is *a rationalization* of) a choice function C on $(X, \mathcal{S})$ if and only if, for every $S \in \mathcal{S}$, $C(S)$ consists on the greatest points of S:

$$G(S, \geq) := \{x \in S : x \geq y \text{ for all } y \in S\}$$

Using this notation, we now offer a definition.

Definition 10.1 A binary relation $\geq$ on a universal set X *G-rationalizes* (or is a *G-rationalization* of) a choice function C on a choice space $(X, \mathcal{S})$ if for every $S \in \mathcal{S}$, $C(S) = G(S, \geq)$.

We also say that a choice function C is *G-rational* if there is a binary relation $\geq$ that G-rationalizes C. This formalization of a rational choice function is what Sen and others often call an *optimizing* notion of rationality.

The second formalization of a rational choice function captures a less stringent notion of rationality, demanding only that a choice function selects the *maximal* elements from each set $S \in \mathcal{S}$. Again, we require some notation. The set of maximal elements of a set S with respect to a binary relation $>$ is defined as follows:

$$M(S, >) := \{x \in S : y > x \text{ for no } y \in S\}$$

With this notation at hand, we again offer a definition.

Definition 10.2 A binary relation $>$ on a universal set X *M-rationalizes* (or is a *M-rationalization* of) a choice function C on a choice space $(X, \mathcal{S})$ if for every $S \in \mathcal{S}$, $C(S) = M(S, >)$.

As with G-rationality, we say that a choice function C is *M-rational* if there is a binary relation $>$ that M-rationalizes C. Sen and others call this a *maximizing* notion of rationality. We will assume this notion of rationality in this article.

In the following, let $\mathcal{P}_{fin}(X)$ denote the family of all finite subsets of a collection of objects X. The next proposition establishes that any reasonable binary relation that M-rationalizes a choice function on a choice space $(X, \mathcal{P}_{fin}(X))$ is unique.

Proposition 10.1 *Let C be a choice function on a choice space $(X, \mathcal{P}_{fin}(X))$. Then if $>$ is a irreflexive binary relation on X that M-rationalizes C, $>$ uniquely M-rationalizes C.*

There has been some work on the project of characterizing the notion of rationality axiomatically using so-called *coherence constraints*. One salient coherence constraint is the condition Sen calls *Property* α (Sen 1971, p. 313), also known as *Chernoff's axiom* (Suzumura 1983, p. 31).

Property α. For all $S, T \in \mathcal{S}$, if $S \subseteq T$, then $S \cap C(T) \subseteq C(S)$.
Sen's Property $\alpha 2$. For all $S \in K$ such that $\{x, y\} \in K$ for every $y \in S$, if $x \in C(S)$, then $x \in \bigcap_{y \in S} C(\{x, y\})$.

There are two lines of argument for the characterization of rationality, one proposed by Sen (1971) (and reconsidered in Sen (1997)) and another proposed by Kotaro Suzumura in Suzumura (1983). Both use the notion of *base preference* (Sen 1971, p. 308, Sen 1997, p. 64, Suzumura 1983, p. 28). We modify Sen's argument in terms of maximization following the presentation offered in Arló-Costa and Pedersen (2009a).

Definition 10.3 (Base Preference) Let C be a choice function on a choice space $(X, \mathcal{P}_{fin}(X))$. We define (*strict*) *base preference* by setting

$$>^C := \{(x, y) \in X \times X : x \in C(\{x, y\}) \text{ and } y \notin C(\{x, y\})\}$$

Observe that $>^C$ must be asymmetric and so irreflexive.

We need an additional axiom in order to present Sen's argument:

Property γ. For every nonempty $I \subseteq \mathcal{S}$ such that $\bigcup_{S \in I} S \in \mathcal{S}$,

$$\bigcap_{S \in I} C(S) \subseteq C\left(\bigcup_{S \in I} S\right)$$

(Sen's γ^*) For all $S, T \in K$ such that $S \cup T \in \mathcal{S}$, $C(S) \cap C(T) \subseteq C(T \cup S)$.

(Sen's $\gamma 2$) For all $S \in K$ such that $\{x, y\} \in K$ for every $y \in S$, if $x \in \bigcap_{y \in S} C(\{x, y\})$, then $x \in C(S)$.

Condition γ entails the following coherence constraint:

(γ^*) For every $S, T \in \mathcal{S}$ such that $S \cup T \in \mathcal{S}$, $\gamma(S) \cap \gamma(T) \subseteq \gamma(S \cup T)$. (*Sen's Property* γ^*)

If $\mathcal{S}$ is finitely additive and compact, then condition γ is equivalent to condition γ^*.

With these new conditions we can state the following theorem:

Theorem 10.1 *A choice function C on a choice space $(X, \mathcal{P}_{fin}(X))$ is acyclic M-rational if and only if it is regular and satisfies Property α and Property γ.*

The main proofs in this section appear in Arló-Costa and Pedersen (2009a). We present them here for the sake of completeness.

Proof It suffices to show that a regular choice function C on a space $(X, \mathcal{P}_{fin}(X))$ is M-rational if and only if it satisfies Property α and Property γ.

($\Rightarrow$) Suppose C is M-rational, and let $>$ be a M-rationalization of C. It suffices to show that C satisfies Property γ. Let $I \subseteq \mathcal{P}_{fin}(X)$ be such that $\bigcup_{S \in I} S \in \mathcal{P}_{fin}(X)$, and suppose $x \in \bigcap_{S \in I} C(S)$. Then for each $S \in I$, $y > x$ for no $y \in S$, so $x \in C\left(\bigcup_{S \in I} S\right)$.

($\Leftarrow$) Suppose C satisfies Property α and Property γ. We must show that $>^C$ M-rationalizes C. Again, in light of the proof of Theorem 10.1, we only show that $M(S, >^C) \subseteq C(S)$ for all $S \in \mathcal{P}_{fin}(X)$. So let $S \in \mathcal{P}_{fin}(X)$, and suppose $x \in M(S, >^C)$. Then $y >^C x$ for no $y \in S$ and therefore by Regularity $x \in \bigcap_{y \in S} C(\{x, y\})$, so by Property γ, $x \in C(S)$.

Corollary 10.1 (cf. Suzumura 1983, p. 28) *A regular choice function C is M-rational if and only if $>^C$ uniquely M-rationalizes C.*

Proof The direction from right to left is trivial. For the other direction, observe that by Theorem 10.1, if C is M-rational, then C satisfies Property α and Property γ, so by the proof of Theorem 10.1, $>^C$ M-rationalizes C, whence by Proposition 10.1 $>^C$ uniquely M-rationalizes C.

Theorem 10.1 can be generalized to a larger class of choice functions. Indeed, it is possible to show that this result holds for choice functions that *fail* to be regular (see Arló-Costa and Pedersen (2009a) for an analysis of this case and some example of heuristics where regularity fails).

Many important results involving the role of Chernoff's axiom presuppose that the choice functions used are regular. As Sen points out, Fishburn, Blair, and Suzumura seem to think that Property α guarantees that the base relation is acyclic. But it is easy to see that this is incorrect, for it is Regularity that corresponds to acyclicity of the base relation, and Property α is independent of Regularity.

10.3.1 Pseudo-rationality

One of the main issues treated by the theory of choice concerns the functional characterization of different notions of rationality. One interesting notion of rationality is the notion of *pseudo-rationality* characterized in Aizerman and Malishevski (1981) and presented later on with a slightly different perspective in Moulin (1985).

Definition 10.4 (Pseudo-rationality) The choice function C is *pseudo-rationalized* by the orderings $R_1, \ldots, R_n$ if C can be written as:

$$C(B) = \bigcup_{1 \leq i \leq n} M(B, R_i), \text{ for all } B$$

As Moulin points out, not all pseudo-rationalizable choice functions are rationalizable. To see this take A = {a, b, c}, and consider the orderings $R_1 = (a, b, c)$, and $R_2 = (c, b, a)$. Then we have that $C(A)$ = {a, c}, yet $C(a, b)$ = {a, b}, and $C(b, c)$ = {b, c}, which leads to a violation of γ.

Aizerman and Malishevski proposed a functional characterization of pseudo-rationality in terms of α and the following condition named after the first author of Aizerman and Malishevski (1981):

Property Aizerman. For all $S, S' \in \mathcal{S}$, if $C(S') \subseteq S \subseteq S'$, then $C(S) \subseteq C(S')$.

We can state the theorem explicitly as follows:

Theorem 10.2 *(Aizerman and Malishevski 1981) A choice function is pseudo-rationalizable if and only if it satisfies α and $Aizerman$.*[2]

At this point it should be clear to the reader the formal connection between pseudo-rationalizability and the first-tier notion of admissibility used by Isaac Levi. If we construct a choice function for this notion of admissibility, the choice function will be pseudo-rationalized by the permissible orderings used in Levi's first-tier decision rule. Now we have therefore a functional characterization of a choice function for Levi's decision rule. So, if we treat indeterminacy in belief change by appealing to Levi's decision rule we now know what are the abstract properties that a choice function for this decision rule would have. What we need now is a bridge between the abstract properties of choice functions and properties of belief revision operators. Such bridges have been provided by the work of Rott (2001). We will focus now on this issue.

[2]Paul Pedersen presents a clear proof of the result in Pedersen (2009). In addition, Pedersen (2009) provides several extensions of the result to cases in which the universal set is infinite and the collection of menus satisfies certain closure conditions.

10.4 Belief Revision

Belief change has been formalized in several frameworks. In this article, the general framework of belief change under discussion is based on the work of Alchourrón, Gärdenfors, and Makinson (AGM) Alchourrón et al. (1985). We will presume familiarity with the AGM framework, but here we will review some of the basic ideas.[3,4]

In the AGM framework, an agent's belief state is represented by a logically closed set of sentences K, called a *belief set*. The sentences of K are intended to represent the *beliefs* held by the agent. belief change then comes in three flavors: *expansion, revision,* and *contraction*.

In expansion, a sentence φ is added to a belief set K to obtain an expanded belief set $K + \varphi$. This expanded belief set $K + \varphi$ might be logically inconsistent. In revision, by contrast, a sentence φ is added to a belief set K to obtain a revised belief set $K * \varphi$ in a way that preserves logical consistency. To ensure that $K * \varphi$ is consistent, some sentences from K might be removed. In contraction, a sentence φ is removed from K to obtain a contracted belief set $K \dot{-} \varphi$ that does not include φ. In this article we will be primarily concerned with *belief revision*.

10.4.1 Postulates for Belief Revision

For a fixed belief set K, the following are the six *basic postulates* of belief revision (Alchourrón et al. 1985, p. 513, Hansson 1999, p. 212):

($*1$) $K * \varphi$ is a belief set. (*Closure*)
($*2$) $\varphi \in K * \varphi$. (*Success*)
($*3$) $K * \varphi \subseteq \mathrm{Cn}(K \cup \{\varphi\})$. (*Inclusion*)
($*4$) If $\neg\varphi \notin K$, then $\mathrm{Cn}(K \cup \{\varphi\}) \subseteq K * \varphi$. (*Vacuity*)
($*5$) If $\mathrm{Cn}(\{\varphi\}) \neq For(\mathcal{L})$, then $K * \varphi \neq For(\mathcal{L})$. (*Consistency*)
($*6$) If $\mathrm{Cn}(\{\varphi\}) = \mathrm{Cn}(\{\psi\})$, then $K * \varphi = K * \psi$. (*Extensionality*)

Let us henceforth call a function $*_K \colon For(\mathcal{L}) \to \mathbb{K}$ a *revision function* over K if it satisfies postulates ($*1$), ($*2$), and ($*6$). Of course, we write $K * \varphi$ instead of $*_K \varphi$.

The six basic postulates are elementary requirements of belief revision and taken by themselves are much too permissive. Invariably, several postulates are added to the basic postulates to rein in this permissiveness and to add structure to belief change. Such postulates are called *supplementary postulates*. Among the various postulates added to the mix, the following postulate – or some equivalent or stronger version of it – never fails to be set forth Gärdenfors (1988):

[3] A comprehensive introduction to theories of belief change is Hansson (1999). A brief introduction to belief change may be found in Gärdenfors (1992).

[4] Much of the presentation here follow the one introduced in Arló-Costa and Pedersen (2009b). In both cases the section introduces the reader to material that is well known in the belief revision literature.

$(*7g)\ K * \varphi \cap K * \psi \subseteq K * (\varphi \vee \psi)$.

In Hansson (1999, p. 217), $(*7g)$ is called *Disjunctive Overlap*.[5] It encodes the intuitive idea that if an agent believes δ whether it revises its beliefs K by φ or by ψ, then the agent ought to believe δ if the agent revises its beliefs K by $\varphi \vee \psi$. In Gärdenfors (1988, pp. 211–212), Peter Gärdenfors shows that in the presence of postulates $(*1)$–$(*6)$, postulate $(*7g)$ is equivalent to the following postulate:

$(*7)\ K * (\varphi \wedge \psi) \subseteq \mathrm{Cn}((K * \varphi) \cup \{\psi\})$.

In fact, an examination of the proof in Gärdenfors (1988) reveals that this equivalence holds even in the presence of only postulates $(*1)$, $(*2)$, and $(*6)$, i.e., if $*$ is revision function over K.

The AGM theory adds as well the following strong postulate:

$(*8)$ If $\neg\varphi \notin K * \varphi$, then $K * \varphi \subseteq K * (\varphi \wedge \psi)$.

Often another postulates are often considered:

$(*8r)\ K * (\varphi \vee \psi) \subseteq \mathrm{Cn}(K * \varphi \cup K * \psi)$.
$(*8c)$ If $\psi \in K * \varphi$, then $K * \varphi \subseteq K * (\varphi \wedge \psi)$.

We can combine postulates $(*7g)$ and $(*8r)$ into one postulate:

$(*\mathrm{R})\ K * \varphi \cap K * \psi \subseteq K * (\varphi \vee \psi) \subseteq \mathrm{Cn}(K * \varphi \cup K * \psi)$.

A first approach to belief revision in conditions of indeterminacy was presented in Arló-Costa (2006). In this case the recommendation was the adoption of the postulate $(*8r)$ and $(*8c)$, instead of $(*8)$ (as well as the basic postulates and $(*7)$). The postulate $(*8c)$ will play a crucial role in the approach presented here.

10.4.2 Selection Functions in Belief Revision

The major innovation in Alchourrón et al. (1985) is the employment of selection functions to define operators of belief change. In Alchourrón et al. (1985), selection functions take *remainder sets* as arguments.[6] In this article we utilize selection

[5] The "g" in $(*7g)$ is for "Gärdenfors" (Rott 2001, p. 110).

[6] For a belief set K and a sentence φ, a *remainder set* $K \perp \varphi$ is the set of maximal consistent subsets of K that do not imply φ. Members of $K \perp \varphi$ are called *remainders*. Thus, in the AGM framework, a belief set K is fixed, and for every sentence φ such that $\varphi \notin \mathrm{Cn}(\emptyset)$, $\gamma(K \perp \varphi)$ selects a set of remainders of $K \perp \varphi$. The situation in which $\varphi \in \mathrm{Cn}(\emptyset)$ can be handled as a limiting case at the level of the selection function Alchourrón et al. (1985) or at the level of the revision operator.

functions which take *propositions* expressed by formulae as arguments, i.e., selection functions on the choice space $(\mathcal{W}_{\mathcal{L}}, \mathcal{E}_{\mathcal{L}})$. Such selection functions are called *semantic selection functions*. Hans Rott has shown in Rott (2001) that this approach is a fruitful generalization of the AGM approach.

Optimization, called *strong maximization* in Gärdenfors and Rott (1995, p. 65), is put to use in the classical AGM theory of belief change (Alchourrón et al. 1985). There a selection function chooses the remainders of a remainder set that are "best" in the sense that they are most worth retaining according to some non-strict ordering (the so-called "marking-off" relation in Hansson (1999, p. 82)).[7]

It is also possible to apply maximization to study belief change. This notion, called *weak maximization* in Gärdenfors and Rott (1995, p. 65), is explored at length in Arló-Costa (2006), and Rott advocates using this notion in Rott (2001, p. 156). Indeed, there are good reasons to believe that this formalization is superior to the aforementioned formalization.

We point to a simple formal connection between rational choice on the one hand and belief change and non-monotonic reasoning on the other. In rational choice, G-rational and M-rational selection functions are often called *rationalizable*. However, in the study of belief change and non-monotonic reasoning, G-rational (i.e., strongly rationalizable) and M-rational (i.e., weakly rationalizable) selection functions are often called *relational*. Thus, formally speaking, rationalizablity in rational choice is equivalent to relationality in belief change and non-monotonic reasoning.

10.5 Rott's Correspondence Results

In this section we will review Rott's correspondence results linking conditions of belief revision and coherence postulates in the theory of rational choice (following the presentation introduced in Arló-Costa and Pedersen (2009b)). We will present his results in a way that brings out their bearing upon rationalizabillty in belief change . We begin with several definitions .

Definition 10.5 A *semantic selection function* is a selection function on choice space $(\mathcal{W}_{\mathcal{L}}, \mathcal{E}_{\mathcal{L}})$.

Definition 10.6 Let γ be a semantic selection function.

(i) We define a semantic selection function $\overline{\gamma}$ by setting for all $S \in \mathcal{E}_{\mathcal{L}}$,

$$\overline{\gamma}(S) := \begin{cases} [\![\widehat{\gamma(S)}]\!] & \text{if } \gamma(S) \neq \emptyset \\ \emptyset & \text{otherwise.} \end{cases}$$

---

[7] In Alchourrón et al. (1985, pp. 517–518), a relation $\geq$ is defined over remainder sets for a fixed belief set K, and $(\mathrm{Eq}_{\geq})$ is called the *marking off identity*:

$$\gamma(K \perp \varphi) = \{B \in K \perp \varphi : B \geq B' \text{ for all } B' \in K \perp \varphi\}.$$

We call $\overline{\gamma}$ the *completion* of γ.
(ii) We say that γ is *complete* if $\gamma = \overline{\gamma}$.

Observe that for every $S \in \mathcal{E}_{\mathcal{L}}$, $\overline{\gamma}(S) \subseteq S$, so $\overline{\gamma}$ is a selection function. Also observe that for all $S \in \mathcal{E}_{\mathcal{L}}$, $\gamma(S) \subseteq \overline{\gamma}(S)$. Finally, observe that if $\mathcal{L}$ is finite, then every semantic selection function is complete.

We now define choice-based revision functions.

Definition 10.7 Let K be a belief set, and let γ be a semantic selection function. The *semantic choice-based revision function* $*$ *over* K *generated by* γ is defined by setting for every $\varphi \in For(\mathcal{L})$,

$$K * \varphi := \begin{cases} \widehat{\gamma(\llbracket\varphi\rrbracket)} & \text{if } \gamma(\llbracket\varphi\rrbracket) \neq \emptyset \\ For(\mathcal{L}) & \text{otherwise.} \end{cases}$$

We say that γ *generates* $*$ or that $*$ is *generated by* γ.

To bring the ideas concerning rationalizablity to the foreground, we offer the following definition.

Definition 10.8 Let K be a belief set. We call a function $*$ a (*regular, rational, pseudo-rational, G-rational, complete*) *choice-based revision function* over K if there is a (regular, rational, pseudo-rational, G-rational, complete) semantic selection function γ that generates $*$.

Observe that every semantic choice-based revision function over a belief set K satisfies postulates ($*1$), ($*2$), and ($*6$) and so is indeed a revision function over K. It is an easy matter to check that the converse holds as well: If $*$ is a revision function over a belief set K, then $*$ is a semantic choice-based revision function over K.

Also observe that $*$ is a semantic choice-based function over K generated by γ if and only if for every sentence ψ of $\mathcal{L}$,

$$\psi \in K * \varphi \text{ if and only if } \gamma(\llbracket\varphi\rrbracket) \subseteq \llbracket\psi\rrbracket.$$

Intuitively, an agent believes a sentence ψ in the revision of K by φ just in case ψ is true in all the most "plausible" worlds in which φ is true. Of course, the role of a semantic selection function – or any selection function – can be interpreted in various ways in different contexts.

In Rott (2001), Hans Rott discusses a handful of coherence constraints for selection functions, some of which are well-known and others of which Rott debuts. We present two conditions of the latter sort without offering motivation (see Rott 2001, pp. 147–149) for such motivation):

($F1_B$) For every $S \in \mathcal{S}$, if $S \cap B \neq \emptyset$, then $\gamma(S) \subseteq B$ (*Faith 1 respect to B*)
($F2_B$) For every $S \in \mathcal{S}$, $S \cap B \subseteq \gamma(S)$ (*Faith 2 respect to B*)

Let us now see how some of the coherence constraints – especially condition α – are intimately connected with the presumption that selection functions are rationalizable in the study of belief change . Here we turn to Rott's recent correspondence results. Among other things, Rott's recent results establish a connection between condition α and postulate $(*7)$ of belief revision. Presented in a form suitable for this article, the following theorem provides one part of this connection (Rott 2001, p. 197).

Theorem 10.3 (Rott 2001) *Let K be a belief set. For every semantic selection function γ which satisfies*

$$\left\{\begin{array}{c} - \\ F1_{[\![K]\!]} \\ F2_{[\![K]\!]} \\ \text{Regularity} \\ \alpha \\ \text{Aizerman} \\ \gamma^* \text{ and is complete} \end{array}\right\}$$

the semantic choice-based revision function $*$ over K generated by γ satisfies, respectively

$$\left\{\begin{array}{c} - \\ (*4) \\ (*3) \\ (*5) \\ (*7) \\ (*8c) \\ (*8r) \end{array}\right\}$$

Theorem 10.3 is a "soundness" result. Rott also establishes a number of "completeness" results. Also presented in a form suitable for this article, the following completeness result is the other part of the connection between coherence constraints and rationality postulates of belief revision (Rott 2001, p. 198).

Theorem 10.4 (Rott 2001) *Every revision function $*$ over a belief set K which satisfies*

$$\left\{\begin{array}{c} - \\ (*3) \\ (*4) \\ (*5) \\ (*7) \\ (*8c) \\ (*8r) \end{array}\right\}$$

can be represented as the semantic choice-based revision function over K generated by a semantic selection function γ which satisfies, respectively.

$$\left\{\begin{array}{c} - \\ \text{F2}_{[\![K]\!]} \\ \text{F1}_{[\![K]\!]} \\ \text{Regularity} \\ \alpha \\ \text{Aizerman} \\ \gamma^* \end{array}\right\}$$

Observe that the preceding theorems do not presuppose any basic postulates other than (∗1), (∗2), and (∗6). Since $(\mathcal{W}_{\mathcal{L}}, \mathcal{E}_{\mathcal{L}})$ is finitely additive, subtractive, and compact,[8] we can apply the results from the previous section to obtain the following corollary which is of particular relevance for the purposes of this article.

Corollary 10.2 *Let K be a belief set.*

(i) Every rational choice-based revision function ∗ over K is a revision function satisfying (∗7), and every rational complete choice-based revision function ∗ over K satisfies (∗7) and (∗8r).
(ii) Every revision function ∗ over K satisfying (∗7) and (∗8r) is a rational (complete) choice-based revision function over K.
(iii) Assuming that the language is finite, every revision function ∗ over K satisfying (∗7) and (∗8c) is a pseudo-rational (complete) choice-based revision function over K.

The corollary and the previous results give us an idea of the conditions that mirror choice-based revision functions that are, respectively, rational and pseudo-rational. The postulate (∗7) is crucial in both approaches, but in the case of pseudo-rationality the central postulate is (∗8c).

If we face a problem where there is indeterminacy in belief revision this can be represented by the fact that there is a multiplicity of permissible orderings that are epistemically relevant. As we have explained above, a first approximation to this problem is to adopt Levi's decision rule recommending to take as admissible any option that is deemed as maximal by some permissible ordering. We can have a choice function collecting all these maximal points. This choice function is pseudo-rationalizable by the permissible orderings and it obeys the structural conditions α and *Aizerman*. The completeness results deriving from the work of Rott indicate, in turn, that a choice-based revision function of this sort should obey the basic postulates of AGM plus postulates (∗7) and (∗8c).

The approach to indeterminacy presented in Arló-Costa (2006) is characterized in terms of the basic postulates plus (∗7), (∗8c) and (∗8r). So, the approach

[8] Now let $(X, \mathcal{S})$ be a choice space. We call $\mathcal{S}$ *finitely additive* if it is closed under finite unions; we call $\mathcal{S}$ *subtractive* if it is closed under relative complements; and we say that $\mathcal{S}$ is *compact* if for every $S \in \mathcal{S}$ and $I \subseteq \mathcal{S}$, if $S \subseteq \bigcup_{T \in I} T$, then there is some finite $I_0 \subseteq I$ such that $S \subseteq \bigcup_{T \in I_0} T$. If γ is a selection function on $(X, \mathcal{S})$, we call γ *finitely additive* (*subtractive, compact*) if $\mathcal{S}$ is finitely additive (subtractive, compact).

considered here is weaker and more general (it succeeds at giving useful recommendations in cases where the approach considered in Arló-Costa (2006) is unable to produce solutions – see the introduction for an example).

10.6 Choosing What to Believe

The formal results linking the theory of choice and the theory of belief revision are robust but we have to keep in mind that the two theories related by the mappings are quite different. Isaac Levi presents in Levi (2004) various reasons to think that the formal results should not be interpreted as providing decision theoretic foundations for belief revision. We use the mappings here as a heuristic device that helps us to find the right postulates for belief revision in conditions of indeterminacy.

In any case, Rott's approach or other approaches that appeal to rational choice or decision theory in order to provide foundations to belief change presuppose that it makes sense to study a model where one is able to choose beliefs. An anonymous referee pointed out that this idea might not work in general. There are indeed many things that we cannot choose to believe, like that the Pope is female, for example. And most probably we cannot choose our perceptual beliefs. Isaac Levi analyzed these philosophical issues in detail in Chapter 3 of Levi (1991). The main point is that the connection between rational choice and decision theory on the one hand and belief revision on the other has severe limitations. Levi proposes that we change our perceptual beliefs via *routine expansions* which have a very different structure than the belief revision functions used in the literature on belief change (see Levi 1991, for details).

Still one might ask the question: is it possible at all to choose what to believe? If the changes of belief in question are construed as responses to stimuli or as dispositions to such responses the answer is negative. Such beliefs are not fully under the control of the deliberating agent. But one can argue that a rational agent is indeed capable of changing his or her *doxastic commitments* (see Levi 1991, p. 71). A scientist who endorse the axioms of a scientific theory is committed to the truth of the logical consequences of these axioms, for example. The main idea is that the deliberating agent has rational control over these doxastic commitments and that therefore he is able to choose how to revise them. In this restricted sense the rational agent we are focusing on is able to deliberatively change her beliefs. Most of the examples we will consider below concern theory choice in science and other situations where one can argue that there is deliberate change of doxastic commitments.

10.7 Admissibility

When we considered the example at the beginning of the chapter we had two orderings R and R'. According to R a is preferred to b and b is preferred to c. According to R' c is preferred to b and b is preferred to c. The decision criteria we considered indicated that in this situation both a and c are equally admissible. But in many situations we would like to determine a unique option as admissible. One way of doing this is by deploying a secondary ordering capable of discriminating between a and c.

A two-tier decision rule of this sort has been proposed by Isaac Levi in various writings. Unfortunately its functional characterization is an open problem. We will present below some preliminary results about the two-tier rule in order to give the reader a preliminary idea of its structural properties.

10.7.1 Admissibility by a Two-Tier Rule

Let Y be a feasible set and let O be a set of permissible orderings on X, where X is the universal set of a choice space $(X, \mathcal{S})$. For $Y \in \mathcal{S}$, $y \in Y$ is *first-tier admissible* in Y iff there is at least one ordering R_i in O according to which there is no $z \in Y$ such that zR_iy. If R_S is an ordering on X, the intended interpretation of which is some notion of security, then, in keeping with the earlier presentation of Levi's criterion, we can formulate admissibility as follows: $y \in Y$ is admissible in Y iff it is first-tier admissible in Y and there is no z that is first-tier admissible in Y such that zR_Sy.

If we construct a choice function for this notion (with $O = \{R_1, \ldots, R_n\}$) we have the following:

$$C(B) = M\left(\bigcup_{1\leq i\leq n} M(B, R_i)\right), <_S, \text{ for all B.}$$

This is a more sophisticated choice function, and a functional characterization for it is not available. Some axioms are sound with respect to the two-tier rule, but it is not difficult to construct counterexamples for some of the best known choice conditions in the literature.

Remark 10.1 α is violated by the two-tier decision rule. To see this consider a set T = {a, b, c} and a subset S = {b, c}. Consider in addition two orderings R_1 = (a, c, b)[9] and R_2 = (b, a, c). Finally consider a security ordering R_S = (c, b, a). According to Levi's rule $C(T)$ = {b} and $C(S)$ = {c}. So, we have that $S \cap C(T) \nsubseteq C(S)$, violating α.

Amartya Sen proposed some weaker versions of α in Sen (1977). The weakest of these properties is the following one:

> *Property* $(\alpha - -)$. If Y is a three-element subset, then there is some $x \in Y$ such that $x \in C(\{x, y\})$ for all $y \in Y$.

It is not difficult to see that our counterexample to α is also a counterexample to $(\alpha - -)$ and therefore to the property $(\alpha -)$, which is stronger than $(\alpha - -)$.

---

[9] The parenthetical notation used here indicates that a is preferred to c and c to b.

Levi's decision rule violates various additional axioms for choice functions. Since α and *Aizerman* are the two central conditions for the first tier decision rule, we will present in addition an explicit counterexample against *Aizerman*.

Remark 10.2 To see a violation of *Aizerman* consider a set $S' = \{a, b, c\}$ and a subset $S = \{a, b\}$. Consider in addition two orderings $R_1 = (c, a, b)$ and $R_2 = (b, a, c)$. Finally consider a security ordering $R_S = (a, b, c)$. $C(B') = \{b\}$, so the antecedent of *Aizerman* is satisfied. But, since $C(B) = \{a\}$ the consequent of *Aizerman* does not hold.

Finally we have as well a violation of γ^*:

Remark 10.3 To see a violation of γ^* consider a set $S' = \{a, b\}$ and a set $S' = \{b, c\}$. Consider in addition two orderings $R_1 = (c, b, a)$ and $R_2 = (a, b, c)$. Finally consider a security ordering $R_S = (b, c, a)$. $C(S) = C(S') = \{b\}$, but $C(S \cup S') = \{c\}$.

It is not difficult to see that similar counterexamples can be constructed to other central axioms of choice that matter for the application we are considering, like β or weaker versions of β considered by Sen (1977).

So, is there any condition satisfied by the rule? Jeff Helzner recently proved in Helzner. (2008) that Fishburn's condition A5 is indeed satisfied by the rule:

> Fishburn Condition A5. If $Y \subseteq X$ has more that two elements and $y \in C(Y)$, then $y \in C(\{y, z\})$ for some $z \in Y$ such that $y \neq z$.

This condition is quite weak and it is unlikely that it suffices to characterize the rule. Most probably non-standard functional constraints would be needed to do so. Preliminary work in this direction has been carried out by Ruth Poproski in her Master thesis at CMU (Poproski 2008). In any case, the challenge here is at the level of the theory of choice functions required to characterize the two-tier decision rule. All elements concerned with belief revision properly are guaranteed by an adequate translation once the functional constraints that characterize the rule have been isolated. Taking into account the shape of postulates that are sound with respect to the decision rule (like Fishburn's A5) it is likely that the resulting constraints on belief revision will be less elegant than the ones required to reflect the first-tier decision rule. But the two-tier rule seem to have advantages with respect to the first-tier rule that might compensate for the lack of elegance of the resulting axiomatic base.

10.8 Applications: Philosophy of Science

Sidney Morgenbesser and Arnold Koslow wrote circa 1960 an important paper (Morgenbesser and Koslow 2008) explaining how to aggregate different dimensions of epistemic utility in order to construct a total utility function used to measure the worth of theories. Their proposal is formulated in terms of utility functions but it is not difficult to determine its qualitative implications. The paper has been corrected and edited by Arnold Koslow who plans to publish it now with notes added by various scholars interested in this issue. In a way the paper is of vital interest today and it is very relevant for the topic of this chapter.

The main problem that they consider is similar to the problems we have considered here, only that they focus on a particular application. Also, they do not consider directly the problem of theory change or even the problem of theory choice, but they do consider problems where there are various dimensions of the worth of (scientific) theories, the challenge being how to aggregate these different dimensions in a total utility function. They also consider the problem where different theories are ranked in accordance with different dimensions of epistemic value and the problem is to determine a total worth functional yielding an aggregate ordering of theories. We can see this problem as a particular case where there is indeterminacy regarding epistemic value. One dimension of epistemic worth could be simplicity, and another explanatory power, for example, and theories can be ranked differently according to each dimension of epistemic value. In cases of this sort we have considered decision rules that tell us what to choose (like Levi's first-teir decision rule). Koslow and Morgenbesser consider instead how to construct an aggregate ordering and an aggregate (total) utility function. Presumably one would choose by taking what is maximal (optimal) according to the aggregate ordering (utility function).

Let me present first the basis of the theory of total utility functions proposed in Morgenbesser and Koslow (2008). We have two theoretical constructs: a Total-Utility functional and Total-Worth functional. A Total-Utility functional is a function which assigns a utility function to (say) any n -tuple of utility functions $u_1, u_2, \ldots, u_n$, where each u_i is a function which has as its domain the consequences of scientific acts. A Total-Worth functional on the other hand, is a function which assigns an ordering relation to (say) any n-tuple of ordering relations $W_1, W_2, \ldots, W_n$, where each W_i is an ordering relation which ranks theories. A (TUF) aggregates specific utility functions, and a (TWF) aggregates various kinds of merit or worth.

Let's focus now on a quasi-formal description of Total-Utility functionals. We would be more interested in Total-Worth functionals but both notions share important features and the authors only offer an explicit presentation of Total-Utility functionals.

Let C be the set of consequences of various scientific acts. A utility function will be a real-valued function with domain C. If u_1 is a utility function on C, let $[u_1]$ denote the class of all those real-valued functions φ on C, such that

$$\varphi(a) \geq \varphi(b) \text{ if and only if } u_1(a) \geq u_1(b),$$

for all elements a, b, of C. Let $U = [u_1] \times [u_2] \times \ldots \times [u_n]$, that is, the set of all ordered n – tuples $\langle f_1, \ldots, f_n \rangle$ where f_i is a member of $[u_i]$, for $i = 1, 2, \ldots, n$.

A Total-Utility Functional Φ is any function with domain U, which satisfies the following conditions:

(1) Φ is a mapping of U into the set of all real-valued functions defined on C. That is, if $\lambda = \langle f_1, \ldots, f_n \rangle$ is an element of U, then Φ_λ is a real-valued function defined on the set C.

(2) If λ and μ are elements of U, then for any two members a and b of C,

$$\Phi_\lambda(a) \geq \Phi_\lambda(b) \text{ if and only if } \Phi_\mu(a) \geq \Phi_\mu(b).$$

That is, although different elements λ and μ of U may have different values Φ_λ and Φ_μ associated with them by the total-utility functional, nevertheless the functions Φ_λ and Φ_μ rank the elements of C in exactly the same order. Since the rankings which Φ_λ induce on C are independent of the particular λ of U, we shall omit the index when we compare the values of the Total-Utility functional on elements of C.

Finally there is a third condition that the authors impose on the functional:

(3) If the utility functions $\varphi_1, \ldots, \varphi_n$ in $[u_1], \ldots, [u_n]$ respectively each rank a higher than b, then the Total-Utility functional cannot rank b higher than a. Specifically, if $u_i a \geq u_i b$ for i = 1, 2, ..., n, then Φ_λ (a) $\geq$ Φ_λ (b), where λ is any element of U.

This rule establishes a weaker constraint on possible aggregations than the one considered in Arló-Costa (2006). Finally the authors impose a constraint that will be important for some of the philosophical discussions that follow. The idea of the constraint is that explanatory and predictive power count for something via their influence in explanatory and predictive utilities.

(EP) If any Total-Utility functional ranks a at least as high as b, then either $(u_{exp}a \geq u_{exp}b)$ or $(u_{pr}a \geq u_{pr}b)$.

The principle **EP** establishes a necessary, but not a sufficient condition, on the Total-Utility. The Total-Utility might rank a greater than b even though b is ranked higher than a both with respect to predictive and explanatory value. Other aspects of value involved in Total-Utility might justify this aggregation.

It is immediate to see that there are functionals which are Total-Utility functionals. With U defined as above, set $\Phi \langle f_1, \ldots, f_n \rangle = f_1$, where f_1 is any function in the set $[u_1]$. Conditions (1) and (2) are trivially satisfied. Further, if we set u_1 to be the utility function u_{exp}, then condition **EP** is satisfied also.

Cases of this sort where Total-Utility "collapses" to one of the utilities used in its construction are labeled by the authors as functionals that are simple or of the projective type. The theory is sufficiently expressive to represent ordering relations among theories (or consequences) that are not projective (or imposed). One such account would be *qualified pragmatism*. If this account is formulated as a thesis about the worth of theories we have that: one theory S has higher (qualified) pragmatic worth than theory T if and only if either S is true and T is false, or S and T have the same truth value, but S has greater predictive power than T. It is clear that qualified pragmatism is not of the projective type and that it can accommodate non trivial claims about the worth of theories. So, the theory has an interesting expressive power.

Total-Utility functionals should not be conflated with the Social Welfare Functions (SWF) used by K. Arrow in his celebrated essay (Arrow 1951). The theory of Total-Utility is too weak to obey the axioms that Arrow imposes on SWFs.

Coming back to our main concern, the problem of belief revision in conditions of indeterminacy, it is clear that the use of the account that we sketched above can provide a solution to this problem. For every time that we have a set of orderings representing relevant dimensions of epistemic value, we would have a function yielding an aggregate ordering that we can then use to determine a revision. It seems, nevertheless, that the theory is too permissive, if one is ultimately interested in theory choice. In other words, if one is interested in the aggregate ordering as a tool to decide what to believe next, it is natural that one can maximize the aggregate ordering. But many permissible aggregate orderings do not seem adequate for this task.

Consider the example that we presented at the beginning. Say that we have two weak orders representing different dimensions of epistemic value, R and R'. Suppose, for simplicity, that there are three salient options: a, b and c. According to R a is preferred to b and b is preferred to c. According to R' c is preferred to b and b is preferred to c. If these orderings are the orderings among consequences induced by two utility theories U and U', a possible aggregation of utility can yield a utility function that ranks the consequences as U (R) does. Or the aggregation can yield a utility function that ranks the consequences as U' (R') does. It is unclear why one should consider these aggregations as permissible. At least this is so if the two dimensions of epistemic worth represented by U and U' are the only two dimensions of epistemic value or worth that matter for the problem. It seems that as long as these two dimensions of value are the only that matter for the problem under consideration, the aggregation pattern that ranks a and c at the same level and above of b has a salient role. For it seems that this ordering has a maximal set that yields the options that one would like to choose in the given conditions of indeterminacy. So, some aggregations seem more preferable than others, and the ones that are preferred seem to be the ones that cohere with the adoption of a decision rule of the sort we have explored in previous sections.

In view of the previous remarks it is not clear whether the use of an aggregation function of the sort proposed by Morgenbesser and Koslow constitute a real improvement with respect to the solution considered in previous sections. It seems that in cases of unresolved conflict of the sort we just considered, ultimately the aggregated ordering is either superfluous (because it gives the same recommendations as the decision rule) or is of a trivial type (projective) or gives an arbitrary solution that can only be justified if there are other dimensions of epistemic worth that we have not considered explicitly in the representation (one permissible Total-Utility in the case of the example can rank the three options at the same level, for example). But as long as we represent each dimension of epistemic worth via a corresponding ordering it seems that the aggregation rule that yields the same maximal set as the first-tier (two-tier) decision rule used by Levi has a salient role. At least this is so if we consider the problem of theory choice as the main issue motivating the use of aggregation rules.

10.8.1 An Interesting Challenge: Theories-of and Theories-for

Most of the second part of Morgenbesser and Koslow (2008) is devoted to distinguish between two types of theories, theories-for and theories-of. Examples of theories-for are the Germ Theory of diseases, or the Kinetic Theory of Gases. These theories seem to behave as research programs rather than scientific theories in the usual sense. They are theories-for the systematization of their instances.

> That Yellow-fever b is caused by an Arbor Virus in Group B (Arthropoid borne animal virus) is not explained by pointing out that Yellow-fever is a communicable disease and that for every communicable disease there is *some* organic causative agent. The statement about the Arbor Viruses isn't even a deductive consequence of these two statements. (p. 21)

The Germ Theory is constituted by this existential statement: that for every communicable disease there is *some* organic causative agent. But this is too weak to explain any of the instances of the theory.

> We can generalize and say that the Germ Theory is not used in an explanation or prediction of its instances. It is not, as we shall call it, a theory-of, of any of its instances, since it is not used in their explanation. On the other hand such a theory seems to be useful for explaining, or used for explaining. We shall say that the Germ Theory of Disease is a theory-for, but it is not a theory-of any of the specific instances we have discussed. It is, at the very least, a theory for the systematization of these instances, though it is not a theory of any of those instances. (p. 22)

Now, theories-for have a peculiar behavior if utility is assigned to them. It seems that they have minimal worth compared with any other arbitrary theory. For assume that one of these theories-for K, ranks above another theory T. It would follow that K ranked higher than T either in explanatory or predictive power. But K does neither, given that the theory in question is not used in explanation or prediction of its instances, and therefore its explanatory or predictive power should be lower than the explanatory or predictive power than any other theory. As a result:

> [...] no use of K, no scientific act involving K, will ever be preferred to the use of any other theory, an extremely counterintuitive conclusion about a class of theories whose scientific merit is admittedly high. (pp. 17–18)

The postulate **EP** was essentially used in the derivation of this counterintuitive conclusion. If one wants to block the counterintuitive result it seems that there are two salient moves available. One can question **EP** or one can refuse to consider theories-for as scientific theories that can be evaluated in terms of epistemic utility.

Let's first focus on **EP**. Notice that the postulate depends essentially on the idea that indeterminacy should be resolved by implementing an aggregation procedure. If one adopts a decision rule of Levi's type the postulate is not longer needed. In fact, **EP** puts constraints on Total-Utility functions.

And if one insists on aggregation as a way of treating indeterminacy it is not clear that **EP** needs to be adopted. Why not to consider simplicity as one of the deciding dimensions in the postulate? Perhaps an argument can be concocted in order to show that simplicity is not one of the main dimensions (perhaps a pragmatist argument can accomplish that). So, the main point here seems to challenge the very fact of using an aggregation rule to deal with indeterminacy. The method seems to create unnecessary problems that can be circumvented via the use of decision rules.

But perhaps the problem presented by Mogenbesser and Koslow reappears if one uses a decision rule. Here is a possible principle that applies to theory choice (rather than to aggregation):

> (EPD) If theory K is admissible and theory T is not in a pairwise choice between K and T, with respect to a set of orderings O that includes explanatory and predictive power as rankings, then K ranks higher than T both in explanatory and predictive power.

This is not a constraint on aggregation, but a constraint on theory choice. Unlike **EP** this is a logical constraint that is a consequence of the use of a first-tier admissibility rule. For if we have that in a pairwise choice between K and T, K is admissible and T is not, then T cannot be selected as admissible by any of the orderings representing different dimensions of epistemic value. In particular the following scenarios are ruled out: K and T are tied or T is preferred to K according to either explanatory or predictive power. T would immediately be admissible in either case.

So, the puzzle about theories-for can apparently be reconstructed in terms of **EDP**. Only that in this case one concludes that we can never have any theory T such that in a pairwise choice between K and T, K is selected and T is not. So, Morgenbesser and Koslow's puzzle seems to have a wider range of applicability than one might have expected. It seems that the problem is also a problem for the type of decision rules we have used in this article.

The second point is related to the fact that perhaps theories-for should not have the status of scientific theories at all (and therefore they should not be the bearers of utility in this application). Here the central issue is what is the underlying understanding of what counts as a scientific theory. Morgenbesser and Koslow seem to appeal to a syntactic test for determining whether a given body of information counts as a scientific theory. But there are many competing views of what counts as a scientific theory in the literature of philosophy of science. For example, one can adopt a model-theoretical view of theories. It is unclear whether theories-for pass the test as scientific theories in one of these rival accounts. Denying theories-for the status of bearers of utility seems nevertheless the most direct way out of the puzzle we have been considering.

It seems that Morgenbesser and Koslow's puzzle applies across a wide spectrum of plausible solutions of the problem of indeterminacy. It also seems that the puzzle deserves more attention in the literature devoted to this issue. Our goal here was to show that the puzzle arises not only for views that aggregate utility (rankings) but also for views of the sort defended here, which are based on the use of decision rules.

Acknowledgment I would like to thank specially Arnold Koslow for sharing with me an unpublished version of a paper he wrote in collaboration with Sidney Morgenbesser (discussed in last section of this chapter). I understand that this paper is currently under submission. So, perhaps the published version will be slightly different from the manuscript I read. In any case, I think that most of the discussed aspects of the original version of the paper are quite important and deserve a philosophical discussion.

I would like to thank also Arnie Koslow for an enlightening philosophical discussion of the contents of the unpublished manuscript.

Finally I would like to thank Paul Pedersen who read the entire manuscript and made useful comments as well as an anonymous referee.

References

Aizerman M, Malishevski A (1981) General theory of best variants choice: Some aspects. In: IEEE Transactions of Automatic Control, vol 26, pp 1030–1040

Alchourrón CE, Gärdenfors P, Makinson D (1985) On the logic of theory change: partial meet contraction and revision functions. Journal of Symbolic Logic 50:510–530

Arló-Costa H (2006) Rationality and value: The epistemological role of interdeterminate and agent-dependent values. Philosophical Studies 128(1):7–48

Arló-Costa H, Pedersen A (2009a) Bounded rationality: Models for some fast and frugal heuristics. In: Proceedings of the Second Indian Conference on Logic and its Relationship with Other Disciplines

Arló-Costa H, Pedersen A (2009b) Social norms, rational choice and belief change. In: Olsson E (ed) Science in Flux, Springer, Berlin

Arrow K (1951) Social choice and individual value. Cowles Foundations and Wiley, New York

Gärdenfors P (1988) Knowledge in flux: Modeling the dynamics of epistemic states. MIT Press, Cambridge

Gärdenfors P (1992) Belief revision: An introduction. In: Gäardenfor P (ed) Belief revision, Cambridge University Press, Cambridge pp 1–28

Gärdenfors P, Rott H (1995) Belief revision. In: Gabbay DM, Hogger CJ, Robinson JA (eds) Handbook of logic in artificial intelligence and logic programming, vol 4, Oxford University Press, Oxford pp 35–132

Hansson SO (1999) A textbook of belief dynamics, applied logic series, vol 11. Kluwer Academic Publishers, Dordrecht

Helzner J (2008) Indeterminacy in additive models of choice. In: Proceedings of *Foundations of the Formal Sciences VI: Reasoning about Probabilities and Probabilistic Reasoning*

Levi I (1986) Hard choices. Cambridge University Press, Cambridge

Levi I (1991) The fixation of belief and its undoing: Changing beliefs through inquiry. Cambridge University Press, Cambridge

Levi I (2004) Mild contractions: Evaluating loss of information due to loss of belief. Oxford University Press, Oxford

Morgenbesser S, Koslow A (2008) Theories and their worth. Tech. Rep., CUNY-Graduate Center, part I-II

Moulin H (1985) Choice functions over a finite set: A summary. Social Choice and Welfare 2:147–160

Pedersen A (2009) Pseudo-rationalizability over infinite choice spaces. Tech. Rep., Carnegie Mellon University

Poproski R (2008) The rationalizability of two-step choices. Master's thesis, Carnegie Mellon University [Published on-line in the *Journal of Philosophical Logic*, 20 August 2010]

Rott H (2001) Change, choice and inference: A study of belief revision and non-monotonic reasoning. Oxford Science Publications, Oxford

Sen A (1971) Choice functions and revealed preference. The Review of Economic Studies 38(3):307–317

Sen A (1977) Social choice theory: A re-examination. Econometrica 45(1):53–89

Sen A (1997) Maximization and the act of choice. Econometrica 65:745–779

Suzumura K (1983) Rational choice, collective decisions, and social welfare. Cambridge University Press, Cambridge

Chapter 11
Perspectival Act Utilitarianism

John F. Horty

11.1 Introduction

This chapter works within a particular framework for reasoning about actions – sometimes known as the framework of "stit semantics" – originally due to Belnap and Perloff, based ultimately on the theory of indeterminism set out in Prior's indeterministic tense logic, and developed in full detail by Belnap et al. (2001). The issues I want to consider arise when certain normative, or decision theoretic, notions are introduced into this framework: here I will focus on the notion of a *right action*, and so on the formulation of act utilitarianism within this indeterministic setting. The problem is simply that there are two different, and conflicting, ways of defining this notion, both well-motivated, and both carrying intuitive weight.

This problem was first pointed out in my Horty (2001), but here I address what I now think of as a mistake in that treatment. In that earlier book, in order to explain our conflicting judgments about right actions, I set out two substantially different accounts of the notion, which I labeled as the "dominance" and "orthodox" accounts. But here, there is only one account, only one theory of right actions, and our conflicting intuitions are instead explained by showing how this theory yields different results when actions are evaluated from different perspectives. In effect, a semantic explanation, which postulates an ambiguity in the notion of a right action, is replaced by a pragmatic explanation.

The chapter is structured as follows. In the next section, I review Prior's indeterministic framework as well as the structures underlying stit semantics. Although these structures were originally introduced for the purpose of interpreting formal languages containing special modal operators – tense operators, agency operators – there is none of that here. The concepts I am concerned with in this chapter are

J.F. Horty (✉)
Philosophy Department and Institute for Advanced Computer Studies, University of Maryland, College Park, MD 20742, USA
e-mail: horty@umiacs.umd.edu

P. Girard et al. (eds.), *Dynamic Formal Epistemology*, Synthese Library 351,
DOI 10.1007/978-94-007-0074-1_11,

defined entirely in terms of the underlying structures themselves; there is no need to introduce or interpret any formal language. In the third and fourth sections, I motivate the two ways of understanding the notion of a right action, and define the corresponding dominance and orthodox act utilitarian theories. Finally, in the fifth section, I show how these two theories can be unified, and how our conflicting intuitions about right actions can then be explained as resulting from the different perspectives from which actions might be evaluated. An appendix shows how the account can be generalized to group as well as individual actions, and how the relation between the right actions available to a group and to the individuals belonging to that group can then be seen to depend on the perspective from which these actions are evaluated.

11.2 Background

11.2.1 Individual Actions

Prior's theory of indeterminism, set out in his Prior (1967) and developed in more detail by Thomason (1970), is based on a picture of moments as ordered into a treelike structure, with forward branching representing the openness or indeterminacy of the future and the absence of backward branching representing the determinacy of the past.

This picture can be represented as a nonempty set Tree of moments together with an ordering $<$ on Tree that is transitive and irreflexive, and that satisfies the treelike property according to which, for any m_1, m_2, and m_3 in Tree, if $m_1 < m_3$ and $m_2 < m_3$, then either $m_1 = m_2$ or $m_1 < m_2$ or $m_2 < m_1$. A maximal set of linearly ordered moments from Tree is a *history*, representing some complete temporal evolution of the world. If m is a moment and h is a history, then the statement that $m \in h$ can be taken to mean that m occurs at some point in the course of the history h, or that h passes through m. Of course, because of indeterminism, a single moment might be contained in several distinct histories. We let $H_m = \{h : m \in h\}$ represent the set of histories passing through m, those histories in which m occurs; and when h belongs to H_m, we speak of a moment/history pair of the form m/h as an *index*.

In this framework, it is the histories themselves that represent possibilities, or "possible worlds." The set of possible worlds accessible at a moment m can thus be identified with the set H_m of histories passing through that moment; those histories lying outside of H_m are taken to represent worlds that are no longer accessible. We can therefore identify the *propositions* at m with the subsets of H_m, where of course, H_m itself is the least informative of these propositions.

These various ideas can be illustrated as in Fig. 11.1, where the upward direction represents the forward direction of time. This diagram depicts a branching time structure containing five histories, h_1 through h_5. The moments m_1 through m_4 are highlighted; and we have, for example, $m_2 \in h_3$ and $H_{m_4} = \{h_4, h_5\}$.

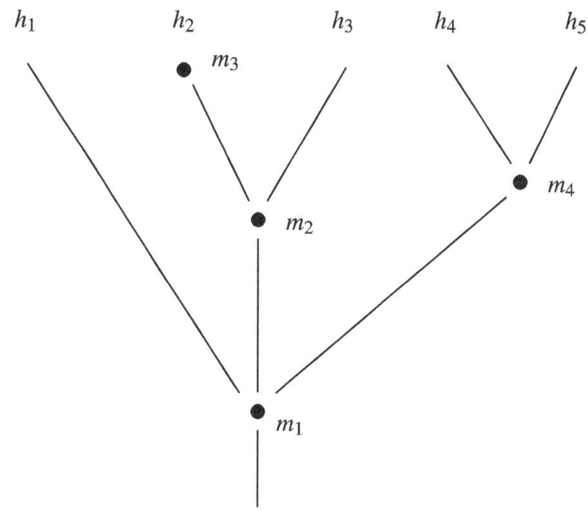

Fig. 11.1 Branching time

We now turn to the treatment of agency. The goal is to represent the notion that an agent, through its action, guarantees the truth of some proposition.[1] We must therefore be able to speak of individual agents, and also of their actions or choices; and so the basic framework of branching time is supplemented with two additional primitives.

The first is simply a set Agent of agents, individuals thought of as making choices, or acting, in time.

Now what is it for one of these agents to act, or choose, in this way? We idealize by ignoring any intentional components involved in the concept of action, by ignoring vagueness and probability, and also by treating actions as instantaneous. In this rarefied environment, acting can be thought of simply as constraining the course of events to lie within some definite subset of the possible histories still available. When an agent α butters the toast, for example, the nature of its action is to constrain the history to be realized so that it must lie among those in which the toast is buttered. Of course, such an action still leaves room for a good deal of variation in the future course of events, and so cannot determine a unique history; but it does rule out all those histories in which the toast is not buttered.

Our second additional primitive, then, is a device for representing the possible constraints that an agent is able to exercise upon the course of events at a given moment, the actions or choices open to the agent at that moment. These constraints are encoded formally through a function *Choice*, mapping each agent α and moment

[1] In an effort to find language that is both gender neutral and unobtrusive, I assume here that the agents are impersonal acting devices, such as robots, which it is appropriate to refer to using the pronoun "it".

m into a partition $Choice_{\alpha}^{m}$ of the set of histories H_m through m.[2] The idea behind this formalism is that, by acting at m, the agent α selects a particular one of the equivalence classes, or choice cells, from $Choice_{\alpha}^{m}$ within which the history to be realized must then lie, but that this is the extent of the agent's influence. If K is such a choice cell, an equivalence class from $Choice_{\alpha}^{m}$, we speak of K as an *action* available to the agent α at the moment m, and we speak of the histories belonging to K as the possible *outcomes* that might result from this action.

These various concepts relating to choice functions are illustrated in Fig. 11.2, which depicts a structure containing six histories, and in which the actions available to the agent α at three moments are highlighted. The cells at the highlighted moments represent the actions available to α at those moments. For example, there are three actions available to α at m_1 – $Choice_{\alpha}^{m_1} = \{K_1, K_2, K_3\}$, with $K_1 = \{h_1, h_2\}$, $K_2 = \{h_3\}$, and $K_3 = \{h_4, h_5, h_6\}$. If the agent selects K_3, then the histories h_4, h_5, and h_6 are the possible outcomes of its action.

11.2.2 Group Actions

To see how this account can be extended to group actions, it is best to begin with an example; so consider the multiple agent situation depicted in Fig. 11.3. Here, the actions open to the agent α at the moment m are depicted by the vertical partitions of

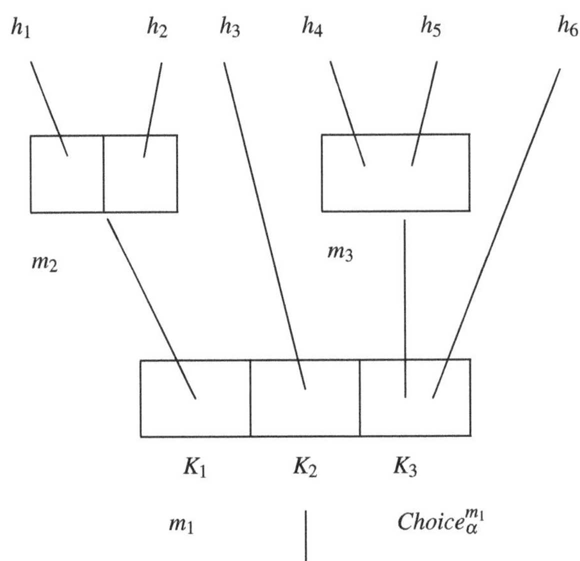

Fig. 11.2 An agent's choices

[2] The *Choice* function is subject to two technical constraints of "no choice between undivided histories" and "independence of actions," which I will not go into here. The constraints can be found in my Horty (2001), and are described in authoritative detail in Belnap et al. (2001).

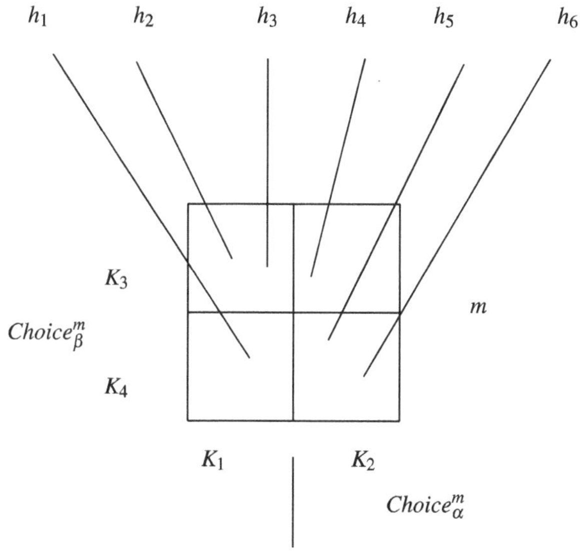

Fig. 11.3 Group actions

H_m; that is, $Choice^m_\alpha = \{K_1, K_2\}$, with $K_1 = \{h_1, h_2, h_3\}$ and $K_2 = \{h_4, h_5, h_6\}$. The actions open to the agent β are depicted by the horizontal partitions; $Choice^m_\beta = \{K_3, K_4\}$, with $K_3 = \{h_2, h_3, h_4\}$ and $K_4 = \{h_1, h_5, h_6\}$.

Now consider the proposition $X = \{h_2, h_3, h_6\}$. It should be clear that, in this situation, neither the agent α nor the agent β acting alone has the ability to guarantee the truth of X. Each action available to each of these agents allows for a possible outcome in which X fails. Still, it seems that the group of agents $\{\alpha, \beta\}$ acting together does have the ability to guarantee the truth of X. If α performs the action K_1 and β performs the action K_3, the group $\{\alpha, \beta\}$ can be said to perform the action $K_1 \cap K_3$, and X holds at each possible outcome of this group action.

As this example suggests, group actions can usefully be defined as patterns of individual actions: an action available to a group of agents can be defined as an intersection of the actions available to the individual agents belonging to that group, one action for each agent.

In order to develop this suggestion, it is convenient to reify patterns of action by defining an *action selection* function at a moment m as a function assigning to each agent some action available to that agent at m – that is, a function s mapping each agent α into some member of $Choice^m_\alpha$. Each of these action selection functions represents a possible pattern of action at the moment m, a selection of an available action for each agent. These patterns of action can be collected together into the set $Select_m$, containing the various action selection functions at m. And where Γ is a group of agents, the set $Choice^m_\Gamma$ of action available to the group at the moment m – the patterns of action available to the members of the group – can then be defined as follows:

$$Choice^m_\Gamma = \{\bigcap_{\alpha \in \Gamma} s(\alpha) : s \in Select_m\}.$$

It should be clear that this definition says what it should: the set of actions available to the group Γ is identified with the set of intersections of actions available to the agents belonging to that group, one action for each agent.

11.3 The Dominance Account

11.3.1 Our Question

With this much of the framework in place, we now add one final primitive: a function Value mapping each history into a real number representing the overall value of that history, however that is conceived. This new primitive is illustrated in Fig. 11.4, where the numbers written beside histories indicate the values assigned to those histories, so that, for example, Value$(h_1) = 10$.

Now that values have been assigned to the various histories consistent with an agent's actions – the various possible outcomes of those actions – we can turn to the central question of this chapter: How, in this indeterministic setting, can we characterize the act utilitarian notion of a *right action* for the agent to perform?

According to the standard formulation of act utilitarianism, an action is defined as right if there is no action among the available alternatives with better consequences, and wrong otherwise.[3] In the present framework, it is easy enough to define the

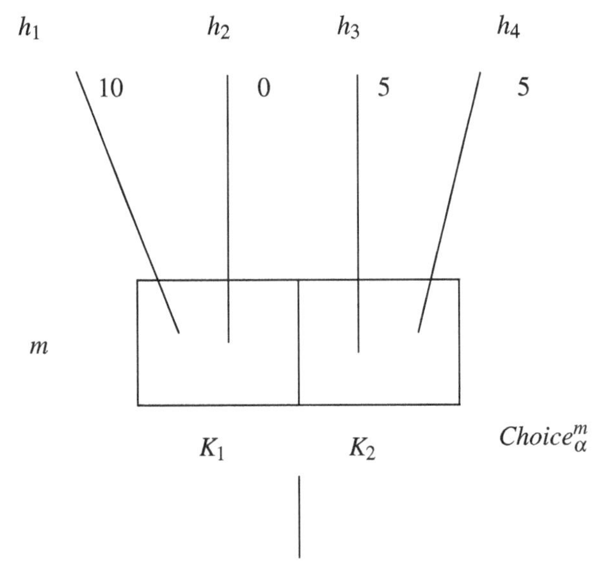

Fig. 11.4 Histories with values

[3] Perhaps the most careful formulation of act utilitarianism can be found in Bergström (1966); for work along similar lines, see Carlson (1995).

alternatives available to an agent α at a moment m; these are simply the actions from $Choice_{\alpha}^{m}$. And our Value function, of course, provides a straightforward ranking of possible outcomes. But in a setting that is genuinely indeterministic, how can we define the notion of an action's consequences?

The problem that a robust indeterminism presents for the characterization of an action's consequences – and so for a definition of act utilitarianism – was noted some time ago by Prior, in his contribution to a symposium on the topic:

> Suppose that determinism is *not* true. Then there may indeed be a number of alternative actions which we could perform on a given occasion, but none of these actions can be said to have any "total consequences," or to bring about a definite state of the world which is better than any other that might be brought about by other choices ... it's not merely that one cannot calculate the totality of what will happen if one decides in a certain way; the point is rather that there *is* no such totality. (Prior 1956, pp. 91–92)

And the general point is clear enough. In the case of Fig. 11.4, for example, the agent must choose between two available actions. The choice of K_2 leads invariably to an outcome whose value is 5, while the choice of K_1 leads to an outcome whose value is either 10 or 0, depending on whether things evolve along the lines of h_1 or h_2. But since, if K_1 is selected, it is then indeterminate whether h_1 or h_2 will be realized, how can we possibly say which of the two actions, K_1 or K_2, has the better consequences?

In response to this problem, Prior himself offers the standard suggestion of appealing to probabilistic information, such as a probability distribution on the histories that might result from an action. Using this information, we could assign an expected value to each of the actions available to an agent, and the ordering of actions based on their expected values would then allow us to define a form of act utilitarianism that did not, in fact, rely on some definite notion of an action's consequences: an action could be defined as right whenever there is no alternative with greater expected value.

This approach – leading to a theory that might be described as *expected value act utilitarianism* – is, of course, very natural when the required probability distribution can be found. But there are many situations in which this information is either unavailable or meaningless; this is true, particularly, when the outcome resulting from an agent's action depends, not simply on a roll of the dice, but on the independent choice of another free agent. In the literature on decision theory, a situation in which the actions available to an agent might lead to their various possible outcomes with known probability is described as a case of *risk*, while a situation in which the probability with which the available actions might lead to their various possible outcomes is either unknown or meaningless is described as a case of *uncertainty*.[4]

---

[4] A discussion of this terminology can be found, for example, in Sections 2.1 and 13.1 of Luce and Raiffa (1957). Of course, the legitimacy of the distinction between uncertainty and risk is itself an issue: following Ramsey (1931) and Savage (1954), many writers in the Bayesian tradition assume that an agent's assessment of the possible outcomes in a given situation can always be represented through a probability measure, so that uncertainty always reduces to risk. However, there is an

Our concern here is with situations involving uncertainty, rather than risk, and we proceed by adapting a standard treatment of these situations from decision theory: since an ordering based on expected value is not possible, we instead define a notion of dominance that can be used to order the actions available to an agent.

11.3.2 Dominance Act Utilitarianism

We begin with a preference ordering on propositions, arbitrary sets of histories through a moment.

PREFERENCES ORDERING ON PROPOSITIONS: Let X and Y be propositions at a moment. Then $X \leq Y$ (Y is *weakly preferred* to X) if and only if Value$(h) \leq$ Value(h') for each $h \in X$ and each $h' \in Y$; and $X < Y$ (Y is *strongly preferred* to X) if and only if $X \leq Y$ and it is not the case that $Y \leq X$.

The idea is that, if Y is weakly preferred to X, each history from Y is at least as valuable as any history from X, so that we are sure to do at least as well in a history at which Y holds as we would in a history at which X holds. If Y is strongly preferred to X, then not only is each history from Y at least as valuable as any history from X, but some history from Y is actually more valuable than some history from X, so that we are not only sure to do at least as well with Y as with X, we might do better.

In the current framework, the actions available to an agent at a moment are reified as sets of histories through that moment. Each action is therefore a proposition, and so it is tempting to imagine that the dominance relations among actions might be identified with the preference orderings defined for propositions more generally. This idea is plausible, and there are a number of examples in which it seems to yield the correct results, including the earlier Fig. 11.4, where it tells us that neither of the two actions, K_1 or K_2, dominates the other. However, the suggestion of simply identifying the dominance orderings over an agent's actions with the preference orderings on propositions fails in more complicated cases.

To see this, consider Fig. 11.5, depicting a situation of simultaneous choice by two agents, and interpreted as follows. We suppose that the agent α is holding a nickel in its hand, and that at the moment m, the agent is faced with a choice between two actions: placing this nickel on a certain table either heads up, performing the action K_1, or tails up, performing the action K_2. At the same moment, the agent β must likewise choose between placing a dime on the table either heads up or tails up, performing either the action K_3 or the action K_4. If α places the nickel on the table heads up, then the resulting utility is 9 if β places the dime heads up and 4 if β places the dime tails up; but if α places the dime on the table tails up, the resulting utility is 10 if β places the dime heads up and 5 if β places the dime tails up.

---

important tradition of resistance to the assimilation of uncertainty and risk in a single numerical measure. A classic paper in this tradition is Ellsberg (1961); for more recent work on decision theory in situations that mix elements of risk and uncertainty, see the papers contained in Parts II and IV of Gärdenfors and Sahlin (1988).

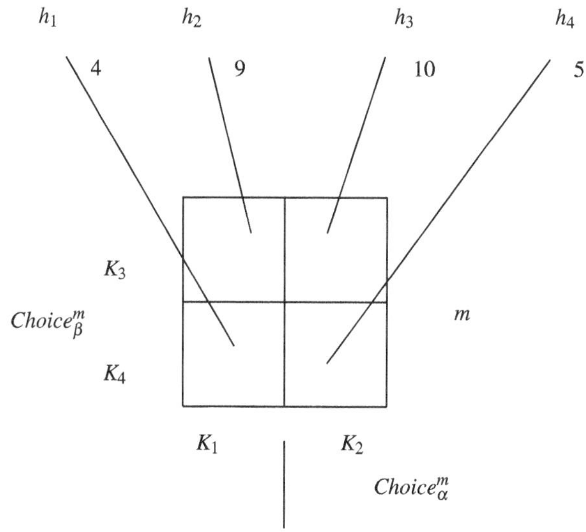

Fig. 11.5 The coin example

In this situation, neither of the two actions open to α is even weakly preferred to the other in the sense of the propositional ordering, since each contains an outcome more valuable than some outcome belonging to the other. Nevertheless, there is a persuasive argument in favor of the conclusion that K_2 is a better action than K_1 for α to perform: The agent β must place the dime on the table either heads up or tails up, performing either K_3 or K_4. So suppose, first, that β places the dime heads up, performing K_3. In that case, it is clearly better for α to place the nickel on the table tails up, performing K_2 rather than K_1, since the unique history h_3 belonging to $K_2 \cap K_3$ is more valuable than the unique history h_2 belonging to $K_1 \cap K_3$. Next, suppose that β places the dime tails up, performing K_4. Then it is again better for α to place the nickel on the table tails up, again performing K_2 rather than K_1, since the unique history h_4 belonging to $K_2 \cap K_4$ is more valuable than the unique history h_1 belonging to $K_1 \cap K_4$. In each of these two cases, then, it is better for α to perform K_2 rather than K_1, and since these cases exhaust the possibilities, a pattern of reasoning sometimes described as the *sure-thing principle* suggests that K_2 is simply a better action than K_1 for α to perform.[5]

The key to applying sure-thing reasoning in a given situation lies in identifying an appropriate partition of the possible outcomes into a set of *states* (sometimes called "states of nature" or "conditioning events"), against the background of which

[5] This pattern of reasoning is first explicitly characterized as the "sure-thing principle" in Savage (1954), but the principle appears already in some of Savage's earlier work, such as (Savage 1951, p. 58), where he writes concerning situations of uncertainty that "there is one unquestionably appropriate criterion for preferring some act to some others: If for every possible state, the expected income of one act is never less and is in some cases greater than the corresponding income of another, then the former act is preferable to the latter."

the actions available to an agent can then be evaluated through a state-by-state comparison of their results. This is often a difficult task, but we simplify in the current setting, not only by supposing that probabilistic information is unavailable, but also by imagining that the only sources of causality present are the actions of the various agents.

Given these assumptions, it is natural to identify the set of states confronting an agent α at the moment m – here abbreviated as $State_\alpha^m$ – with the possible patterns of action that might be performed at that moment by all other agents. In the case of Fig. 11.5, for example, if we assume that α and β are the only two agents – that is, Agent $= \{\alpha, \beta\}$ – then $State_\alpha^m$ can be identified with $Choice_\beta^m$, the set $\{K_3, K_4\}$ of actions available to β. Although we concentrate in this chapter on simple cases like this, with two agents at most, the definition of a state is more general. Where Agent contains an arbitrary group of agents, the set of agents other than α is Agent $- \{\alpha\}$, of course, and we can then define the set of states confronting α at m by stipulating that:

$$State_\alpha^m = Choice^m_{\text{Agent}-\{\alpha\}}.$$

Given this treatment of the states facing an agent, we can now define a dominance ordering on the actions available to the agent through a state-by-state comparison of their results. As an initial step, we must first specify a standard for comparing the possible results of two actions against the background of a particular state. The example depicted in Fig. 11.5 is deceptively simple in this regard, for in this situation, once a particular state from $State_\alpha^m$ is fixed, each action available to α then determines a unique outcome, so that these actions can simply be ranked along with their outcomes.

In the more general case, of course, even against the background of a fixed state, the actions available to an agent may determine only sets of outcomes, or propositions, rather than unique outcomes – but here, we can compare the results of different actions in a state by appealing to the preference ordering defined earlier on propositions. Where S is a state belonging to $State_\alpha^m$, and where K and K' are actions available to α at m, we can say that the results of K' are at least as good as those of K in the state S whenever $K \cap S \leq K' \cap S$ – whenever, that is, the proposition $K' \cap S$, determined by performing the action K' in the state S, is weakly preferred to the proposition $K \cap S$, determined by performing K in S.

With these various concepts in place, we are now in a position to define a dominance ordering on the actions available to an agent at a moment.

DOMINANCE ORDERING ON ACTIONS: Let α be an agent and m a moment, and let K and K' be members of $Choice_\alpha^m$. Then $K \preceq K'$ (*K' weakly dominates K*) if and only if $K \cap S \leq K' \cap S$ for each state $S \in State_\alpha^m$; and $K \prec K'$ (*K' strongly dominates K*) if and only if $K \preceq K'$ and it is not the case that $K' \preceq K$.

The idea is that, K' weakly dominates K, then the results of performing K' are at least as good as those of performing K in every state, so that, no matter which state is realized, the agent is sure to do at least as well with K' as with K. If K' strongly dominates K, then not only are the results of performing K' at least as good as those of performing K in every state, but there is some state in which K' yields better results, so that the agent is sure to do at least as well with K' as with K, and might do better.

Let us now return to our central question: how, in this indeterminist setting, can we define the utilitarian notion of a right action? The dominance account provides an answer that is both precise and intuitively plausible.

We begin by defining the set $Optimal_\alpha^m$ containing the *optimal actions* available to an agent α at a moment m, those actions available to the agent that are not strongly dominated by any others:

$$Optimal_\alpha^m = \{K \in Choice_\alpha^m : \neg\exists K' \in Choice_\alpha^m \,(K \prec K')\}.$$

It is then natural formulate a theory that might be characterized as *dominance act utilitarianism* simply by identifying the right actions available to an agent at a moment with the optimal actions.

> DOMINANCE ACT UTILITARIANISM: Let α be an agent and m a moment, and suppose $K \in Choice_\alpha^m$. Then the action K is *right* at the moment m if and only if $K \in Optimal_\alpha^m$, and *wrong* otherwise.

The theory can be illustrated with our earlier examples. In the case of Fig. 11.4, we have $Optimal_\alpha^m = \{K_1, K_2\}$, so that both actions available to the agent at the moment m are right. In the case of Fig. 11.5, we have $Optimal_\alpha^m = \{K_2\}$, so that K_2 is right and K_1 is wrong.

11.4 The Orthodox Account

11.4.1 An Example

This theory of dominance act utilitarianism is, I suspect, not too surprising. It is perhaps even obvious. The underlying ideas of dominance and optimality are familiar from decision theory, generalized only slightly to allow for the fact that an action in a state yields a proposition, rather than a unique outcome.

What may be surprising, however – and particularly if the dominance theory does seem to be obvious – is the fact that the treatment of utilitarianism within the ethical literature does not follow this dominance account at all, but is based on an entirely different approach, which I will refer to, in deference to the literature, as the *orthodox* account.

In order to illustrate this orthodox account, let us consider an example that has figured prominently in the discussion of different forms of utilitarianism. Although the example was first introduced by Gibbard (1965), and was elaborated on shortly thereafter by Sobel (1968), I take the later but more extensive discussion by Regan (1980) as my primary source:

> Suppose that there are only two agents in the moral universe, called Whiff and Poof. Each has a button in front of him which he can push or not. If both Whiff and Poof push their buttons, the consequences will be such that the overall state of the world has a value of ten units. If neither Whiff nor Poof pushes his button, the consequences will be such that the overall state of the world has a value of 6 units. Finally, if one and only one of the pair pushes his button (and it does not matter who pushes and who does not), the consequences will be such that the overall state of the world has a value of 0 (zero) units. Neither agent, we assume, is in a position to influence the other's choice. (Regan 1980, p. 19)

In the present framework, this example can be depicted as in Fig. 11.6, where α represents Whiff, β represents Poof, and m is the moment at which each of these two agents must choose whether or not to push his button.[6] The action K_1 represents Whiff's option of pushing his button, and K_2 his option of refraining; likewise, K_3 and K_4 represent Poof's options of pushing or refraining; and the possible outcomes resulting from the choices by these agents are represented by the histories h_1 through h_4, which are assigned the values indicated in Regan's description.

Now, when the example is set out in this way, it is easy to see that both agents will satisfy our previous theory of dominance act utilitarianism no matter what they do. Neither action available to either agent is dominated, and so we have both $Optimal^m_\alpha = \{K_1, K_2\}$ and $Optimal^m_\beta = \{K_3, K_4\}$. Since both of the actions K_1 and K_2 available to Whiff are optimal, both are right according to the dominance theory; and both of the actions K_3 and K_4 available to Poof are right as well.

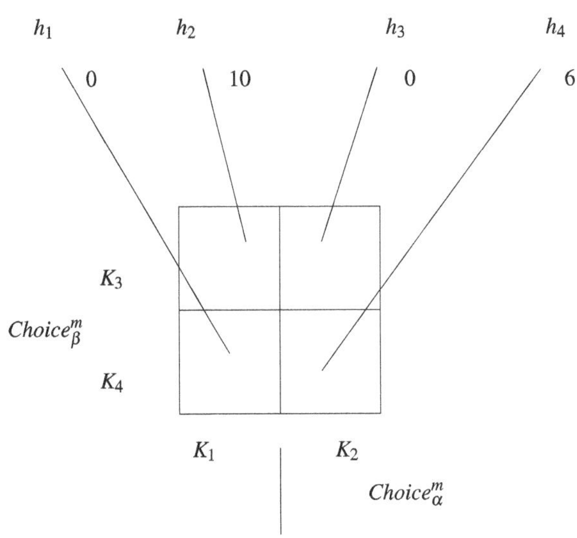

Fig. 11.6 Whiff and Poof

[6] Regan does not actually require that these choices must be simultaneous (though simultaneity is part of Gibbard's earlier description), but he does require the choices to be independent, and we guarantee independence through simultaneity.

The theory of dominance act utilitarianism, then, yields results that are at least definite in this case, even if not particularly constraining: each of the two agents can satisfy the theory by selecting either of the available actions. However, Regan's own conclusions – based on his own theory of act utilitarianism or, as he calls it, AU – are strikingly different:

> Now, if we ask what AU directs Whiff to do, we find that we cannot say. If Poof pushes, then AU directs Whiff to push. If Poof does not push, then AU directs Whiff not to push. Until we specify how Poof behaves, AU gives Whiff no clear direction. The same is true, *mutatis mutandis*, of Poof. (Regan 1980, p. 18)

In saying that act utilitarianism gives Whiff no clear direction, Regan does not mean only that this theory, like the dominance theory, classifies multiple actions as right, allowing the agents to choose among them. Instead, he means that, on the basis only of the information provided so far, the theory is simply unable to generate any results at all: no actions can be classified either as right or as wrong. In order to arrive at a situation in which act utilitarianism is able to yield definite results, Regan feels that it is necessary to supplement the description of the example provided so far, and depicted in Fig. 11.6, with additional information concerning the actions actually performed by the individuals involved:

> If we shift our attention to patterns of behavior for the pair, we can decide whether each agent satisfies AU in any specified pattern. (Regan 1980, p. 18)

And he illustrates the kind of reasoning allowed by this additional information as follows:

> Suppose, for example, Whiff and Poof both push their buttons. The total value thereby achieved is ten units. Does Whiff satisfy AU? Yes. The only other thing he might do is not push his button. But under the circumstances, which include the fact that Poof pushes his button, Whiff's not pushing would result in a total utility of zero. Therefore Whiff's pushing his button has at least as good consequences as any other action available to him under the circumstances. Therefore, it is right according to AU (Regan 1980, pp. 18–19)

11.4.2 Orthodox Act Utilitarianism

Evidently, Regan is unwilling to classify actions as right or wrong absolutely, but only as right or wrong in particular circumstances. That is fair enough. But Regan, following Gibbard and Sobel, also takes the further, and more contentious step of supposing that an agent's circumstances must include whatever actions are simultaneously performed by other agents – so that he is unwilling to classify the actions available to Whiff and Poof as either right or wrong absolutely, but only as right or wrong under the circumstances determined by the actions of the other.[7]

---

[7] Gibbard adopts a similar viewpoint in his original discussion of this example, evaluating each agent's selection only under an assumption about the action selected by the other (Gibbard 1965, p. 215). And Sobel defends Gibbard's strategy as follows: "It is perhaps natural to feel that Gibbard's first case is objectionable just because it includes assumptions concerning what agents will

How can we represent the theory of act utilitarianism that guides Regan's judgments? In my Horty (2001), I adopted a strategy, which still seems reasonable to me, and which I review here, of first introducing a concept of conditional optimality, and then conditionalizing on a proposition that represents the agent's circumstances.

The concept of conditional optimality is introduced in three steps. First, taking X as a proposition, the set of actions available to an agent α at m under the condition that X holds – expressed here as $Choice^m_\alpha/X$ – is simply the set containing those actions open to α at m that are consistent with X:

$$Choice^m_\alpha/X = \{K \in Choice^m_\alpha : K \cap X \neq \emptyset\}.$$

The next step is to generalize our earlier treatment of dominance to include conditional dominance.

CONDITIONAL DOMINANCE ORDERING ON ACTIONS: Let α be an agent and m a moment, and let K and K' be members of $Choice^m_\alpha$, and X a proposition. Then $K \preceq_X K'$ (K' *weakly dominates* K *under the condition* X) if and only if $K \cap X \cap S \leq K' \cap X \cap S$ for each state $S \in State^m_\alpha$; and $K \prec_X K'$ (K' *strongly dominates* K *under the condition* X) if and only if $K \preceq_X K'$ and it is not the case that $K' \preceq_X K$.

This conditional analysis follows the pattern of the absolute treatment set out earlier, except that, in comparing the results of two actions K and K' in a given state S, our attention is now restricted only to those outcomes that are consistent with the background proposition X.

Finally, having generalized both choice and dominance to the conditional setting, we can now combine these ideas to arrive at a concept of conditional optimality. Again taking X as a proposition, we define the set of optimal actions available to α at m under the condition X – expressed as $Optimal^m_\alpha/X$ – to be the set of those actions available to α at m under the condition X that are not strongly dominated under this condition by any other such action:

$$Optimal^m_\alpha/X = \{K \in Choice^m_\alpha/X : \neg\exists K' \in Choice^m_\alpha/X\ (K \prec_X K')\}.$$

It is easy to verify, but worth noting explicitly that the conditional notions of choice, dominance, and optimality introduced here are, in fact, generalizations of our earlier concepts. When the background condition X is identified with the trivial proposition H_m – that is, when $X = H_m$ – each of these three conditional notions coincides with its absolute counterpart. In particular, we have

$$Optimal^m_\alpha/H_m = Optimal^m_\alpha;$$

the actions available to α at m that are optimal under the condition that the trivial proposition holds are simply the optimal actions.

---

and would do. But this can be no objection since it is obvious that such assumptions are essential to the application of AU; without such assumptions the dictates of AU could not be determined ..." (Sobel 1968, p. 152).

Now that the notion of conditional optimality has been introduced, it remains only to define the propositions on which we conditionalize.[8] Just as $Choice^m_\alpha/X$ represents the set of actions available to α at m that are consistent with X, we can likewise define

$$State^m_\alpha/X = \{K \in State^m_\alpha : K \cap X \neq \emptyset\}$$

as the set of states confronting α at m that are consistent with X. And in this case, it is also convenient to represent the proposition formed by taking the union of these states – the proposition, that is, according to which one of these states holds – written $State^m_\alpha(X)$ and defined as follows:

$$State^m_\alpha(X) = \bigcup State^m_\alpha/X.$$

To illustrate this notation, suppose in the case of Fig. 11.6, the Whiff and Poof example, that the proposition $X = \{h_2, h_4\}$. Then $State^m_\alpha/X = \{K_3, K_4\}$ is the set of states confronting α at m that are consistent with this proposition, and $State^m_\alpha(X) = K_3 \cup K_4$ therefore represents the proposition that one of these states obtains.

In the special case in which $X = \{h\}$ is a maximally specific proposition, containing only a single history, we write $State^m_\alpha/h$ and $State^m_\alpha(h)$ for convenience; and here, $State^m_\alpha/h$ is a unit set containing the unique state consistent with that history, and $State^m_\alpha(h)$ is simply this unique state itself. Thus, for example, again in the case of Fig. 11.6, we have $State^m_\alpha/h_2 = \{K_3\}$ and so $State^m_\alpha(h_2) = K_3$.

With these concepts before us, we can now, as in Horty (2001), define a form of act utilitarianism designed to model the orthodox notion found in the work of Gibbard, Sobel, Regan, and others.

ORTHODOX ACT UTILITARIANISM: Let α be an agent and m a moment, and suppose $K \in Choice^m_\alpha$. Then the action K is *right* at the index m/h if and only if $K \in Optimal^m_\alpha/State^m_\alpha(h)$, and *wrong* otherwise.

What the definition tells us, then, is simply that the action K is right at the index m/h whenever K is optimal under the condition specified by the state containing the history h.

Returning to the Whiff and Poof example, let us consider, for example, the index m/h_2, where both Whiff and Poof push their buttons. At this index, the situation confronting Whiff, determined by Poof's action, is K_3; that is, $State^m_\alpha(h_2) = K_3$. We therefore have $Optimal^m_\alpha/State^m_\alpha(h_2) = Optimal^m_\alpha/K_3$. And it is easy to verify also that $Optimal^m_\alpha/K_3 = \{K_1\}$, so that the action K_1 is classified as right at m/h_2. In the same way, however, we can see that $Optimal^m_\alpha(h_1) = \{K_2\}$, so that the action K_1 is classified as wrong at the index m/h_1.

As this example shows, the orthodox classification of actions as right or wrong – in contrast to the dominance account – depends on a full index, not just a moment.

[8] These definitions may seems to be needlessly general, but please bear with me; the generality will help us later on.

Here, the same action, K_1, is classified as right at the index m/h_2 but wrong at the index m/h_1; although Whiff performs the same action at each of these two indices, this agent satisfies orthodox act utilitarianism at the first, performing an action that is classified as right, but not at the second. It is as Regan says: we cannot define which of an agent's actions are right or wrong until we know the circumstances under with the action is performed – that state confronting that agent, here defined as the actions simultaneously performed by the other agents involved.

11.5 The Perspectival Account

11.5.1 A Problem

At this point, we have before us two accounts of right action, dominance and orthodox. In order to compare these accounts, I now want to introduce yet another example, which I have found to be especially helpful in highlighting their differences.[9]

Imagine that two drivers are traveling toward each other on a one-lane road, with no time to stop or communicate, and with a single moment at which each must choose, independently, either to swerve or to continue along the road. There is only one direction in which the drivers might swerve, and so a collision can be avoided only if one of the drivers swerves and the other does not; if neither swerves, or both do, a collision occurs. This example is depicted in Fig. 11.7, where α and β repre-

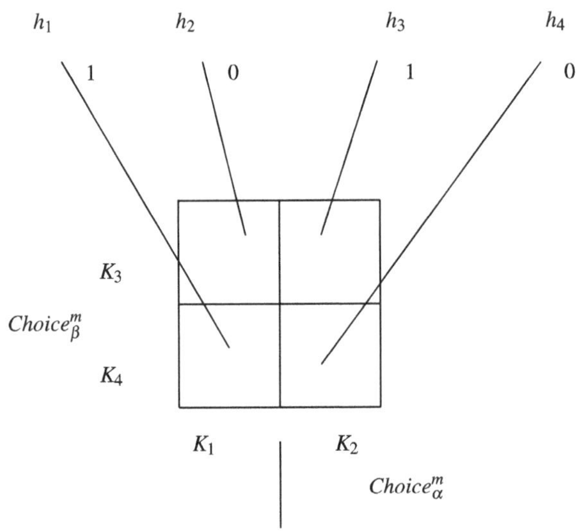

Fig. 11.7 The driving example

[9] The example is due to Goldman (1976), but also discussed by Humberstone (1983), a paper that sets out in a different context some of the fundamental ideas underlying the orthodox account.

sent the two drivers, K_1 and K_2 represent the actions available to α of swerving or continuing along the road, K_3 and K_4 likewise represent the swerving or continuing actions available to β, and m represents the moment at which α and β must make their choice. The histories h_1 and h_3 are the ideal outcomes, resulting when one driver swerves and the other does not; collision is avoided. The histories h_2 and h_4, resulting either when both drivers swerve or both continue along the road, are nonideal outcomes in which a collision occurs.

Now imagine that what actually happens is that both agents continue along the road, so that the resulting outcome is the history h_4, in which there is a collision. Suppose that, looking back at the situation from some later moment belonging to h_4 – perhaps while recovering in the hospital – the agent α says to itself: I performed the wrong action; it would have been right to swerve. And let us ask: is what the agent says correct, or not? The answer, I think, is that we can legitimately understand this statement either as correct or as incorrect, and that the contrast between these two different readings can be captured by appeal to our distinction between the orthodox and dominance accounts of right action.

On the one hand, it is clear from the standpoint of the later moment that, if the agent had swerved, there would have been no collision. Things would have gone much better for everyone had the agent swerved, and therefore, from a utilitarian point of view, the agent was wrong not to. This way of understanding the agent's statement is captured by the orthodox account, according to which $\mathit{Optimal}^m_\alpha/\mathit{State}^m_\alpha(h_4) = \{K_1\}$, so that the action K_1 is classified as right and the action K_2, which the agent actually performed, as wrong at the index m/h_4. On the other hand, if we consider the situation from the standpoint of the earlier moment m, when the agent's action was actually performed, it is hard to see how we could have said at this moment that it would be right for the agent to swerve and wrong not to. Surely there is nothing in the situation as it appears at this moment – with the four histories each lying ahead as future possibilities – that could justify such a judgment. This way of evaluating the agent's statement is captured by the dominance account, according to which $\mathit{Optimal}^m_\alpha = \{K_1, K_2\}$, so that both actions are classified as right at the moment m.

The situation pictured in Fig. 11.7, then, seems to support two different evaluations of the agent's decision not to swerve – that it was wrong, or right – which can then be captured by our two theories of right and wrong actions, orthodox and dominance. This idea, originally set out in Horty (2001), of analyzing examples of this kind by appeal to two separate utilitarian theories carries some distinct advantages. It does not force us into the artificial position of classifying the agent's action either as unequivocally right or as unequivocally wrong, ignoring the pull of the opposite intuition. However, in allowing us this freedom, it also does not lead us into the muddled position of describing the action as somehow both right and wrong. What we can say, instead, is that the agent's action is right in one definite sense and wrong in another – that it is right in the dominance sense, but wrong in the orthodox sense.

Although I do not have space (or time) to justify this claim here, I believe that the contrast apparent in this example between the two different ways of evaluating the agent's action can be seen as underlying many of the debates in utilitarian theory that commanded so much attention during the 1970 s and 1980 s. One of these, about which I will say nothing here at all, is the debate over the "actualist" and "possibilist" positions regarding the relations between an agent's present obligations and future choices.[10] Another, about which I will say only a bit more in the appendix to this chapter, is the problem that occupied Gibbard, Sobel, Regan, and others concerning the relation between individual and group utilitarian theories.[11] The current proposal therefore has the real benefit of providing a rigorous explication of two different ways of understanding our normative evaluation that can be felt both in the example presented here and also, I believe, in other cases from the literature on utilitarian theory.

This benefit, however, comes with a cost. The cost is that the way in which the current proposal allows us to treat the same action as both right and wrong, respecting our conflicting intuitions, is by offering two different theories of right and wrong action. In effect, the proposal treats the words "right" and "wrong" as carrying two different senses, two different meanings. Of course, philosophy often proceeds like this, by discovering hidden ambiguities in items of ordinary language, which are then teased apart and provided with different formal explications. But in this case the idea simply seems wrongheaded. It is hard to think of these words as *semantically* ambiguous.

I now want to show that there is a better way. We can preserve the benefits of the account presented here, allowing appeal to both the orthodox and dominance perspectives in evaluating an agent's actions, without postulating semantic ambiguity, by relying instead on a pragmatic difference.

11.5.2 Perspectival Act Utilitarianism

The basic idea is that an action performed by an agent at one moment is to be evaluated as right or wrong from the perspective of another moment, which may or may not be identical to the first. The key component of this idea – the appeal to "double time reference" – was first set out systematically by Belnap (2001), with an emphasis on the assessment of speech acts, particularly the speech act of assertion.[12] It was later developed in a somewhat different way by MacFarlane (2003) and elsewhere, who is concerned with the role of perspective in the assessment of a statement's content: what is said, rather than the act of saying it.

---

[10] This problem was originally presented in a trio of papers: Goldman (1976), Sobel (1976), and Thomason (1981). Further discussion can be found, for example, in Bergström (1977), Carlson (1995), Feldman (1986), Goldman (1978), Greenspan (1978), Humberstone (1983), Jackson (1985), Jackson (1988), Jackson and Pargetter (1986), McKinsey (1979), and Zimmerman (1990).

[11] In addition to the work by Gibbard, Sobel, and Regan cited earlier, further discussion of this issue can be found in Carlson (1995), Feldman (1986), Jackson (1987), Jackson (1988), and of course Parfit (1984).

[12] Further discussion can be found at various points throughout Belnap et al. (2001) (see index entries under "double time reference"), and an informal presentation appears in Belnap (2004).

Let us take m as the *moment of action* and m' as the moment from which the action selected at m is evaluated – the *moment of evaluation*, which we can sensibly assume to be comparable to m in the treelike ordering of moments: either later than, earlier than, or identical with m. In that case, $State^m_\alpha/H_{m'}$ – the set of states consistent with $H_{m'}$, the trivial proposition at m' – can be taken to represent the states confronting the agent at m, as judged from the standpoint of m'. As we have seen, $State^m_\alpha(H_{m'})$ is simply the proposition that one of these states holds. And so the set $Optimal^m_\alpha/State^m_\alpha(H_{m'})$ contains those actions available to the agent at m that are optimal under the conditions in which the agent finds itself, where these conditions are themselves judged from the standpoint of m'.

Using these ideas, we can therefore define *perspectival act utilitarianism* as the theory according to which an action available to the agent α at the moment m is right from the standpoint of the moment m' just in case that action is optimal given the states that the agent is confronting at m, as judged from the standpoint of m'.

PERSPECTIVAL ACT UTILITARIANISM: Let α be an agent and m and m' moments such that either $m < m'$ or $m' < m$ or $m = m'$, and suppose $K \in Choice^m_\alpha$. Then the action K is *right* at m from the standpoint of m' if and only if $K \in Optimal^m_\alpha/State^m_\alpha(H_{m'})$, and *wrong* otherwise.

This perspectival account allows us to capture the intuitions underlying the orthodox approach, as we can see by returning to the driving example. Suppose, again, that neither driver swerves, the crash occurs, and we are considering the incident from the standpoint of some later moment – call it m_1 – lying on the history h_4. Since m_1 lies on the history h_4 at some time later than m, and h_4 itself belongs to the state K_4, it follows that each history from H_{m_1}, the set of histories passing through m_1, must likewise belong to K_4.[13] From this is follows that K_4 is the only state confronting α at m that is consistent with H_{m_1} – that is, $State^m_\alpha/H_{m_1} = \{K_4\}$; the set of states confronting the agent at m, as judged from the standpoint of m_1, contains K_4 alone. From this is follows that $State^m_\alpha(H_{m_1}) = K_4$. We therefore have

$$\begin{aligned} Optimal^m_\alpha/State^m_\alpha(H_{m_1}) &= Optimal^m_\alpha/K_4 \\ &= \{K_1\}, \end{aligned}$$

so that, from the standpoint of m_1, we reach the orthodox judgment that the action K_2 chosen by the agent was wrong and K_1 would have been right, optimal under the circumstances in which the agent found itself.

On the other hand, suppose that, at the crucial moment, both drivers swerve, another crash occurs, things proceed along the history h_2, and that we are now reflecting on the incident from some later moment – say m_2 – lying on that history. Parallel reasoning thus gives us $State^m_\alpha(H_{m_2}) = K_3$, from which we can conclude, just as before, that $Optimal^m_\alpha/State^m_\alpha(H_{m_2}) = \{K_2\}$. From the standpoint of m_2,

[13] Although this point is "visually obvious," it actually relies on the technical constraint of "no choice between undivided histories," not discussed in this chapter, according to which histories that are still undivided at a given moment cannot be separated at that moment by the *Choice* partition.

then, we conclude that the action K_1 was wrong and K_2 would have been right. A different point of evaluation leads to a different result.

The perspectival approach, then, allows us to recover the intuitions underlying orthodox act utilitarianism, but interestingly, it subsumes the dominance account as well. This can be seen to hold quite generally. Suppose that the actions available to an agent at the moment m are evaluated from the standpoint of a moment m' that is either identical with or earlier than the moment m itself: $m' = m$ or $m' < m$. Then it is easy to see that each member of $State_\alpha^m$ contains some history from $H_{m'}$, so that the set of states confronting the agent at m, judged from the standpoint of m', is simply $State_\alpha^m$ itself: $State_\alpha^m / H_{m'} = State_\alpha^m$. From this it follows, since $State_\alpha^m$ partitions the set H_m, that $State_\alpha^m(H_{m'}) = H_m$. As noted earlier, the set $Optimal_\alpha^m / H_m$, containing those actions available to α at m that are optimal under the conditions specified by the trivial proposition, coincides with the set $Optimal_\alpha^m$ itself. It therefore follows that

$$\begin{aligned} Optimal_\alpha^m / State_\alpha^m(H_{m'}) &= Optimal_\alpha^m / H_m \\ &= Optimal_\alpha^m, \end{aligned}$$

so that the set of actions available at m that are right from the standpoint of m' coincides with the set of actions available at m that are right according to the dominance account.

This general point can be illustrated with our driving example, Fig. 11.7, if we suppose that m' is some moment of evaluation identical with or earlier than the moment m of action. In that case, we have

$$\begin{aligned} Optimal_\alpha^m / State_\alpha^m(H_{m'}) &= Optimal_\alpha^m / H_m \\ &= Optimal_\alpha^m \\ &= \{K_1, K_2\}. \end{aligned}$$

Taking such a moment m' as our moment of evaluation, then, we arrive at the dominance intuition that either action available to the agent at m is right.

11.6 Conclusion

The theory of perspectival act utilitarianism set out here allows us see how we can say, in the driving example, for instance, that the agent's actions at the crucial moment might legitimately be viewed as both right and wrong. The theory thus preserves the advantages of my earlier account, from Horty (2001), by allowing us to respect our conflicting intuitions in cases like this. But it does not do so by postulating two separate senses of the words "right" and "wrong" – an orthodox and a dominance sense – captured by two separate utilitarian theories. These words can now be taken as semantically unambiguous.

When we say that an agent's action is right, from the standpoint of some moment of evaluation, we always mean exactly the same thing: the action is optimal under the conditions in which the agent finds itself at the moment of action, where these conditions are themselves judged from the standpoint of the moment of evaluation. Our conflicting intuitions about right and wrong can now be provided with a pragmatic, rather than a semantic, explanation, shifting with the relation between moment of action and moment of evaluation, and reflecting different evaluative judgments about the conditions confronting the agent at the moment of action. If the moment if evaluation is strictly later than the moment of action, then the perspectival theory agrees with orthodox act utilitarianism. But if the evaluation takes place at the very moment of action, or earlier, the perspectival theory agrees with dominance act utilitarianism. The difference between our orthodox and dominance intuitions is not, therefore, a substantial difference that needs to be explained by postulating two separate utilitarian theories, but only a matter of perspective.

Appendix: Act Utilitarianism for Groups

This appendix shows how perspectival act utilitarianism can be extended from individual actions to group actions, and how the relation between the right actions available to groups and individuals can then be seen to depend on the standpoint from which these actions are evaluated.

The extension of perspectival act utilitarianism to group actions is straightforward, involving nothing more than a generalization of several of our previous notions. We have already seen, in the text, how the set $Choice^m_\Gamma$ of actions available to the group Γ at the moment m can be defined, with each group action identified as a pattern of actions available to the individuals from that group. The states confronting the group Γ at m can then be defined as the patterns of actions available at m to all agents except those from that particular group:

$$State^m_\Gamma = Choice^m_{\text{Agent}-\Gamma}$$

And where X is some proposition, weak and strong dominance relations under the condition X can be defined among the actions available to a group in a way exactly parallel to the definition for individual actions.

> CONDITIONAL DOMINANCE ORDERING ON GROUP ACTIONS: Let Γ be a group of agents and m a moment, and let K and K' be members of $Choice^m_\Gamma$, and X a proposition. Then $K \preceq_X K'$ (*K' weakly dominates K under the condition X*) if and only if $K \cap X \cap S \leq K' \cap X \cap S$ for each state $S \in State^m_\Gamma$; and $K \prec_X K'$ (*K' strongly dominates K under the condition X*) if and only if $K \preceq_X K'$ and it is not the case that $K' \preceq_X K$.

The set of actions available to the group Γ under the condition X can be defined as those among the available actions that are consistent with this condition:

$$Choice^m_\Gamma/X = \{K \in Choice^m_\Gamma : K \cap X \neq \emptyset\}.$$

And the optimal actions available to the group under this condition can then be defined as the actions available under this condition that are not dominated under this condition by any other such actions:

$$Optimal^m_\Gamma/X = \{K \in Choice^m_\Gamma/X : \neg\exists K' \in Choice^m_\Gamma/X\ (K \prec_X K')\}.$$

Finally, the set of states confronting Γ at m that are consistent with the proposition X can be represented just as before:

$$State^m_\Gamma/X = \{K \in State^m_\Gamma : K \cap X \neq \emptyset\}.^{14}$$

And likewise the proposition that one of these states holds:

$$State^m_\Gamma(X) = \bigcup State^m_\Gamma/X.$$

Given these materials, we can now introduce a form a perspectival act utilitarianism for groups, according to which an action available to the group Γ at a moment m is right from the standpoint of the moment m' just in case that action is optimal under the conditions in which the group Γ finds itself at m, where these conditions are judged from the standpoint of m':

PERSPECTIVAL ACT UTILITARIANISM FOR GROUPS: Let Γ be a group of agents and m and m' moments such that either $m < m'$ or $m' < m$ or $m = m'$, and suppose $K \in Choice^m_\Gamma$. Then the action K is *right* at m from the standpoint of m' if and only if $K \in Optimal^m_\Gamma/State^m_\Gamma(H_{m'})$, and *wrong* otherwise.

As with individual actions, this perspectival account supports the orthodox intuitions concerning group actions when the moment m' of evaluation is later than the moment m of action, while the dominance intuitions are supported when m' is earlier than or identical with m.

Now that the perspectival account has been extended from individuals to groups, let us turn briefly to two of the most central questions concerning the relation between individual and group act utilitarianism. First, if each individual belonging to a group performs a right action, does that entail that the group itself performs a right action? And second, if a group performs a right action, does that entail that the individuals belonging to the group do so?

The answer to the first question is No. This fact is well-known and can be illustrated with the Whiff and Poof example from Fig. 11.6, which was originally formulated to make exactly this point. Still, it is useful to consider the question separately from the dominance and orthodox perspectives, since the contours of this negative answer differ.

---

[14] In the group case, this fact actually follows from the previous definitions of $State^m_\Gamma$ as the set of states confronting Γ at m and $Choice^m_\Gamma/X$ as the actions available to Γ that are consistent with X. However, it is set out separately here in order to conform to our treatment of the individual case, where the corresponding notion must be introduced through a definition.

Suppose, first, that we evaluate the actions available at the moment m in Fig. 11.6 from the standpoint of m itself, adopting the dominance perspective. Then it is easy to verify that each action available to either agent is classified as right from the standpoint of m. So suppose that Whiff pushes his button, performing the action K_1, while Poof refrains, performing K_4 – each agent therefore performing an action that is classified as right. Then the group $\Gamma = \{\alpha, \beta\}$ containing both Whiff and Poof performs the action $K_1 \cap K_4$, which is clearly non-optimal, leading to a utility of 0 while 10 is possible, and so classified as wrong from the standpoint of m. Indeed the group action $K_1 \cap K_4$ is not even in equilibrium: each agent would be better off with a different choice, given the action chosen by the other. Individual satisfaction of dominance act utilitarianism, then, not only fails to guarantee group satisfaction, but has the even more depressing consequence that the pattern of actions chosen, each right from an individual perspective, may not be an equilibrium pattern.

Next, suppose Whiff and Poof both refrain from pushing their buttons, performing the individual actions K_2 and K_4. The outcome of this pair of actions is the history h_4. So let us evaluate these actions from the standpoint of some later moment along this history, thus adopting the orthodox perspective. It is easy to see that both of these actions are then classified as right from the standpoint of this later moment, and also that the pair of actions is in equilibrium: each agent is performing a best available action, given the actions performed by the other.

This example illustrates the general rule: whenever each individual member of a group of agents performs an action that is right from the standpoint of a later moment – and so right from the orthodox perspective – the pattern of actions performed by the entire group is in equilibrium. However, this does not mean that the group action is itself right. In this case, the group action $K_2 \cap K_4$ is non-optimal, and so wrong, since it yields a utility of 6 while the available group action $K_1 \cap K_3$ yields a utility of 10. If each member of a group performs an action that is right from the standpoint of a later moment, then, the overall pattern of actions will be in equilibrium, but it still may not be a right action for that group to perform, since there may be better equilibrium patterns.

Now to the second question: if a group action is right, does it follow from this that the actions of the individuals belonging to that group are also right? The standard answer to this question is Yes. Regan, for example, writes that "for any group of agents in any situation, any pattern of behaviour by that group of agents in that situation which produces the best consequences possible is a pattern in which the members of the group all satisfy AU" (Regan 1980, p. 54). And Jackson, that "if the right group action is actually performed, then that group action's constituent individual actions must be right" (Jackson 1988, p. 264). In the case of this question, however, the dominance and orthodox perspectives yield different answers.

Both Regan and Jackson adopt the orthodox perspective, evaluating actions from the standpoint of a later moment, and from that perspective what they say is right. In our current language, it can be put like this: if a group action performed at m is right from the standpoint of a later moment m', then the actions performed by the individual members of that group are also right from the standpoint of m'.

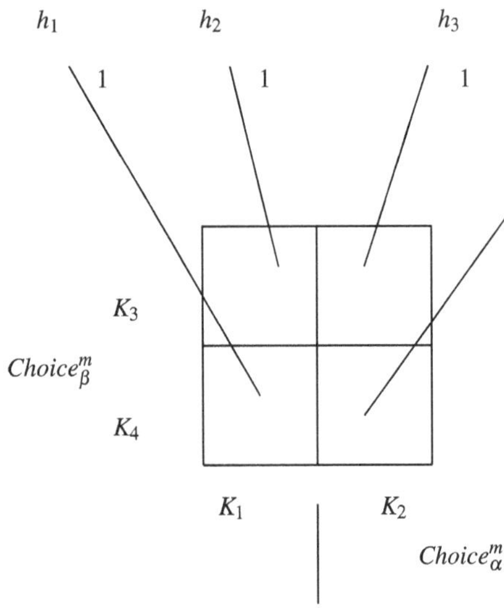

Fig. 11.8 Group action right, individual action wrong

However, the implication fails if we consider the matter from the dominance perspective, evaluating actions from the standpoint of a moment at or before the moment of their performance: where m' is identical with or earlier than m, it might well be possible that a group action performed at m is right from the standpoint of m', while the individual action of some member of that group is wrong from the standpoint of m'. This possibility is illustrated in Fig. 11.8. Here, it is easy to see that the action $K_2 \cap K_3$ performed at the moment m by the group $\Gamma = \{\alpha, \beta\}$ is right from the standpoint of the moment m itself, since this group action leads to an outcome of utility 1, the highest available, and is therefore optimal. But the component action K_2 by the agent α is wrong from the standpoint of m, since it is dominated by K_1. Of course, from the standpoint of some future moment along the history h_3, we can see the action K_2 by α was performed under circumstances in which β performed the action K_3, so that an outcome of utility 1 was achieved; from this later standpoint, the action K_2 is therefore right. But at the moment m itself, while it is still unclear which action β will perform, the choice of K_2 allows for an outcome of utility 0, and is therefore dominated by K_1, which guarantees an outcome of utility of 1.

References

Belnap N (2001) Double time references: speech act reports as modalities in an indeterministic setting. In: Wolter F, Wansing H, de Rijke M, Zakharyaschev M (eds) Advances in modal logic, vol 3, CSLI Publications, Stanford pp 1–21

Belnap N (2004) Future contingents and the sea battle tomorrow, manuscript, Philosophy Department, University of Pittsburgh

Belnap N, Perloff M, Xu M (2001) Facing the future: Agents and choices in our indeterminist world. Oxford University Press, Oxford
Bergström L (1966) The alternatives and consequences of actions, Stockholm Studies in Philosophy, vol 4. Almqvist and Wiksell, Stockholm
Bergström L (1977) Utilitarianism and future mistakes. Theoria 43:84–102
Carlson E (1995) Consequentialism reconsidered, theory and decision library, series A: Philosophy and methodology of the social sciences, vol 20. Kluwer Academic Publishers, Dordrecht
Ellsberg D (1961) Risk, ambiguity, and the Savage axioms. Quarterly Journal of Economics 75:643–669
Feldman F (1986) Doing the best we can: An essay in informal deontic logic. D. Reidel Publishing Company, Dordrecht
Gärdenfors P, Sahlin NE (eds) (1988) Decision, probability, and utility: Selected readings. Cambridge University Press, Cambridge
Gibbard A (1965) Rule-utilitarianism: merely an illusory alternative? Australasian Journal of Philosophy 43:211–220
Goldman H (1976) Dated rightness and moral imperfection. The Philosophical Review 85:449–487
Goldman H (1978) Doing the best one can. In: Goldman AI, Kim J (eds) Values and morals, D. Reidel Publishing Compamy, Dordrecht, pp 185–214
Greenspan P (1978) Oughts and determinism: a response to Goldman. Philosophical Review pp 77–83
Horty J (2001) Agency and deontic logic. Oxford University Press, Oxford
Humberstone IL (1983) The background of circumstances. Pacific Philosophical Quarterly 64:19–34
Jackson F (1985) On the semantics and logic of obligation. Mind 94:177–195
Jackson F (1987) Group morality. In: Pettit P, Sylvan R, Norman J (eds) Metaphysics and morality: Essays in honour of J. J.C. Smart, Basil Blackwell Inc., Oxford pp 91–110
Jackson F (1988) Understanding the logic of obligation. In: Proceedings of the Aristotelian Society, Supplementary Volume 62, Harrison and Sons
Jackson F, Pargetter R (1986) Oughts, options, and actualism. Philosophical Review 99:233–255
Luce RD, Raiffa H (1957) Games and decisions. John Wiley and Sons, New York
MacFarlane J (2003) Future contingents and relative truth. Philosophical Quarterly 53:321–336
McKinsey M (1979) Levels of obligation. Philosophical Studies 35:385–395
Parfit D (1984) Reasons and persons. Oxford University Press, Oxford
Prior A (1956) The consequences of actions. In: Proceedings of the Aristotelian Society, Supplementary Volume 30, Harrison and Sons
Prior A (1967) Past, present, and future. Clarendon Press, Oxford
Ramsey F (1931) Truth and probability. In: Braithwaite RB (ed) The foundations of mathematics and other logical essays, Routledge and Kegan Paul, Oxford pp 156–191, originally published in 1926
Regan D (1980) Utilitarianism and co-operation. Clarendon Press, Oxford
Savage L (1951) The theory of statistical decision. Journal of the American Statistics Association 46:55–67
Savage L (1954) The foundations of statistics. John Wiley and Sons, New York second revised edition published by Dover Publications, 1972
Sobel JH (1968) Rule-utilitarianism. Australasian Journal of Philosophy 46:146–165
Sobel JH (1976) Utilitarianism and past and future mistakes. Nous 10:195–219
Thomason R (1970) Indeterminist time and truth-value gaps. Theoria 36:264–281
Thomason R (1981) Deontic logic and the role of freedom in moral deliberation. In: Hilpinen R (ed) New Studies in Deontic Logic, D. Reidel Publishing Company, Dordrecht pp 177–186
Zimmerman M (1990) Where did I go wrong? Philosophical Studies 59:55–77

Chapter 12
Real Change, Deontic Action

Krister Segerberg

12.1 Real Change

Let U be a given universe (environment) of (possible total) states. Then a real change in U can be represented as a sequence of states. Thus a certain sequence

$$u_0 u_1 u_n \tag{12.1}$$

may be seen as a path instantiating or exemplifying a certain event (but it is also possible that it is not instantiating or exemplifying any particular event or action at all).

But when we give this model a deontic dimension, as we do in the following section, a more complicated representation is needed. If h is a given past history, we write cont(h) for the set of possible future histories (possible continuations) of h. An actual situation may then be represented as a pair hh, $\langle cont(h)\rangle$, where h is the *past* and cont(h) is the *open future*. The same real change $u_0 u_1 u_n$ just considered in (12.1) (but now situated after a certain past history h with last element $h(\#) = u_0$) may now be given the following more complicated representation:[1]

$$\langle hh, \ cont(h)\rangle, \langle hhu_1, cont(hu_1)\rangle, , \langle hhu_1 \dots u_n, cont(hu_1 \dots u_n)\rangle \tag{12.2}$$

A further complication is that for some purposes it is not enough to consider what we have just termed the actual situation. Sometimes it is important to be able to take a more general view. By a *(possible) situation* let us mean a pair $\langle hh, S\rangle$, where h (the *real past*) is a past history and S (the *real prospect*) is a subset of cont(h) (the focal future, the set of possible futures on which for some reason we wish to focus). Then the real change is represented as:

K. Seserberg (✉)
Filosotiska Institution, Uppsala Universitet, Box 627, 75126 Uppsala, Sweden
e-mail: krister.segerberg@filosofi.uu.se

[1] Here hu_1 is the path consisting of h followed by u_1. In general $hu_1 \dots u_n$ is the path consisting of h followed by $u_1 \dots u_n$.

P. Girard et al. (eds.), *Dynamic Formal Epistemology*, Synthese Library 351,

DOI 10.1007/978-94-007-0074-1_12,

$$\langle hh, S_0\rangle, \langle hh_1, S_1\rangle, , \langle hhu_1 \ldots u_n, S_n\rangle \tag{12.3}$$

where

$u_0 = h(\#)$, for some element u_0 in U,
$S_0 = S$,
$S_i = \{f \in cont(h_1 \ldots{}_i)\colon u_0 u_1 \ldots u_{i1} f \in S\}$, for all positive $i \leq n$.

12.2 Pure Deontic Actions

In order to represent deontic actions we need the concept of a norm – all deontic actions take place within a norm. (Two complications are ignored here. One is that in real life we are confronted by many norms, some of which may clash. The other is that norms may change, for example by what may be called higher order or legislative actions.[2] But here the norm is unique and unchanging.)

A complete norm is supposed to deliver answers to all questions of the following kind: Given a past history h, which of the elements of a certain prospect S are legal (normal, in accordance with the norm)? Trying to make this notion precise, let us define a norm as a function N on the set of situations that satisfies the following conditions:

$N(h, S) \subseteq S$ (INCLUSION),
if $S \subseteq S'$ then $N(h, S) \neq \emptyset$ only if $N(h, S) \neq \emptyset$ (MONEYS),[3]
if $S \subseteq S'$ then $S \cap N(h, S') \neq \emptyset$ only if $N(h, S) = S \cap N(h, S')$ (ARROW),
if $f = pf'$, for any finite path p in U, then $f \in N(h, S)$ only if $f' \in N(hp, S')$, where $S' = \{g \in cont(hp)\colon\ pg \in S\}$ (COHERENCE).

We are now in a position to represent both real change and deontic action. The three best-known, pure simple deontic actions are (simple) ordering, permitting and forbidding. There are different possible conceptions of these generic concepts. If we follow Ross (1941) and Hans (1974), we arrive at the following definitions: a situation $\langle hh, S\rangle$ is directly replaced by the situation $\langle hh, S\rangle$, where, in the case of a being ordered,

$$S' = \bigcup\nolimits_{q \in a} N\{f \in S\colon\ q \text{ occurs in } f\};$$

in the case of a being permitted,

[2] Cf. (There is also a third kind of change: legislative change. This occurs when what is here called the norm is modified, for example, when a new law is enacted or an old one is repealed. But that kind of change is not considered here.)
[3] MOnotonicity for NonEmptY Segments.

$$S' = N(h, S) \cup \bigcup_{q \in a} N\{f \in S : q\ hboxoccursin\ f\};$$

in the case of a being forbidden,

$$S' = N\{f \in S \colon \neg\exists q \in a(q \text{ occurs in } f)\};$$

For the sake of symmetry we might add a fourth (nontraditional) condition: in the case of a being made omissible,

$$S' = N(h, S) \cup N\{f \in S \colon \neg\exists q \in a(q \text{ occurs in } f)\};$$

12.3 Mixed Deontic Actions

Real change is change in the real state of the world. Deontic actions change the normative position. But sometimes real actions do too. You sign you name on a piece of paper, a real action. By doing so, you achieve some immaterial effect: for example, if the circumstances are right, you are contracting to sell a house of which you are the owner. The logician who wishes to formalize what is going on may elect to leave out of the formalism the real action (producing the signature) and concentrate on what is the point of the action: the sale of the house. But it may also be that he wishes to include both the real action and the deontic action in the formal representation. Then what?

Let us limit ourselves to what may be called atomic change: a minimal change from a situation $\langle h, S\rangle$ to a situation $\langle hh', S'\rangle$. There are three possibilities, mutually exclusive and jointly exhaustive:

(i) $h = h'$ and $S \neq S'$,
(ii) $hv = h'$, for some element v in U, and $S' = \{f \in \text{cont}(h) \colon vf \in S\}$,
(iii) $hv = h'$, for some element v in U, and $S' \neq \{f \in \text{cont}(h) \colon vf \in S\}$.

In case (i), the real past history does not change, only the normative position does; this is a case of *pure deontic action*. In case (ii), the real state changes, and the normative position also changes but only the in way described by (COHERENCE); this we may call *real change with ordinary deontic import*. But in case (iii), in which again the real state changes, the normative position changes in a way not described by (COHERENCE) but evidently as the result of some deontic action. In the last case we may term the change *real change with extraordinary deontic import*.

In the formalism developed in **?**, case (iii) could not be captured. In that respect the formalism provided here seems like an improvement.

Acknowledgements This chapter was written while the author was a Killam Visiting Scholar at the University of Calgary.

References

Hans K (1974) Free choice permission. Proceedings of the Aristotelian Society, vol. 74, pp 57–74.
Ross A (1941) Imperative and logic. Theoria 7:53–71.

Chapter 13
Neither Logically Omniscient nor Completely Irrational Agents: Principles for a Fine-Grained Analysis of Propositional Attitudes and Attitude Revision

Daniel Vanderveken

13.1 Introduction

Contemporary logic is confined to a few paradigmatic attitudes such as belief, knowledge, desire and intention. My purpose in this chapter is to present a general approach of propositional attitudes of any cognitive or volitive mode. In my view, one can recursively define the set of all psychological modes of attitudes. As Descartes anticipated, the two primitive modes are those of belief and desire. Complex modes are obtained by adding to primitive modes special cognitive and volitive ways or special propositional content or preparatory conditions.

According to standard logic of attitudes, for instance Hintikka's epistemic logic Hintikka (1971), human agents are either perfectly rational or totally irrational. I will proceed to a finer analysis of propositional attitudes that accounts for our imperfect but minimal rationality Cherniak (1986). For that purpose I will use a non standard predicative logic which distinguishes propositions with the same truth conditions that have different cognitive values.

In recent year much attention has been given to the theory of attitude change, for instance in belief revision theory or in the theory of preference change. At the end of this chapter I show that the logic I propose can easily be extended to deal with attitude dynamics. More precisely, I show how minimally rational agents dynamically revise propositional attitudes of any mode.

This chapter aims at presenting the *principles* that underlie the logic I propose. The reader interested in the details of the logical analysis, i.e. the syntax and model theoretical semantics, can consult the companion Vanderveken (2011) paper available on: http://www.vanderveken.org

D. Vanderveken (✉)
Département de Philosophie, Université du Québec, Trois-Rivières, QC G9A 5H7, Canada
e-mail: daniel.vanderveken@uqtr.ca

P. Girard et al. (eds.), *Dynamic Formal Epistemology*, Synthese Library 351,
DOI 10.1007/978-94-007-0074-1_13,

13.2 Compositional Analysis of Propositional Attitudes

In this Section I present the general principles behind my logic of propositional attitudes. More details can be found in Vanderveken (2008).

Propositional attitudes are directed at objects and facts of the world and they have logically related conditions of possession and of satisfaction. Beliefs and other cognitive attitudes are satisfied whenever they are true, desires and wishes whenever they are realized and intentions and plans whenever they are executed. Whoever possesses such an attitude is in principle able to determine what has to happen in the world in order that his or her attitude is satisfied. Propositional attitudes consist of a *psychological mode M* with a *propositional content P*. They are the simplest kinds of individual attitudes directed at facts.

Many philosophers tend to reduce all propositional attitudes to sums of beliefs and desires. However, our intentions are much more than a desire to do something with a belief that we are able to do it. Of course, cognitive attitudes (e.g. anticipation, conviction, faith, confidence, knowledge, certainty, presumption, pride, and so on) are types of beliefs. Similarly, volitive attitudes (e.g. wish, will, intention, ambition, project, hope, aspiration, satisfaction, pleasure, enjoyment, delight, gladness, joy, elation, etc.) are desires. Psychological modes, however, divide into other components than the basic categories of *cognition* and *volition:* complex modes have a *proper way* of believing or desiring, proper *conditions on their propositional content* or proper *preparatory conditions*. Many modes require a special *cognitive or volitive way* of believing or desiring. Thus, *knowledge* is a belief based on strong evidence that gives confidence and guarantees truth. Whoever has an *intention* feels such a strong desire that he or she is disposed to *act* in order to satisfy that desire.

Formally, a *cognitive or volitive way* is a function $f_{\bar{\omega}}$ which restricts basic psychological categories. Like illocutionary forces Searle and Vanderveken (1985), modes also have *propositional content* and *preparatory conditions*. *Previsions* and *anticipations* are directed towards the future. *Intentions* are desires to carry out a present or future action. From a logical point of view, a *condition on the propositional content* is a function f_θ that associates which each agent and moment a set of propositions. The one who holds an attitude or who performs an illocution *presupposes* certain propositions. His or her attitude and illocution would be *defective* if these propositions were then false. Thus *promises* and *intentions* have the preparatory condition that the agent is then able to do the action represented by their propositional content. No agent can lie to himself. Whoever has an attitude both believes and presupposes that its preparatory conditions are fulfilled. A preparatory condition is a function f_Σ associating with each agent, moment and propositional content a set of propositions that the agent would presuppose and believe if he had then an attitude with that preparatory condition and propositional content. The sets of cognitive and volitive ways, of propositional content and of preparatory conditions are *Boolean algebras*. They contain a *neutral* mode, preparatory and propositional condition and they are closed under the operations of meet *and* joint.

Such a compositional analysis can distinguish formally different modes of attitudes, e.g. fear, regret and sadness, which apparently reduce to the same sums of beliefs and desires. Identical psychological modes have the same components. Possession conditions of propositional attitudes are entirely determined by components of their mode. By definition, an agent *a possesses a cognitive* (*or volitive*) *attitude of the form M(P) at a moment m* when he or she then *believes* (or *desires*) its propositional content *P*, he or she feels that belief or desire that *P in the cognitive or volitive way* $\bar{\omega}_M$ proper to psychological mode *M*, the *proposition P then satisfies propositional content conditions* $\theta_M(a, m)$ and finally that *agent then presupposes and believes all* propositions determined by *preparatory conditions* $\Sigma_M(a,m,P)$ of mode *M* with respect to the content *P*. Thus an agent *intends that P* at a moment when proposition *P* then represents a present or future action of that agent, he or she desires so much that action that he or she is committed to carrying it out and moreover the agent presupposes and believes to be able to carry it out. An *attitude strongly commits an agent to another attitude at a moment* when he or she could not then have that attitude without having the second. Thus whoever believes that it will rain tomorrow then foresees rain tomorrow. The day after tomorrow the same belief wont be a prevision. It will be a belief about the past. An attitude *contains another* when it strongly commits any agent to that other attitude at any moment. There are *strong and weak psychological commitments* just as there are strong and weak illocutionary commitments. One must distinguish between the overt possession of an attitude and a simple psychological commitment to that attitude. Whoever believes that every man is mortal is weakly committed to believing that Nebuchadnezzars is mortal, even if he has not Nebuchadnezzars concept in mind and if he or she does not then possess the second belief. No one could simultaneously believe the first universally quantified proposition and the negation of the second.

Psychological modes, however, are not a simple sequence of a basic psychological category, a cognitive or volitive way, a propositional content condition and a preparatory condition, for their components are not logically independent. Certain components *determine* others of the same or of another kind. Thus the volitive way of the mode of *intention* determines its propositional content and preparatory conditions: the content of an intention must represent a present or future action of the agent and that agent must presuppose and believe that he or she is then able to carry out that action. The two primitive modes of *belief* and *desire* are the simplest cognitive and volitive modes. They have no special cognitive or volitive way, no special propositional content or preparatory condition.[1] Complex modes are obtained by adding to primitive modes special cognitive or volitive ways, propositional content conditions or preparatory conditions. Thus the mode of *prevision* M_{foresee} is obtained by adding to the mode of belief the propositional content condition θ_{future} that associates with each agent and moment the set of propositions that are future with respect to that moment ($M_{\text{foresee}} = [\theta_{\text{future}}]Belief$). The mode of *expectation* is obtained from that of prevision by adding the special cognitive way that the agent is then in a state of expectation ($M_{\text{expect}} = [\bar{\omega}_{\text{expectation}}]M_{\text{foresee}}$). The mode of *hope* is obtained from that of *desire* by adding the special cognitive way that the agent is

[1] For the logic of primitive modes see Vanderveken (2009).

then uncertain as regards the existence and the inexistence of the represented fact and the preparatory condition that that fact is then possible. The mode of *satisfaction* is obtained from that of *desire* by adding the *preparatory condition* that the desired fact exists. The mode of *pleasure* has, in addition, the *volitive way* that the satisfaction of the desire puts the agent in a state of pleasure and the preparatory condition that it is good for the agent. Because all operations on modes add new components, they generate stronger modes. Attitudes $M(P)$ with a complex mode *contains* attitudes $M(P)$ whose modes have less components.

13.3 Neither Logically Omniscient nor Completely Irrational Agents

Unlike many logics for cognitive or volitive attitudes, my approach avoids the problem of logical omniscience, and can deal with para-consistent beliefs. This is important because propositions with the same truth conditions are not necessarily the contents of the same attitudes. Moreover, agents ignore the necessary truth of many propositions that they understand. They have to learn a lot of essential properties of objects. By *essential property* of an object I mean a property that it *really* possesses in any possible circumstance. An essential property of each agent is to have certain parents. Some do not know their parents. Others are wrong about their identity. In my approach, all *circumstances* remain *possible*. So objects keep their essential properties (each of us keeps his real parents) and necessarily true propositions remain true in all circumstances. In order to account for our human inconsistency and minimal rationality, I advocate a *predicative* non classical propositional logic which takes into account acts of predication and reference that we make in apprehending propositions.

In my view, reference is indirect and each proposition has a finite *structure of constituents*. It predicates *attributes* (properties or relations) of *objects subsumed under concepts*. We understand a proposition when we understand which attributes our objects of reference must possess in a possible circumstance in order that this proposition be true in that circumstance. We understand most propositions without knowing in which possible circumstances they are true, because we ignore *real denotations* of their attributes and concepts in many circumstances. One can refer to a colleague's mother without knowing who she is. However we can always in principle think of persons who could be that mother. So in any possible use of language, there are a lot of *possible denotation assignments to attributes and concepts* in addition to the standard *real denotation assignment* which associates with each propositional constituent its actual denotation in every possible circumstance. They are functions of the same type that associate with each individual concept a unique individual or no individual at all in every possible circumstance. According to the real denotation assignment, my colleague's mother is the woman who gave birth to him. According to other possible denotation assignments, his mother is another person. However, all possible denotation assignments respect *meaning postulates*, e.g. any mother is a female parent. Most of the time we ignore the real denotation of concepts and attributes in certain circumstances, but we can nevertheless think of denotations that they could have. When we have concepts and attributes in mind, only some possible denotation assignments to these concepts and attributes *are then compatible with* our beliefs. Suppose that according to you my mother is French or Belgian. In that case, possible denotation assignments according to which she is neither French nor Belgian are then incompatible with your beliefs. Possible denotation assignments rather than possible circumstances are compatible with the beliefs of agents.

In my logic, the truth definition is relative to possible circumstances and denotation assignments. Most propositions have therefore *a lot of possible truth conditions*. However, in order to be *true in a circumstance* a proposition has of course to be *true in that circumstance according to the real denotation assignment*. So among all possible truth conditions of a proposition, there are its *real Carnapian truth conditions* that determine the set of possible circumstances where it is true according to the real denotation assignment.

Identical propositions make the same predications and they are true in the same circumstances according to the same possible denotation assignments. Necessarily true propositions with a different structure of constituents are therefore different. So are propositions that are not true according to the same possible denotation assignments to their constituents. We do not understand them to be true in the same possible circumstances. Few necessarily true propositions are *pure tautologies* that we know *a priori*. A proposition is *necessarily true* when it is true in every possible circumstance according to the real denotation assignment. In order to be *tautologically true*, a proposition has to be true in every circumstance according to every possible denotation assignment. Unlike the proposition that my mother is a mother, the necessarily true proposition that my mother is Gabrielle Louise Albertine Charron is not a pure tautology. It is false according to some denotation assignments. A proposition is *subjectively possible* when it is true in a circumstance according to a possible denotation assignment. In order to be *objectively possible* it must be true in a circumstance according to the real denotation assignment.

13.4 Agency, Time and Indeterminism

The logic of attitudes I propose is cast in a ramified conception of time, compatible with indeterminism. The attitudes and actions of agents are not determined; they could have thought or acted differently. In other words, their future is open.

In branching time, a *moment* is a complete possible state of the actual world at a certain instant and the *temporal relation* of *anteriority* between moments is partial. There is a single causal route to the past. However, there are multiple future routes. Consequently, the set of moments of time is a *tree-like frame* (Fig. 13.1).

A maximal chain h of moments of time is called a *history*. It represents a *possible course of history of our world*. Following Belnap's approach Belnap et al. (2001), I consider that each *possible circumstance* is a pair of a moment m and of a history

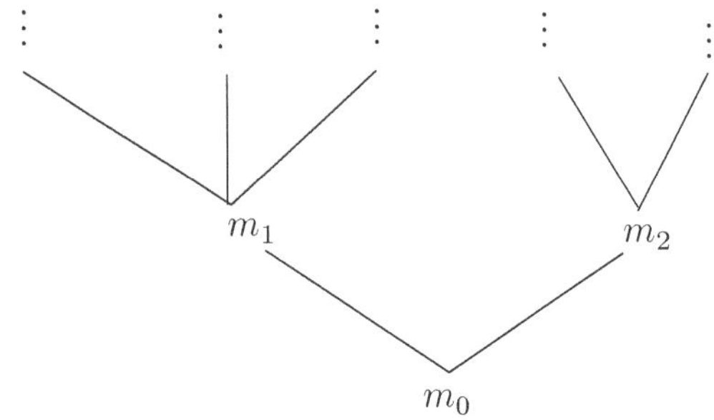

Fig. 13.1 A tree-like frame

h to which that moment belongs. Thanks to histories temporal logic can analyze important modal notions like settled truth and historic necessity. Certain propositions are true at a moment according to all histories.Their truth is then *settled at that moment* no matter how the world continues. So are propositions which attribute attitudes to agents. Whoever desires something at a moment then desires that thing no matter what happens later. Contrary to the past, however, the future is open, and so the truth of future propositions is not settled at non final moments. It depends on which historical continuation of such moments is under consideration. When a moment can continue in different possible ways, its actual future continuation is not then determined.

I adopt an occamist Prior (1967) point of view for the evaluation of propositions about future events: if the world continues after a moment, it will continue in one way. The actual historic continuation of any moment is unique even if it is still undetermined at that moment. So each moment m has a *proper history* h_m in each model. A proposition is *true at a moment* m according to a denotation assignment when it is true according to that assignment at that moment in its proper history h_m. Coinstantaneous moments which belong to the same instant are on the same horizontal line in each tree-like frame. Logic analyzes *historic necessity* by quantifying over coinstantaneous moments.[2]

13.5 Satisfaction as a Generalization of Truth

Like elementary illocutions, propositional attitudes are directed towards facts of the world represented by their propositional content. Most often agents establish a correspondence between their ideas and things to which they refer. According to them their ideas correspond to represented things in the world or represented things have to correspond to their ideas. Their attitudes and illocutions have for that reason *satisfaction conditions*. At each moment where agents think and act they ignore

[2] The proposition that P is then necessary (in symbols $\Box P$) is true at a moment when P is true at all coinstantaneous moments according to all histories.

how the world will continue. However, their attitudes and actions are intrinsically directed toward the real historic continuation of that moment. In order that a present desire directed at the future is realized, it is not enough that things will be at a posterior moment as the agent now desires. They must be so later in the real future. So the *satisfaction* of propositional attitudes and elementary illocutions of an agent at an indeterminist moment requires the *truth* of their propositional content *at that moment* in its real historic continuation.

The notion of *satisfaction* is a generalization of the notion of *actual truth* that covers most attitudes and elementary illocutions. Like assertive illocutions, cognitive attitudes have the *mind-to-things direction of fit*. They are satisfied when their propositional content is true at the moment under consideration. The agents ideas correspond to things as they are then in the world. In the cognitive case, when the agent realizes that there is no correspondence, he immediately changes his beliefs. This is why satisfaction amounts to truth in the case of cognitive attitudes.

However, this is not the case for volitive attitudes whose direction of fit goes from things to ideas. For the world and not the agent is at fault in the case of dissatisfaction of volitive attitudes. The agent can keep his desires and remains dissatisfied. Most often agents having a *volitive* attitude desire the fact *represented* by the propositional content no matter how that fact turns to be existent in the world. So most volitive attitudes are satisfied when their content is then true, no matter for which reason. Things are then such as the agent desires them to be, no matter what is the cause of their existence.

Volitive attitudes like will, intentions, projects, plans and programs have a special volitive way which requires that things fit the agents ideas because he or she wants them in that way. Such attitudes as well as illocutions like orders, commands, pledges and promises that express these volitive attitudes have *self-referential satisfaction conditions*. Their satisfaction requires more than the actual truth of their propositional content. It requires that the represented fact turns to be existent in order to satisfy the agent's attitude. In order to execute a prior intention, an agent must later carry out the intended action because of that previous intention. An agent who would later be obliged to act would not then execute that prior intention. My logic of attitudes explains such cases of self-referential satisfaction by relying on the notion of *intentional causation*. The agent's attitude must then be a *practical reason* why the represented fact turns to be existent.

Volitive modes like joy, gladness, pride, pleasure regret, sadness, sorrow, and shame have the *empty direction of fit*. Agents of such attitudes do not want to establish a correspondence between their ideas and things in the world. They just take for granted either correspondence or lack of correspondence. In the case of *joy*, *gladness*, *pride* and *pleasure*, the agent believes that the desired fact exists. In the case of *regret*, *sorrow* and *shame*, he believes the contrary. The first attitudes have the preparatory condition that their propositional content is then true, the second attitudes have the opposite preparatory condition that their content is then false. Volitive attitudes with such special *preparatory condition* have the *empty direction of fit* and no proper *satisfaction conditions*. Instead of being satisfied or dissatisfied, they are just *appropriate* or *inappropriate*.

13.6 A Closer Look at the Basic Categories of Cognition and Volition

Attitudes of agents are about objects that they represent under concepts. No agent can have a propositional attitude without having in mind consciously or potentially all attributes and concepts of its content. Whoever has a conscious propositional attitude has in mind consciously its attributes and concepts. In case of unconscious attitudes the agent can in principle express their senses thanks to his or her language. Otherwise, that agent would be unable to determine under which conditions his or her attitudes are satisfied.

Several denotation assignments to propositional constituents are compatible with the satisfaction of agents attitudes. There corresponds to each agent *a* and moment *m* a unique set *Belief(a,m)* of possible denotation assignments to attributes and concepts that are compatible with the truth of beliefs of that agent at that moment. *Belief(a,m)* is the set *Val* of all possible denotation assignments when agent *a* has no specific attribute or concept in mind at moment *m*. That agent has then no attitudes. Otherwise, *Belief(a,m)* is a *non empty proper* subset of *Val*. Similarly, to each agent *a* and moment *m* there corresponds a unique non empty set *Desire(a,m)* of possible denotation assignments to attributes and concepts that are compatible with the realization of all desires of that agent at that moment.

Any agent who has in mind propositional constituents respects meaning postulates governing them in determining truth conditions of propositions. Otherwise he could not even express them. Because meaning postulates are entirely determined by meaning, agents necessarily internalize such postulates in learning their language. Consequently there always are possible denotation assignments compatible with what an agent believes or desires.

In my approach, an agent *a believes a proposition* at a moment *m* when he or she has then in mind all its concepts and attributes and that proposition is true at that moment according to all possible denotation assignments belonging to *Belief (a,m)*. Most logicians have a logical analysis of desire similar to that of belief. There is however the following important difference between desire and belief that one needs to take into account. We often believe that objects are so and so without believing that they could be otherwise. This is not the case for desires, for any desire contains a *preference*. Whoever feels a desire distinguishes two different ways in which represented objects could be in the actual world. In a first preferred way, objects are in the world as the agent desires, in a second way, they are not. The agent's desire is realized in the first case, it is unrealized in the second case. Consequently in order that an agent *a* desires the fact represented by a proposition at a moment *m*, it is not enough that he or she has then in mind all its constituent senses and that the proposition is true at that moment according to all possible denotation assignments of *Desire(a,m)*. That proposition must moreover be false in a circumstance according to that agent.

This account of belief and desire is in line with major work in the philosophy of mind (Goldman 1970, Searle 1983), which distinguishes conscious and unconscious attitudes and accounts for the fact that human agents are not perfectly rational. We do not have in mind all concepts and attributes. So we ignore logical as well as necessary truths. Our knowledge is limited: we ignore which objects possess many properties in a lot of circumstances. In that case assignments associating different denotations to these properties in these circumstances are then compatible with our beliefs. We have false beliefs and unsatisfied desires. Possible denotation assignments compatible with our beliefs and desires need not assign real denotations to attributes that we have in mind. They can even violate essential properties of objects. In that case we have necessarily false beliefs and insatisfiable desires. My proposal can take such limitations into account.

Human agents cannot be totally irrational, though. They are rather *minimally rational*. Agents cannot believe or desire everything since some possible denotation assignments are always compatible with the satisfaction of their beliefs and desires. Moreover, their beliefs and desires commit them to others. For possible denotation assignments respect meaning postulates. Human agents are therefore logically omniscient, but in a controlled way: they cannot have in mind what I call a pure tautology without knowing for certain that it is necessarily true. Represented objects could not be in another way according to them. Similarly, *pure contradictions* (negations of tautologies) are false in every possible circumstance according to any agent. We can neither believe nor desire contradictory things. Some still hope that arithmetic is complete, but one could never hope for both the completeness and the incompleteness of arithmetic.

My logic also provides a more refined form of *strong* propositional *implication,* much finer than Lewis' strict implication. A proposition *strongly implies* another when whoever expresses that proposition is able to express the other and it cannot be true in a circumstance according to a possible denotation assignment unless the other proposition is also true in that circumstance according to that assignment. Strong implication is finite, tautological, paraconsistent, decidable and *a priori known*.

13.7 Attitudes Revisions

Agents keep many attitudes during an interval of time. Whenever they acquire a new propositional attitude they then lose old attitudes that are incompatible with the new attitude and they acquire others. The nature of these changes can be determined with the conceptual apparatus of my logic.

Minimally rational agents cannot have simultaneously attitudes of certain forms at certain moments. My first principle of update is thus a *principle of revision*: any agent who acquires a new attitude loses *eo ipso* all previous attitudes that are incompatible with that new attitude. There are many reasons of incompatibility between attitudes. Propositional attitudes can be incompatible because they contain incompatible beliefs or desires or incompatible cognitive or volitive ways, propositional content or preparatory conditions. For example agents lose beliefs that they discover to be false and they revoke previous intentions that they discover to be impossible to execute.

My second principle of update is an *expansion principle*: Whenever an agent acquires a new propositional attitude M(P) he or she acquires *eo ipso* many other new attitudes with the same propositional content whose modes can be determined given propositional content conditions that P then satisfies and other attitudes that that agent has at that moment about P. Suppose, for instance, that an agent who sees a lot of heavy low clouds in the sky comes to believe that it will soon rain. Because that new belief is directed towards the future, he or she then *foresees* that it will rain. This is so because the mode of *prevision* is obtained from that of belief by adding the special propositional content condition that the represented fact is future with respect to the moment of the attitude. Suppose now that this new belief brings the agent in a state of expectation, a special cognitive way. He or she then also *expects* rain. Suppose moreover that that agent previously *hoped* and still desires that it would not soon rain. Because hope is a desire whose satisfaction is then possible according to the agent, i.e. special preparatory condition, the agent then loses the previous hope and feels moreover a *disappointment*. Whoever is *disappointed* with a fact desires its inexistence while believing, i.e. special preparatory conditions, that the fact exists and that he or she previously believed that it would not exist.

The expansion principle that I have just stated generates new attitudes having the same propositional content but a new mode, my second expansion principle generates new attitudes having the same mode but a new propositional content. That second expansion principle is based on the fact that most components of psychological modes M are closed under conjunction. Suppose that an agent *a* acquiring a new attitude M(P) at a moment also possesses at that moment another attitude M(Q) with propositional content Q. Then he or she also acquires the new attitude M(P Q) whose content is the conjunction of these propositions. This will happen in case both attitudes M(P) and M(Q) strongly commit that agent to the new attitude M(P & Q). This is always the case when M is a primitive psychological mode. For the categories of cognition and volition are closed under conjunction. Whoever believes (or desires) two things beliefs (or desires) both. Moreover many but not all cognitive and volitive ways and preparatory and propositional content conditions are closed under conjunction. When two propositions are future with respect to a moment so is their conjunction. Consequently, two previsions strongly commit the agent to foreseeing their conjunction. On the other hand, sometimes it can be good to do one thing and it is good to do another thing but it is not good to do both. So one can approve one thing and approve another thing without approving both.

13.8 Conclusion

In this chapter I have presented the general principles that guide my compositional analysis of all propositional attitudes. These can be summarized as follows: (1) All psychological modes of propositional attitudes can be compositionally generated on the basis of the primitive modes of belief (cognition) and desire (volition); (2) A non-standard, predicative logic can account for the limited rationality of agents, avoiding both pitfalls of logical omniscience and complete irrationality;

(3) The logic of attitudes should take indeterminism into account; (4) Satisfaction, a generalization of the notion of truth, is applicable to all psychological modes; (5) The process of attitude revisions is guided by two simple principles, revision and expansion, together with the logical constraints built in my compositional analysis of attitudes of any mode. Of course, to explicate all this one needs a rather rich logical system. The details of this system can be found on the companion paper Vanderveken (2011) on: http://www.vanderveken.org

References

Belnap N, Perloff M, Xu M (2001) Facing the future; agents and choices in our indeterminist world. Oxford University Press, Oxford

Cherniak C (1986) Minimal rationality. MIT Press, Cambridge

Goldman A (1970) A theory of human action. Princeton University Press, Princeton, NJ

Hintikka J (1971) Semantics for propositional attitudes. In: Linsky L (ed) Reference and modality, Oxford University Press, Oxford

Prior A (1967) Past, present and future. Clarendon Press, Oxford

Searle J (1983) Intentionality. Cambridge University Press, Cambridge

Searle J, Vanderveken D (1985) Foundations of illocutionary logic, Cambridge University Press, Cambridge

Vanderveken D (2008) A general logic of propositional attitudes. In: Dégremont C, Keiff L, Ruckert H (eds) Dialogues, logic and other strange things, College Publication, Tributes, vol 7, pp 449–483

Vanderveken D (2009) Beliefs, desires and minimal rationality. In: Johansson L, Österberg J, R S (eds) Logic, ethics and all that jazz: Essays in honour of Jordan Howard Sobel, Uppsala Philosophical Studies

Vanderveken D (2011) Formal Semantics for Propositional Attitudes, forthcoming in Agazzi E, D'Ottaviano I, Mundici D (eds) Science, Truth and Consistency, a special issue in honour of Newton da Costa of Manuscrito: Revista Internacional de Filosofia

Index

P. Girard et al. (eds.), *Dynamic Formal Epistemology*, Synthese Library 351,
DOI 10.1007/978-94-007-0074-1, © Springer Science+Business Media B.V. 2011

www.ingramcontent.com/pod-product-compliance
Lightning Source LLC
Chambersburg PA
CBHW071243230926
51849CB00001B/209
* 9 7 8 9 4 0 0 7 0 0 7 3 4 *